AF789606

Expanded Bed Chromatography

Edited by:

BO MATTIASSON

Department of Biotechnology,
Center for Chemistry and Chemical Engineering,
Lund University,
Lund, Sweden

Reprinted from *Bioseparation*, Volume 8, No. 1/5 (1999)

Springer-Science+Business Media, B.V.

Library of Congress Cataloging-in-Publication Data

ISBN 978-90-481-5380-0 ISBN 978-94-017-1519-5 (eBook)
DOI 10.1007/978-94-017-1519-5

Printed on acid-free paper

All Rights Reserved
© 1999 Springer Science+Business Media Dordrecht
Originally published by Kluwer Academic Publishers in 1999
No part of the material protected by this copyright notice may be reproduced or utilized in any form or by any means, electronic or mechanical, including photocopying, recording or by any information storage and retrieval system, without written permission from the copyright owner.

Contents

Bioseparation **8:** v, 1999.
© 1999 *Kluwer Academic Publishers.*

Preface

Integration is a leading theme in modern process biotechnology. This is done with an aim to reduce the number of processing steps, to simplify the handling, to shorten the time a sensitive target molecule has to be in contact with harsh conditions and also to reduce total processing costs.

The main emphasis has so far been on integration of bioconversions and the initial steps of downstream processing, mainly with regard to harvesting small molecules.

The concept of capturing of biomolecules implies an early enrichment in an isolation step that *per se* does not necessarily increase the purity to high levels. It is rather a matter of reducing the processing volumes and, to some extent, also to remove some contaminants. The use of expanded bed chromatography is such a development. This involves the use of adsorption material which is heterogeneous with regard to both particle size and density such that it forms a stable expanded bed when used in upstream chromatographic processing under conditions that would normally have given a fluidised bed.

The format of expanded bed has many applications in biotechnology. The main area developed so far is the capturing approach. However, this volume also illustrates use of the expanded bed technology in other areas of biotechnology, such as bioanalysis and biocatalysis. It seems safe to predict that these areas also hold interesting potential for future development. One can thus expect that the technology may become a versatile tool in different areas of modern bioprocess technology.

The new material that are being developed for use in expanded bed chromatography will help to broaden the area of applications even more.

The area of expanded bed chromatography is novel and during June 1998 an international meeting was held from which this volume emerges. We have chosen to collect a broad selection of papers illustrating different aspects of the technology, all the way from design of new materials, mathematical simulations of column performances, use of the new adsorbents in processing conditions both from the research laboratories and from the industrial environment. There may sometimes be a lack of some details in a few of the papers that traditionally would have been expected from an academic paper, but in my mind these papers contain in their present shape significant valuable information that makes the volume interesting and useful for the separation science community.

The symposium covered many aspects of expanded bed technology. A large proportion of the contributors have submitted material to the volume and one additional paper is added after the meeting and that makes the proceedings a good coverage of the situation in expanded bed chromatography during 1998.

The scientific committee behind the meeting is gratefully acknowledged for their contribution. The committee had the following members:

Dr. Stuart Builder (Chairman), Genetech, USA;
Dr. Howard Chase, University of Cambridge, UK;
Dr. Rolf Hjort, Amersham Pharmacia Biotech, Sweden;
Prof. Maria-Regina Kula, University of Düsseldorf, Germany;
Prof. Mathias Uhleń, Royal Institute of Technology, Sweden;
Dr. Joachim Walter, Dr. Karl Thomae Gmbh, Germany.

This proceeding is among the first in a line of special issues that are planned within the journal of BIOSEPARATION.

Lund in May 1999
Bo Mattiasson

Bioseparation **8:** 1–9, 1999.
© 1999 *Kluwer Academic Publishers.*

Expanded bed adsorption: elution in expanded bed mode

Rolf Hjorth
Amersham Pharmacia Biotech, SE-751 84 Uppsala, Sweden Tel: +46 18 165814; Fax: +46 18 166396; e-mail: rolf.hjorth@eu.apbiotech.com

Received 19 September 1998; accepted 25 November 1998

Key words: elution, expanded bed adsorption, ion exchange, protein recovery

Abstract

Elution in expanded bed mode has been investigated in the expanded bed adsorption process. Elution was performed at different sample loads and at different liquid velocities using bovine serum albumin as a model. The effect on mixing in the liquid phase and on the volume of the eluted peak were determined. Mixing in the liquid phase was almost unaffected when elution was performed at 100 cm/h, regardless of sample load. However, mixing increased significantly when elution was carried out at high liquid velocities (300 cm/h) at high sample loads. The eluted peak volume increased with liquid velocity and increased sample load. It was approx. 80% higher in expanded bed mode than in packed bed from an adsorbent completely saturated with protein eluted at 300 cm/h.

Introduction

Expanded bed adsorption is an efficient technique for recovery of protein from unclarified feedstocks. It is based on stable fluidisation of an adsorbent in a purposely designed column. Due to the fluidisation process the space between the adsorbent particles is increased, allowing feed material containing cells and/or cell debris to be pumped through the column without blocking it. This makes expanded bed adsorption an attractive technique for recovery of proteins from particle-containing feedstocks. Several examples where biomolecules have been recovered from mammalian cell suspensions (Zapata et al., 1996), yeast cell suspensions (Noda et al., 1996) and *E. coli* (Barnfield Frej et al., 1994; Johansson et al., 1996) in pilot and industrial scale using expanded bed adsorption have been published. Recently, Thatcher et al. (1997) reported the recovery of plasmid DNA from an *E. coli* homogenate by expanded bed adsorption in pilot scale. Expanded bed adsorption has been reviewed in terms of both its basic principles (Chase, 1994, Thömmes, 1997) and its industrial applications (Hjorth, 1997).

Several reports have been published during the last few years describing the adsorption process of expanded bed adsorption (Chang and Chase, 1996b; Hjorth et al., 1995; Karau et al., 1997; Thömmes et al., 1995); however, very little attention has been paid to the elution process of expanded beds. It has been recommended that expanded beds be eluted in packed bed mode (Amersham Pharmacia Biotech). Nevertheless, the few reports describing elution in expanded bed mode (Hjorth et al., 1995; Thömmes et al., 1996) clearly demonstrate that it might be a useful alternative. Hjorth et al. (1995) have shown that elution in expanded bed mode could be performed without adverse effects and that the volume of the eluted peak increased only 40% compared with elution under similar conditions in packed bed mode. Thömmes et al. (1996) described the use of elution in expanded bed mode for recovery of IgG from a protein A containing adsorbent. This mode of operation was reported to result in an increase of the eluted peak volume due to the increased bed voidage in an expanded bed compared to a packed bed.

Selecting elution in expanded bed mode will certainly affect both equipment design and the process itself and therefore it is important to gain more information about this mode of operation. It is under certain conditions (Amersham Pharmacia Biotech) necessary to carry out elution in expanded bed mode in order to avoid excessive back pressure. Furthermore, since elution in expanded bed mode minimises contact between the adsorbent particles it will help prevent bead aggregation and hence make the subsequent cleaning in place procedure more efficient. Finally, elution in expanded bed mode could facilitate a simpler design of the expanded bed column through the use of a fixed upper adapter instead of a movable adapter (Thömmes et al., 1996).

In this paper we report data on elution in expanded bed mode and its effect on the hydrodynamic properties of the expanded bed as well as the functional performance of the expanded bed.

Materials and methods

Expanded bed adsorption experiments were carried out using a STREAMLINE™ 25 column (inner diameter, 2.5 cm) filled with 76 ml STREAMLINE DEAE (Amersham Pharmacia Biotech AB, Sweden), which corresponds to 15.5 cm sedimented bed height. The base matrix in STREAMLINE adsorbents is a highly cross-linked 6% beaded agarose containing crystalline quartz to increase the density of the beads. The average density is 1.2 g ml^{-1}. The particle size range from 100 to 300 μm with an average size of 200 μm. The ion exchange capacity for STREAMLINE DEAE is 0.17±0.04 mmol Cl^- per ml adsorbent. The STREAMLINE column is purpose designed for use in expanded bed adsorption. It is equipped with a liquid distribution inlet consisting of a perforated plate, which generates the pressure drop necessary to obtain an even plug flow in the column as there is no pressure drop over the expanded bed. The set-up of the equipment was as described in Barnfield Frej et al. (1994) except that conductivity was monitored using a conductivity monitor (Amersham Pharmacia Biotech AB, Sweden). All buffer chemicals were of analytical grade. Adsorption was performed in expanded bed mode using bovine serum albumin (BSA), (Intergen, Toronto, Canada) at a concentration of 2 mg ml^{-1} in 50 mm Tris buffer, pH 8, at 300 cm h^{-1} liquid velocity. The amounts of BSA loaded were 30, 300 and 2100 mg. When 2100 mg of BSA were loaded a slight breakthrough of protein could be observed, indicating that the dynamic binding capacity of the adsorbent was fully utilised. Elution was carried out using 50 mM Tris buffer containing 0.5 M NaCl. In certain experiments where the hydrodynamics of the elution process were studied, Tris buffer containing 0.2 or 1 M NaCl was also used. The densities and viscosities of the elution buffers are given in Table I. Elution liquid velocities were nominally 100, 200 and 300 cm h^{-1}; the actual values are given in the results and discussion section. The distance between the bed surface and the upper adapter during elution in expanded bed mode was in the range of 1–6 cm, unless otherwise stated. For comparison, experiments were run in which elution was performed in packed bed mode, at liquid velocities of 100 and 300 cm h^{-1}. In order to ensure proper function of the adsorbent it was cleaned in place (CIP) every third or fourth cycle by washing with 1 M NaOH during 30 min contact time. Some experiments were repeated to check that the adsorbent behaved in a similar manner between experiments, (data not shown).

Bed expansion was evaluated by the Richardson–Zaki equation $u=u_t\epsilon^n$ (Richardson and Zaki, 1954), where u is the liquid superficial velocity, u_t the particle terminal velocity, ϵ the bed voidage and n the Richardson-Zaki coefficient. The value of ϵ was calculated from $H/H_o=(1-\epsilon_o)/(1-\epsilon)$ where H_o is the sedimented bed height, H the expanded bed height and ϵ_o the voidage in the sedimented bed. In this study a value of 0.4 was used for ϵ_o as determined in earlier published work (Draeger and Chase, 1991; Thömmes et al., 1996).

Mixing in the liquid phase was evaluated by determination of axial dispersion. The axial dispersion coefficient, D_A, was calculated using $D_A=uH/2N\epsilon$ (Barnfield Frej et al., 1997), where u is the liquid superficial velocity, H bed height and ϵ the bed voidage. N is the number of theoretical plates calculated as $N = t^2/\sigma^2$, where t is the residence time and σ^2 the variance. The parameters t and σ were determined by residence time distribution (RTD) analysis for expanded bed mode using the positive-step signal from experiments where 0.2, 0.5 and 1 M NaCl were applied as described in Barnfield Frej et al. (1997). Axial dispersion was also determined from the conductivity trace of NaCl during elution of BSA in expanded bed mode.

The Bodenstein number, Bo, was used as a dimensionless number for describing the axial mixing in a fluidised bed system (Sofer and Hagel, 1997). It was calculated as $Bo=uH/\epsilon D_A$, where u is the liquid superficial velocity, H the expanded bed height, ϵ the bed voidage and D_A the axial dispersion coefficient.

The peak volume of the eluted peak was measured at 10% of the maximum peak height.

Table 1. The effect of salt concentration and liquid velocity on expanded bed height, degree of expansion and bed porosity

Buffer	Density (kg m^{-3})	Viscosity (Pa s)	Liquid velocity (cm/h)	Expanded bed height (cm)	Degree of expansion (–)	Bed porosity ϵ (–)
50 mM Tris	999.7	1.015e-03	0	15.5	1	0.40
50 mM Tris	999.7	1.015e-03	107	25.6	1.64	0.63
			208	33.5	2.15	0.72
			309	42.1	2.70	0.78
50 mM Tris+0.2 M NaCl	1006.4	1.021e-03	107	25.7	1.65	0.64
			208	35.2	2.26	0.74
			309	46.7	2.99	0.80
50 mM Tris+0.5 M NaCl	1018.7	1.048e-03	104	25.9	1.67	0.64
			200	35.1	2.26	0.73
			299	48.0	3.10	0.81
50 mM Tris+1.0 M NaCl	1038.6	1.094e-03	104	26.7	1.72	0.65
			208	37.5	2.42	0.75
			302	52.8	3.23	0.81

Results and discussion

The elution of ion exchange adsorbents is commonly performed either by increasing the salt concentration or by changing the pH of the elution buffer. In the case of changing pH, the change in physical parameters, like density and viscosity, which will affect the performance of the expanded bed, are very minor. However, the addition of salt to the elution buffer will significantly change these parameters. To study the effects of different elution buffers on bed expansion and axial dispersion three elution buffers, containing 0.2, 0.5 and 1 M NaCl were evaluated. Elution experiments were carried out using three different loads of BSA in order to study the effects on mixing in the liquid phase and on the volume of the eluted peak in relation to the concentration of the sample being eluted. Elution of BSA from the adsorbent was performed using 0.5 M NaCl only, as that concentration of NaCl was found to desorb BSA quantitatively from STREAMLINE DEAE (data not shown). For certain experiments, the yield of BSA in the eluted fraction was determined and it exceeded 95% in all cases tested.

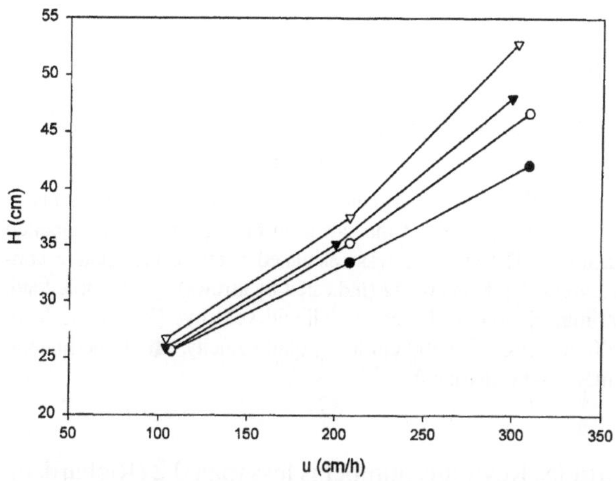

Figure 1. The influence of liquid velocity and salt concentration on expanded bed height. NaCl concentrations: (●) 0 M; (○) 0.2 M; (▼) 0.5 M; (▽) 1 M.

Bed expansion

The results from determining the effect of salt concentration and liquid velocity on expanded bed height (Fig. 1) , degree of expansion and bed porosity are shown in Table 1. These results are in accordance what could be expected from such measurements. The bed porosity increased from 0.63 in buffer at 100 cm h^{-1} liquid velocity to 0.81 in 1 M NaCl at 300 cm h^{-1}. That means that the change in bed porosity under a broad range of operational conditions is relatively small.

One important criterion for the stability of an expanded bed is the relationship between the liquid velocity in the bed and the corresponding void in the bed. This relationship is described by the Richardson–Zaki equation. The stability of the expanded bed is characterised by the value of the exponent n and for a stable bed it is $4.65 + 19.5d_p/D$ where d_p is the particle diameter and D the diameter of the column, if the

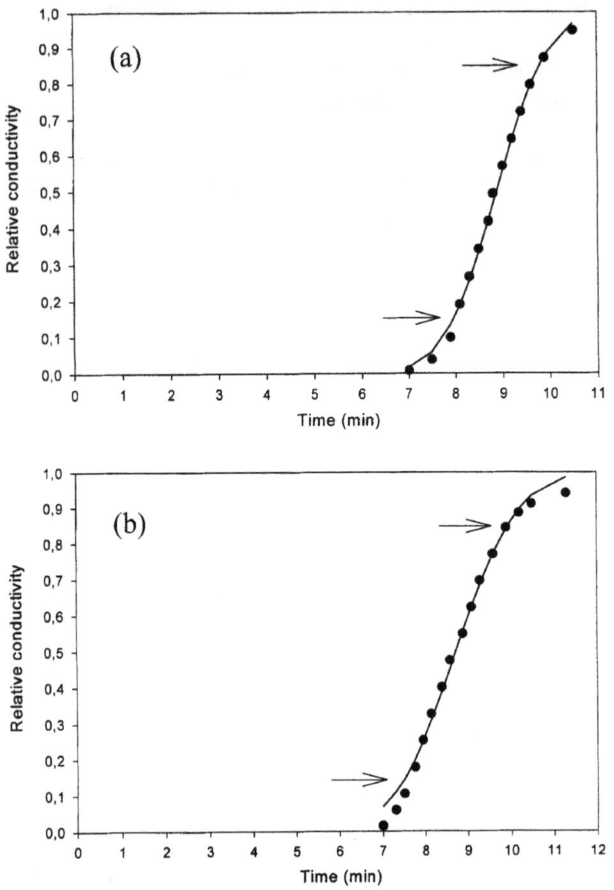

Figure 2. Fitting of experimental data from RTD measurements of the conductivity trace of the salt front during elution in expanded bed mode. The variance was calculated between the relative conductivities 0.158 and 0.842 (indicated by arrows). (A) Sample load, 300 mg; elution at 300 cm h^{-1} liquid velocity; (B) sample load, 2100 mg; elution at 300 cm h^{-1} liquid velocity. (●) Experimental values; (—) Gaussian fit.

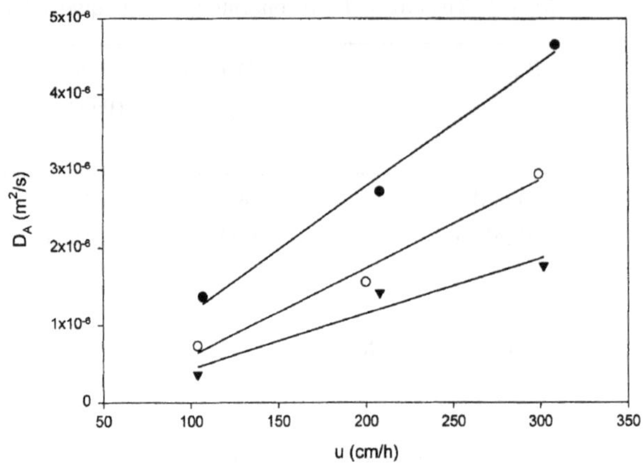

Figure 3. Axial dispersion for buffer containing different concentrations of NaCl at different liquid velocities. The axial dispersion coefficient, D_A, was calculated from the RTD measurements obtained from the conductivity trace of the salt front. NaCl concentrations: (●) 0.2 M; (○) = 0.5 M; (▼) 1 M.

particles Reynolds number is less than 0.2 (Richardson and Zaki, 1954). Under the experimental conditions used in this study an optimal value of *n* should be 4.8. In this investigation the values of *n* varied between 4.5 and 5.7 and there was no significant difference if the measurements were taken in Tris buffer alone or in the presence of a salt-containing buffer. The observed values of *n* are in accordance with values reported earlier for STREAMLINE adsorbents (Chang and Chase, 1996b; Karau et al., 1997). These results show that the bed behaved as a stable expanded bed both in buffer and in the presence of NaCl in concentrations appropriate for elution of proteins from ion exchange adsorbents.

Axial dispersion

The use of the axial dispersion coefficient was two-fold: as a measure of the mixing in the liquid phase of the bed when using different salt concentrations in the eluent; and for determining the mixing during elution of BSA. The formula used for calculation of the axial dispersion coefficient was suggested by Levenspiel (1972) but its use for accurately describing the dispersive mass transport in chromatographic beds has been questioned (Sofer and Hagel, 1997). It should be pointed out that the dispersion coefficients presented might contain contributions from effects, which results from the actual experimental conditions. Such contributions might originate from changes in bed expansion and/or mixing effects due to changed viscosity in the tracer pulse. However, this approach has shown to be valuable for monitoring hydrodynamic properties during scale-up of packed bed chromatography, where the running conditions are quite similar between different experiments (Sofer and Hagel, 1997). The importance of using a correct model for measuring the axial dispersion has been described by Swaine and Daugulis (1988). Fig. 2 shows the fitting of experimental data for the relative conductivity in the salt front in comparison with a Gaussian distribution. The results obtained show a good correlation to the expected theoretical values. It was estimated that σ^2could be calculated with an accuracy of ±5% of the theoretical value.

The effects of viscosity and density of the liquid on axial dispersion at different liquid velocities is shown in Fig. 3. As can be seen, the axial mixing increases with increased liquid velocity, a behaviour which has been reported earlier (Dasari et al., 1993; Goto et al., 1995). It is also clear that increased viscosity and density of the liquid result in a decreased mixing in the liquid phase, e.g., compare the lower axial disper-

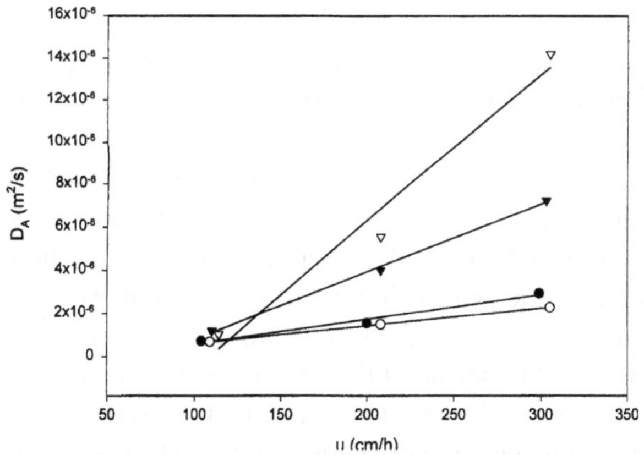

Figure 4. Axial dispersion in the zone of eluted protein during elution in expanded bed mode at different liquid velocities and loads of BSA. The axial dispersion coefficient, D_A, was calculated from RTD measurements of the conductivity trace of the salt front. Elution was performed in 50 mM Tris buffer containing 0.5 M NaCl. Sample loads: (●) = 0 mg; (○) 30 mg; (▼) 300 mg; (▽) 2100 mg.

Table 2. Bodenstein numbers for different sample loadings determined for two ranges of liquid velocities

Sample load (mg)	Liquid velocity (cm h^{-1})	Bo (–)
0	104	161
0	299	167
30	109	175
30	305	204
300	110	103
300	303	61
2100	114	124
2100	305	31

sion for 1 M NaCl with that for 0.2 M NaCl. This is contrary to the observations described by Chang and Chase (1996b) studying mixing in expanded beds in the presence of high concentrations of glycerol, where they found increased mixing with increased viscosity of the liquid. The most possible explanation for these different observations, are differences in how the experiments were performed. In this work, the RTD data were calculated from the salt front when buffer containing salt was introduced into a bed expanded in buffer only. Chang and Chase (1996b) on the other hand used acetone as tracer in the solution used for expansion of the bed. In this work axial dispersion was thus measured when changing from one solution to another, in the case of Chang and Chase (1996b), axial dispersion was measured in a bed already expanded in a solution with high viscosity and density. Values for the axial dispersion coefficient in the range of $1–5\times10^{-6}$ m^2 s^{-1} have also been reported by others using similar conditions (Batt et al., 1995; Chang and Chase, 1996b; Karau et al., 1997; Lindgren et al., 1993). From these data it can be concluded that an increase in viscosity and density of the desorption eluent will not introduce any additional mixing in the expanded bed when elution is performed in expanded mode.

More interesting is how axial dispersion is affected by the presence of protein in the salt front, as in a real situation when a protein is eluted from adsorbent. The results from such experiments are shown in Fig 4. From these data it is clear that both sample load and liquid velocity affect axial dispersion. At low sample loads (30 mg protein) there is no significant change between the axial dispersion coefficient obtained with protein present and that obtained without, i.e., eluent alone, at the liquid velocities tested. That behaviour is expected as the average protein concentration in the eluted peak is low (0.3–0.5 mg ml^{-1}) and the corresponding increase in viscosity and density will not cause any disturbance in the liquid zone. However, at increased sample loadings, and hence increased protein concentration in the eluted peak, there is a significant increase in axial dispersion. This increased mixing is directly dependent on the liquid velocity. At a low liquid velocity (100 cm h^{-1}) there is no difference in axial dispersion for different BSA concentrations in the eluted peak compared to axial dispersion for the eluent alone, even if the average BSA concentration in the eluted peak is 14 mg ml^{-1}. On the other hand, at higher liquid velocities (300 cm h^{-1}) there is a 7-fold increased in the axial dispersion coefficient.

The Bodenstein number, Bo, was calculated in order to compare mixing in the liquid phase in the expanded bed with data obtained by Karau et al. (1997). The results are shown in Table 2. These Bodenstein numbers obtained in this study are higher than those obtained by Karau et al. (1997) where Bo numbers of 80–100 were reported for STREAMLINE DEAE run at liquid velocities ranging from 180 to 540 cm h^{-1} using a bed with a sedimented bed height of 14 cm. One explanation for this difference might be that a column of different design (e.g., liquid distribution inlet) was used in the study by Karau et al. (1997). They also determined the Bo number for an expanded bed containing an adsorbent which was saturated with BSA to 95% and run at 300 cm h^{-1}: the result was a 10-fold reduction of the value of Bo.

A similar decrease in Bo was observed in this work, although the reduction was approximately 5-fold (from 167 to 31). However, it should be pointed out that the data here were obtained during desorption, a situation where both bound and unbound protein are present in the salt front, which is different from the work by Karau et al. (1997) in that the protein was bound to the adsorbent during the RTD measurement. It has been reported that if Bo < 20, the axial dispersion will have an effect on the process performance (Slater, 1991). If the liquid velocities are above 300 cm h^{-1} there is a risk that the Bo number may reach values where mixing in the expanded bed will be significant and affect the peak volume, especially at high loads.

Elution performance

Elution of low loadings (30 mg) of BSA was performed at 100 cm h^{-1} liquid velocity to examine the effect of the distance between the bed surface and the upper adapter. It is important to note that the position of the bed surface is affected by the liquid velocity and the viscosity and density of the liquid used. This distance was varied from 1.1 to 67 cm, and under these conditions the volume of the eluted peak decreased slightly. However, when these experiments were repeated using higher BSA loadings (300 mg) at 100 cm h^{-1} liquid velocity, the volume of the eluted peak increased from 49 ml at a bed surface/adapter distance of 1.8 cm to 76 ml (55%) at a distance of 27 cm. Elution using 1 M NaCl resulted in a similar volume of the eluted peak as noted above. These data show that with increased protein concentration in the eluted peak, the small volume of the eluted peak cannot be mastered when the distance between the bed surface and the upper adapter is enlarged. Furthermore, this behaviour could not be circumvented by using an eluent with increased density and viscosity such as 1 M NaCl. However, it is likely that the use of an additive of much higher density and viscosity such as glycerol could be used for this purpose as it has been shown to be very efficient (i.e. resulting in small volume) in washing out particulate matters from expanded beds (Draeger and Chase, 1991).

A summary of the results for elution of BSA in expanded and packed bed modes at different sample loadings and liquid velocities is given in Table 3. In general, the volume of the eluted peak increased with higher liquid velocities and sample loads as expected. However, it is important to compare the results between elution in expanded and packed bed mode in more detail, in particularly at high sample loads, as that reflects a real process situation. For example, for a sample load of 2100 mg applied at 100 cm h^{-1} liquid velocity, the peak volume during elution in expanded bed mode increased by 43% compared to the peak volume obtained at elution in packed bed mode under the same conditions. This value is in accordance with data published earlier (Hjorth et al., 1995). When the liquid velocity was increased to 300 cm h^{-1}, the volume of the eluted peak increased by 81% in expanded mode compared to that obtained in packed mode. An increase in the volume of the eluted peak in the range of 80% can usually be accepted even if a production scale column is considered. For example, Barnfield Frej et al. (1997) eluted BSA from 42 l of an anion exchange adsorbent in packed bed mode resulting in 92 l of eluted BSA. If we consider that the elution step was performed in expanded bed mode the volume of the eluted fraction would have increased to 166 l. The difference in sample application time for a subsequent chromatography step would approximately be 21 min if we assume a 30-cm diameter column run at 300 cm h^{-1}. Such an increase in time, as well as the extra cost for additional elution buffer, will generally have little impact on the overall cost for the process.

In order to evaluate the full effect of the operational performance on elution in expanded bed mode, a comparison of three different modes of operation were made. All three modes were run at 300 cm h^{-1} liquid velocity with a sample load of 2100 mg BSA. The total cycle was comprised of sample application, washing, elution and re-equilibration. The results are summarised in Table 4. These data show that there is a significant increase in process time and process volume when the process was performed with a large distance between the bed surface and the upper adapter. This increase occurred during the washing, the elution and the re-equilibration steps of the process, but was highest during re-equilibration, due to increased mixing when changing from 0.5 M NaCl to the less dense and less viscous Tris buffer. Operating the expanded bed adsorption process under such conditions results in a largely increased buffer consumption (3.3-fold) and the subsequent increased volumes of waste cannot usually be accepted. An alternative is to perform the entire expanded bed process at constant bed expansion, keeping the liquid velocity

Table 3. Summary of data for elution in expanded and packed bed modes under different conditions

Mode of operation	Sample load (mg)	Liquid velocity (cm h^{-1})	Peak volume (ml)[a]
Expanded bed	30	109	61
	30	208	75
	30	305	90
	300	110	58
	300 [b]	110	76
	300 [b]	109 [c]	83
	300	208	73
	300	303	97
	2100	114	110
	2100	208	134
	2100	305	150
Packed bed	300	101	38
	300	293	57
	2100	104	77
	2100	315	83

[a] Determined at 10% of the maximum peak height.
[b] Adsorbent surface/upper adapter distance 26 cm.
[c] Elution performed using 1 M NaCl.

at a value which maintains the bed surface close to the upper adapter, thereby minimising the mixing volume above the bed surface. This mode of operation has been studied by Chang and Chase (1996b) and it was shown to increase productivity approximately 15% (calculated as mg protein per adsorbent volume per time unit) when lysozyme was used as a model in an expanded bed process. If an expanded bed is operated at constant bed expansion the relative buffer consumption increases from 1 to 1.8 (Table 4). Assuming buffer costs to be \$0.25 l^{-1}, an increase in buffer consumption from 1000 to 1800 l would raise costs per cycle by \$200. One consequence of running at a constant bed expansion is that the liquid velocity during sample application has to be adjusted to the density and viscosity of the feedstock to avoid that the adsorbent pack against the upper adapter. It is usually possible to operate an expanded bed at a constant bed expansion (factor of 2) at approx. 100 cm h^{-1} for most feedstocks. Recently Chang and Chase (1996a) showed that it was possible to apply a 25% wet weight yeast cell homogenate onto a bed expanded by a factor of 2 at a liquid velocity of 66 cm h^{-1}. A more advanced approach keeping an appropriate distance between the bed surface and the upper adapter has recently been described by Thelen et al. (1997). They used an ultrasonic sensor to monitor the position of the bed surface in an expanded bed in the presence of a cell-containing feedstock. Such devices might be useful for optimal operation during elution in expanded bed mode when the process is operated at constant bed expansion.

Conclusions

Different salt concentrations do not have any significant effect on bed expansion, however, a slight decrease in mixing in the liquid phase, measured as axial dispersion, could be observed with increased salt concentration. These findings show that the buffer composition itself will not have any adverse effect on the mixing during elution in expanded bed mode. Mixing was considerable when measured in the zone of the eluted protein, in particular at high liquid velocities (300 cm h^{-1}) and at high sample loads (approx.

Table 4. The effect of different modes of operation in expanded bed adsorption on process time and process volume

Mode of operation	Bed surface/adapter distance (cm)	Total process time (min)	Relative process time	Buffer consumption per cycle (ml)	Relative process volume
Packed	0	74	1	825	1
Expanded	3.5	101	1.4	1500	1.8
Expanded	52	151	2.0	2750	3.3

25 mg BSA/ml adsorbent). This indicates that the liquid velocity should not exceed 300 cm h^{-1} during elution in expanded bed mode as there is a risk that the mixing will reach values which will affect the volume of the eluted peak. The volume of the eluted peak increased by approximately 80% when elution was carried out in expanded bed mode compared to elution in packed bed mode, providing that the liquid volume between the bed surface and the upper adapter was kept small. The economical effect on the process of such an increased volume may in most cases be reasonable. Improved monitoring and automation of the expanded bed process can most probably minimise the dilution effects of elution in expanded bed mode and thereby making the operational advantages of this mode of operation also economically beneficial.

Acknowledgement

I wish to express my gratitude to Dr. Lars Hagel for his valuable comments and suggestions and for his support with statistical analysis of RTD data.

References

Amersham Pharmacia Biotech. Expanded Bed Adsorption. Principles and Methods.

Barnfield Frej A-K, Hjorth R & Hammarström Å (1994) Pilot scale recovery of recombinant annexin V from unclarified *Escherichia coli* homogenate using expanded bed adsorption. Biotechnol. Bioeng. 44: 922–929.

Barnfield Frej A-K, Johansson HJ, Johansson S & Leijon P (1997) Expanded bed adsorption at production scale: scale-up verification, process example and sanitization of column and adsorbent. Bioprocess Eng. 16: 57–63.

Batt BC, Yabannavar VM & Singh V (1995) Expanded bed adsorption process for protein recovery from whole mammalian cell culture broth. Bioseparation 5: 41–52.

Chang YK & Chase HA (1996a) Ion exchange purification of G6PDH from unclarified yeast cell homogenates using expanded bed adsorption. Biotechnol. Bioeng. 49: 204–216.

Chang YK & Chase HA (1996b) Development of operating conditions for protein purification using expanded bed techniques: the effect of the degree of bed expansion on adsorption performance. Biotechnol. Bioeng. 49: 512–526

Chase HA (1994) Purification of proteins by adsorption chromatography in expanded beds. Trends Biotech. 12: 296–303.

Dasari G, Prince I & Hearn MTW (1993) High-performance liquid chromatography of amino acids, peptides and proteins. CXXIV: Physical characterisation of fluidized-bed behaviour of chromatographic packing materials. J. Chromatogr. 631: 115–124.

Draeger NM & Chase HA (1991) Liquid fluidized bed adsorption of proteins in the presence of cells. Bioseparation 2: 67–80.

Goto M, Imamura T & Hirose T (1995) Axial dispersion in liquid magnetically stabilized fluidized beds. J. Chromatogr. 690: 1–8

Hjorth R (1997) Expanded-bed adsorption in industrial bioprocessing: recent developments. Trends Biotech. 15: 230–235.

Hjorth R, Kämpe S & Carlsson M (1995) Analysis of some operating parameters of novel adsorbents for recovery of proteins in expanded beds. Bioseparation 5: 217–223.

Johansson HJ, Jägersten C & Shiloach J (1996) Large scale recovery and purification of periplasmic recombinant protein from *E. coli* using expanded bed adsorption chromatography followed by new ion exchange media. J. Biotechnol. 48: 9–14.

Karau A, Benken C, Thömmes J & Kula MR (1997) The influence of particle size distribution and operating conditions on the adsorption performance in fluidized beds. Biotechnol. Bioeng. 55: 54–64.

Levenspiel O (1972) Chemical Reaction Engineering. Wiley, New York.

Lindgren A, Johansson S & Nyström L-E (1993) BED -Vol 27. In: B. Henon (ed.), Bioprocess Engineering (pp 27–30) The American Society of Mechanical Engineers, New York.

Noda M, Sumi A, Ohmura T & Yokoyama K (1996) European Patent Application EP 069 968 7A2.

Richardson JF & Zaki WN (1954) Sedimentation and fluidization: Part 1. Trans. Inst. Chem. Eng. 32: 35–53.

Slater MJ (1991) The Principles of Ion Exchange Chromatography. Butterworth-Heinemann, Oxford.

Sofer G & Hagel L (1997) Handbook of Process Chromatography: a Guide to Optimization, Scale Up and Validation. Academic Press, San Diego.

Swaine DE & Daugulis AJ (1988) Review of liquid mixing in packed bed biological reactors. Biotechnol. Prog. 4: 134–148.

Thatcher DR, Hitchcock AG, Hanak JA & Varley DL (1997) Patent Application WO 97/29 190.

Thelen TV, Mairal AP, Thorsen CS & Ramirez WF (1997) Application of ultrasonic backscattering for level measurement and process monitoring of expanded-bed adsorption columns. Biotechnol. Prog. 13: 681–687.

Thömmes J (1997) Fluidized bed adsorption as a primary recovery step in protein purification. In: T. Scheper (ed.) Advances in Biochemical Engineering/Biotechnology, Vol 58 (pp. 185–230) Springer-Verlag, Berlin.

Thömmes J, Weiher M, Karau A & Kula MR (1995) Hydrodynamics and performance in fluidized bed adsorption. Biotechnol. Bioeng. 48: 367–374.

Thömmes J, Bader A, Halfar M, Karau A and Kula MR (1996) Isolation of monoclonal antibodies from cell containing hybridoma broth using a protein A coated adsorbent in expanded beds. J. Chromatogr. 752: 111–122.

Zapata G, Lindgren A, Barnfield Frej A-K, Leijon P, Liten AD, Mayes TL, Blank G, Narindray D, Wagner B, Galan W, Beck J, Press M and Builder S (1996) Abstract, Recovery of Biological Products, Tucson, Az., p. 16.

Bioseparation **8:** 11–31, 1999.
© 1999 *Kluwer Academic Publishers.*

Monitoring, modeling, and control strategies for expanded-bed adsorption processes

Travis V. Thelen & W. Fred Ramirez*
University of Colorado, Department of Chemical Engineering, Boulder, CO 80309-0424, U.S.A.

Received 19 August 1998; accepted in revised form 2 May 1999

Key words: fluid dynamics, regulation, IMC-based PID control, two-phase flow theory

Abstract

Expanded-bed adsorption (EBA) is a technique for the purification of proteins from cellular debris in downstream processing. An expanded bed presents the possibility of protein recovery in a single step, eliminating the often costly clarification processing steps such as ultrafiltration, centrifugation, and precipitation. An obstacle to the successful commercialization of this technology is the inability to accurately monitor and control the bed height in these systems.

In this paper, we present an overview of work in our laboratory addressing monitoring, modeling, and control strategies as applied to EBA. First, we present the development of a level measurement technique based upon ultrasonics. It is shown that this technique has great promise for bed-height measurement in EBA systems. Second, we present modeling strategies for bed-height dynamics due to flow rate and fluid property changes, and lastly, we show how monitoring and modeling information can be used for the control and regulation of bed expansion.

Introduction

Most recombinantly derived proteins are intracellular and hence require lysing. For soluble proteins, the feedstock is generally clarified before purification by packed-bed chromatography to eliminate fouling of the resin. The most widely employed techniques for clarification are microfiltration and centrifugation (Lee, 1989). These clarification steps are typified by low product yields, negligible increases in product concentration, and high costs (Hjorth et al., 1995). Expanded-bed adsorption, alternatively, presents the possibility of protein recovery in a single step (e.g., Galliot et al., 1990), eliminating the clarification processing steps.

In the EBA process, the upward flow of the unclarified feedstock fluidizes the resin. In the fluidized state, there is sufficient void volume to allow the free passage of particulate debris. Back-mixing is minimized in these system by using resin with a distribution of diameters. As such, plug flow behavior is achieved with minimal axial dispersion.

In the following sections, we discuss monitoring, modeling, and control strategies for the EBA process.

Ultrasonic monitoring of EBA columns

Theory

Ultrasonic waves behave similar to sound waves, but are characterized by a much shorter wavelength. Typically, ultrasonics denotes vibrations above 20 kHz. Because of this short wavelength, ultrasound is readily reflected off of small surfaces. For this reason, ultrasound can be used to image or detect objects.

The backscattered signal can readily be detected experimentally by the pulse-echo technique (Fay, 1991). This technique involves generating a high-voltage pulse with a pulser-receiver, which is connected to a sending transducer. The transducer converts the excitation pulse into ultrasonic energy. The resulting ultrasonic wave (longitudinal) travels through

* To whom all correspondence should be addressed

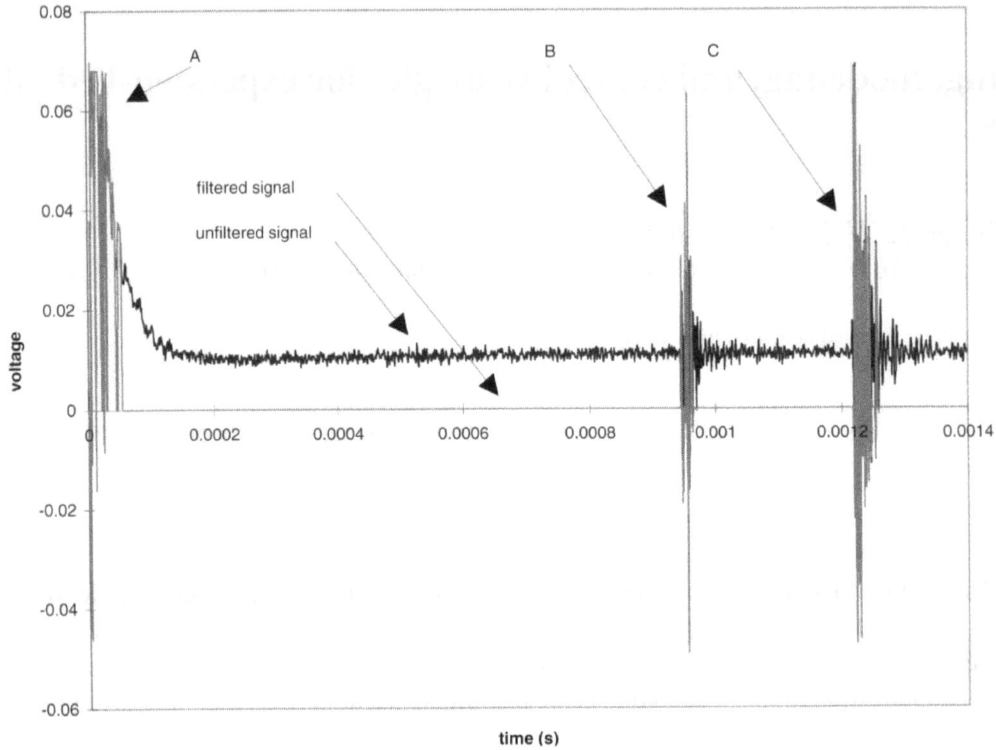

Figure 1. Oscilloscope sweep plot generated by the pulse-echo technique showing unfiltered and filtered signals for particulate-free feedstock. Signal-to-noise ratio of 1.3 dB (unfiltered signal). (A) pulse signal, (B) echo signal from solid-liquid interface, and (C) echo signal from column bottom.

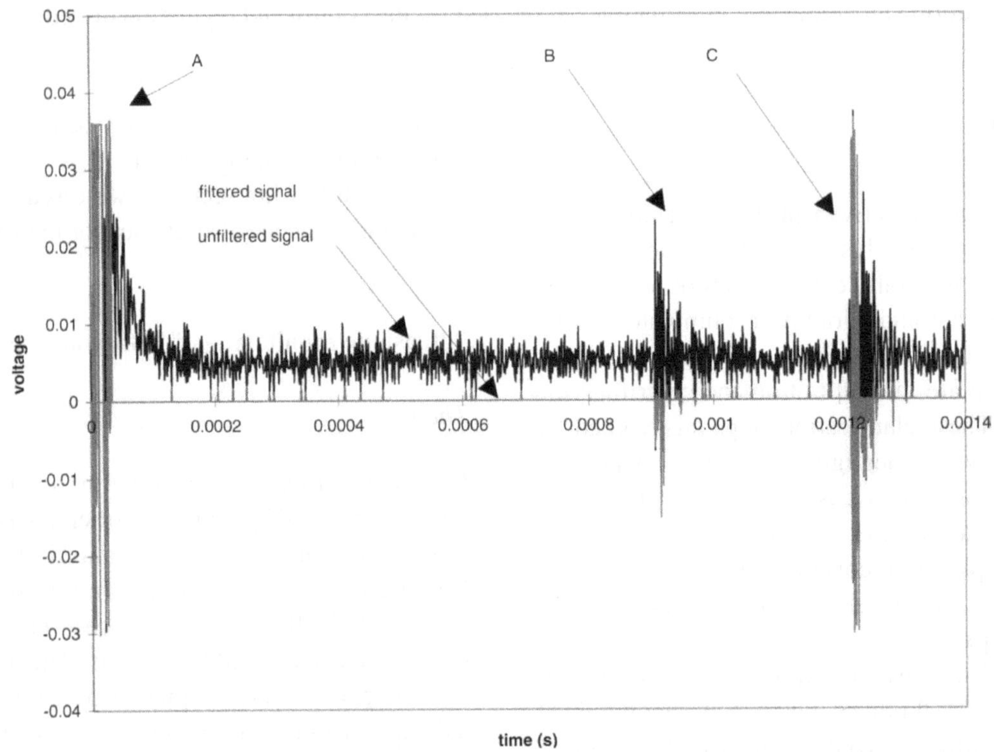

Figure 2. Oscilloscope sweep plot generated by the pulse-echo technique showing unfiltered and filtered signals for particulate-laden feedstock. Signal-to-noise ratio of 0.5 dB (unfiltered signal). (A) pulse signal, (B) echo signal from solid-liquid interface, and (C) echo signal from column bottom.

the column until it is reflected at an interface. The interface also causes some of the acoustic signal to scatter which diminishes the signal strength. The reflected ultrasonic wave returns to the receiver where the transducer converts the ultrasonic energy into electrical energy. This signal is then sent to an oscilloscope for processing. Figures 1 and 2 show typical oscilloscope sweep plots of an EBA column with different signal-to-noise ratios from pulse-echo experiments. In these figures, (A) shows the initial pulse, while (B) and (C) show the echoes from the solid-liquid interface and the bottom of the column, respectively.

The distance from the transducer to a reflecting interface can be determined as follows:

$$d = \frac{v_s t}{2} \tag{1}$$

where d is the distance to the interface, v_s is the velocity of sound in the fluid, and t is the time between the pulse and echo. The longitudinal velocity of acoustic waves in water at 20 °C is 148,000 cm/s (Panametrics, 1995).

Equipment

An expanded-bed column was designed and constructed in our laboratory. The column was made from polycarbonate. It has an internal diameter of 8 cm. There is a 40 μm stainless-steel mesh situated between the main column and the flow distributor regions. The mesh is designed to allow the free passage of cells and cellular debris while retaining the resin. The flow distributor consists of glass shot and Raschig rings. The experimental system is shown schematically in Figure 3.

Poly-methylmethacrylate was used in this study. The resin is spherical in shape and has an approximate density of 1.2 g/cm^3. The resin has a mean diameter of 330 μm. The physical properties of this resin are similar to commercially available adsorbents.

Two feedstocks were used in this study. The particulate-free feedstock used was deionized water. The particulate-laden feedstock used in the experiments included *E. coli* grown in a minimal media consisting of 6 g tryptone (Difco Laboratories), 3 g yeast extract (Difco Laboratories), and 8 g NaCl (Fisher) per liter of deionized (DI) water. The cells were allowed to grow until plateau phase. The cellular concentration was approximately $1x10^9$ cells/mL as measured by a hemacytometer. This cellular concentration is typical of an industrial fermentation (e.g., Bailey and Ollis, 1986; Gailliot et al., 1990).

A Panametrics model 5072PR pulser-receiver was used to excite a piezoelectric transducer and receive the reflected ultrasonic signal. The transducer used in this study was a 500 kHz immersion-style transducer (Panametrics). A 12-bit digital oscilloscope (Nicolet) was used for signal processing.

Methods and results

In order to determine the bed height as a function of time, pulse-echo experiments were performed at regular intervals for bed expansion-contraction experiments. Bed-height transients were introduced by making step changes in the inlet fluidization velocity. From these experiments, bed height can be determined by taking the difference between the total column height (L) and the distance to the scattering interface as measured by (1). The bed height is calculated as

$$H = L - \frac{v_s t}{2} \tag{2}$$

Computer-based signal processing of the oscilloscope sweep plots requires the development of algorithms to determine the time elapsed between pulse and echo signals. In order to determine this time, the following cut-off filter was implemented:

1. Determine the baseline voltage (e.g., Figures 1 and 2 show a baseline voltage of 0.01 V and 0.005 V, respectively).
2. Search in the allowable domain (defined by physical bed-height constraint limits) of the voltage signal for the first value outside a user specified band around the baseline voltage and record time. The mathematical implementation of this filter is as follows:

$$\text{if} \quad V_{bl} - V_c \leq V(t) \leq V_{bl} + V_c \tag{3}$$

$$\text{then} \quad V(t) = 0$$

$$\text{else} \quad V(t) = V(t) \tag{4}$$

   where $V(t)$ is the voltage, V_{bl} is the baseline voltage, and V_c is the cut-off voltage.
3. Calculate the bed height using (2).

The band ($\pm V_c$) around the baseline voltage is a user-defined processing parameter and proper choice is very critical for accurate bed-height determination. In choosing this parameter, there is a trade-off between accurately identifying the echo signal and falsely assigning a noisy baseline signal as the echo signal. False echo identifications are magnified as the signal-to-noise ratio decreases. Figure (1) shows both the filtered and unfiltered signals from a pulse-echo experiment with particulate-free feedstock. The filtered

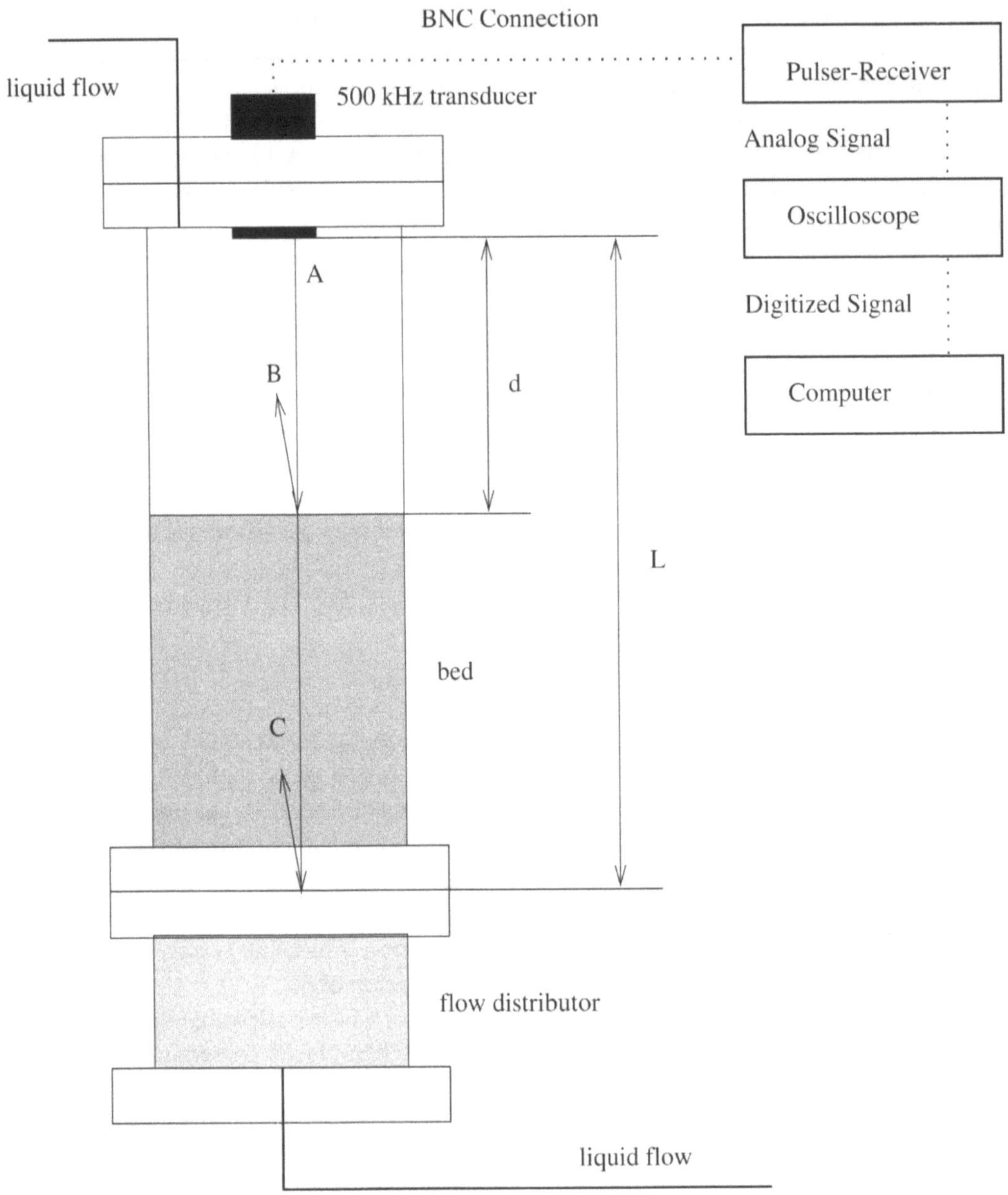

Figure 3. Schematic of experimental set-up showing acoustic signals and relevant distances.

signal was obtained by implementing the above algorithm and setting the voltage to zero if it lied within the band. A relatively high signal-to-noise ratio of 1.3 dB was obtained. Figure 2 shows the analogous plot, but for particulate-laden feedstock. The signal-to-noise ratio was diminished considerably, to a value of 0.5 dB, by the presence of cells. This reduction in the signal to noise ratio makes it much more difficult to accurately identify the echo signal. Figures 4 and 5 show comparisons of bed height as measured by eye and ultrasonically for particulate-free and particulate-laden feedstocks, respectively. The acoustic signal was sampled at a rate of 1 MHz, which corresponds to a bed-height resolution of 0.074 cm. For the particulate-free study, the visual measurements agreed with the ultrasonic measurement to within $\pm$ one sample interval, while for the particulate-laden case the visual and ultrasonic measurements agreed to within $\pm$ two sample intervals.

Measurement resolution is a function of two variables; the sampling rate and echo identification. Sampling rate places a theoretical limit on the measurement resolution, as determined by (1). The resolution can be increased by sampling at a higher rate, but there is a trade-off. To achieve higher sampling rates, one must use a high-speed oscilloscope which is more costly. If sampling is carried out using a personal computer, sampling rates are limited to around 1.0–

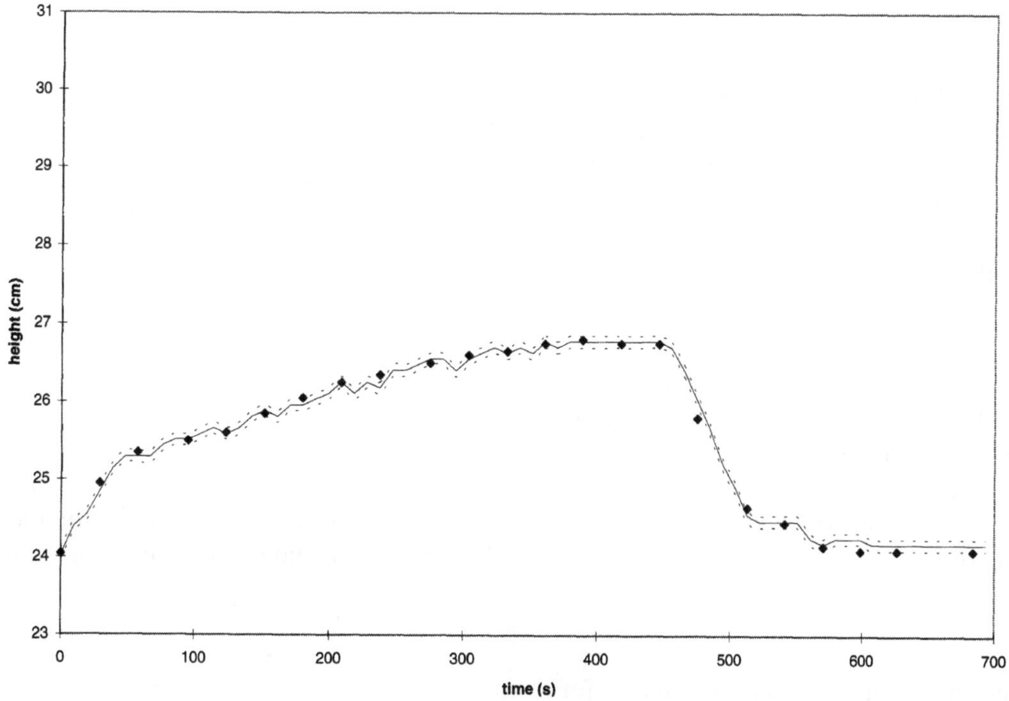

Figure 4. Comparison of bed height as measured by eye and ultrasonically for particulate-free feedstock. Visual measurement (♦), ultrasonic measurement (—), and ultrasonic measurement error bars (– –) at ± one sample interval (0.074 cm).

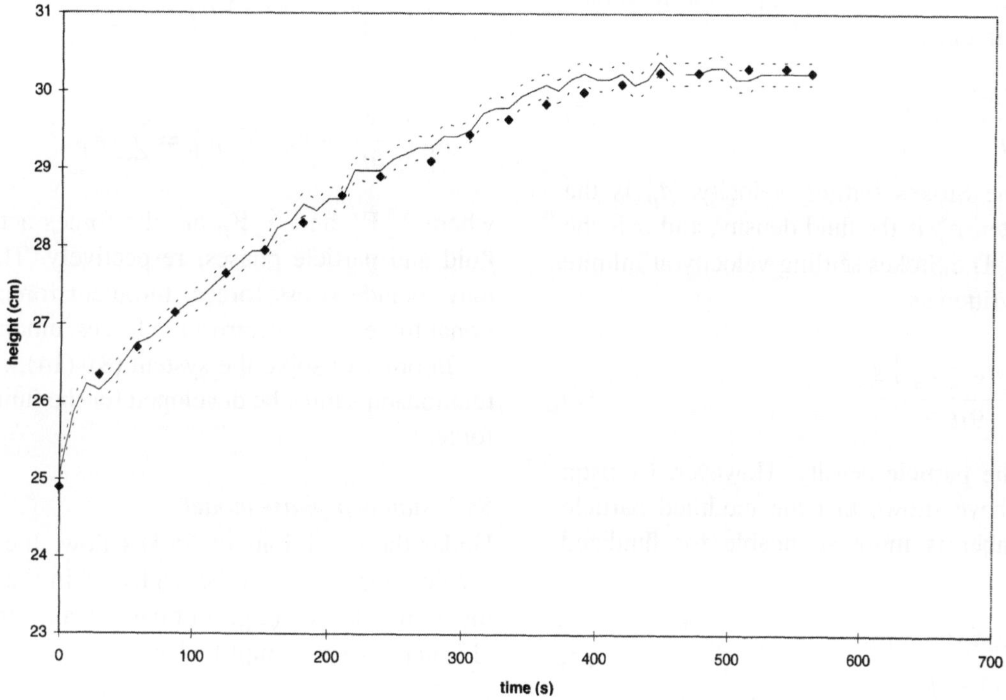

Figure 5. Comparison of bed height as measured by eye and ultrasonically for particulate-laden feedstock. Visual measurement (♦), ultrasonic measurement (—), and ultrasonic measurement error bars (– –) at ± two sample intervals (0.15 cm).

1.25 MHz with an EISA computer bus. The sampling rate attainable with a personal computer provides adequate resolution for most level-detection applications in EBA columns.

Modeling of solid-liquid fluidization

A number of articles have been presented on the modeling of bed-height dynamics of solid-liquid fluidized beds (Slis et al., 1959; Fan et al., 1963; Gibilaro et al.,

1984; Poncelet et al., 1990; Asif et al., 1994; Thelen and Ramirez, 1997); however, these papers have only addressed the bed-height dynamics for changes in the fluidization velocity. As discussed by De Luca et al. (1994), the characteristics of the bed-height response due to changes in the fluidization velocity and that due to changes in the fluid properties differ markedly. In the case of the latter, there is a significant time delay before the bed responds to changes in the fluid properties.

We now present the two-phase flow equations for solid-liquid fluidization. A comparison between experimental and simulation results is then presented for two flow configurations. The first considers changes in the fluidization velocity; the second looks at changes in the fluid properties.

Theory

This section develops the governing equations for solid-liquid fluidization where the inertial effects can be neglected. That is, the analysis is restricted to the Stokes regime, as defined by a particle Reynolds number less than unity

$$Re_p = \frac{v_t d_p \rho_f^*}{\mu} < 1 \tag{5}$$

where v_t is the Stokes settling velocity, d_p is the particle diameter, ρ_f^* is the fluid density, and μ is the fluid viscosity. The Stokes settling velocity at infinite dilution v_t is written as

$$v_t = \frac{2r_p^2\left(\rho_p^* - \rho_f^*\right)g}{9\mu} \tag{6}$$

where ρ_p^* is the particle density. However, Coulson et al. (1991) have shown that the modified particle Reynolds number is more applicable for fluidized systems.

$$Re_{p_m} = \frac{v_o d_p \rho_f^*}{\mu(1-\phi)} \tag{7}$$

where v_o is the superficial fluidization velocity and ϕ is the particle-phase fraction.

General Two-Phase Model

The fluid-phase and solid-phase continuity equations, respectively, are

$$\frac{\partial \rho_f}{\partial t} + \nabla \cdot \left(\rho_f \mathbf{v_f}\right) = 0 \tag{8}$$

$$\frac{\partial \rho_p}{\partial t} + \nabla \cdot \left(\rho_p \mathbf{v_p}\right) = 0 \tag{9}$$

where ρ_f and ρ_p are the in-suspension densities and $\mathbf{v}_f$ and $\mathbf{v}_p$ are the local fluid and particle velocities, respectively. The in-suspension densities ρ_f and ρ_p are related to the intrinsic densities of the fluid and solid as follows

$$\rho_f = \epsilon \rho_f^* \tag{10}$$

$$\rho_p = \phi \rho_p^* \tag{11}$$

where ϵ is the fluid-phase fraction and ϕ is the solid-phase fraction. At any point in the mixture, the phase fractions for the liquid and solid sum to unity.

$$\epsilon + \phi = 1 \tag{12}$$

The linear momentum equations for the fluid and solid phases, respectively, are written as

$$\rho_f\left(\frac{\partial \mathbf{v}_f}{\partial t} + \mathbf{v}_f \cdot \nabla \mathbf{v}_f\right) = \sum \mathbf{F}_f \tag{13}$$

$$\rho_p\left(\frac{\partial \mathbf{v}_p}{\partial t} + \mathbf{v}_p \cdot \nabla \mathbf{v}_p\right) = \sum \mathbf{F}_p \tag{14}$$

where $\sum \mathbf{F}_f$ and $\sum \mathbf{F}_p$ are the forces acting on the fluid and particle phases, respectively. These forces may include stress forces, turbulent forces, gravitational forces, and electrostatic forces, among others.

In order to solve the system (8)–(14), constitutive relationships must be developed for the fluid and solid forces.

Stokesian two-phase model

Under the conditions of Stokes flow, the inertial or acceleration terms can be neglected in the linear momentum balances (e.g., Gibilaro et al., 1984), hence (13) and (14) are simplified as

$$\sum \mathbf{F}_f = \mathbf{0} \tag{15}$$

$$\sum \mathbf{F}_p = \mathbf{0} \tag{16}$$

It is postulated that there are three important forces to consider, namely drag, gravitational, and diffusive.

Under steady-state conditions, the drag force on an individual particle is balanced by its buoyant force.

The steady-state drag force exerted on a particle by the fluid can be written as

$$F_d = \frac{6\pi \mu r_p v_o}{(1-\phi)^n} \tag{17}$$

where r_p is the particle radius and the hindered settling function (Davis, 1995) is given by the Richardson-Zaki expression (Richardson and Zaki, 1954). If (17) is written in terms of the velocity of the fluid relative to the sphere (i.e., $v_r = v_f - v_p$)(Stokes, 1901), the drag force is expressed as

$$F_d = \frac{6\pi \mu r_p v_r}{(1-\phi)^{n-1}} \tag{18}$$

The buoyant force on an individual particle is given by

$$F_b = \frac{4\pi r_p^3 \left(\rho_p^* - \rho_f^*\right) g}{3} \tag{19}$$

where g is the gravitational constant. At steady-state, the drag (17) and buoyant (19) forces are balanced, thus

$$\frac{v_o}{v_t} = (1-\phi)^n \tag{20}$$

Equation (20) is valid for steady-state conditions. For the case where the solid velocity is not zero, the equivalent expression is given by

$$\frac{v_r}{v_t} = (1-\phi)^{n-1} \tag{21}$$

which is a balance of the dynamic drag force (18) and the buoyant force (19).

The parameters, v_t and n, can be determined either experimentally or from theoretical considerations. From steady-state expansion data, a log-log plot of fluidization velocity versus void fraction can be used to regress the parameters. The Stokes velocity can also be determined by substitution of measured values of the physical properties into (6).

Dispersion is also an important factor. Truesdell (1962) has argued that diffusion (dispersion), being a change of motion, arises from forces. And that the motions produced by dispersion forces must conform to the principle of linear momentum.

Peddieson (1974) has shown a parameterization of solid-fluid forces based upon concepts from dusty gas mechanics (Marble, 1970)

$$\sum F_p = \beta_1 \left(v_f - v_p\right) - \beta_2 \nabla \rho_p + \rho_p g \tag{22}$$

The parameters β_1 and β_2 are given by

$$\beta_1 = \frac{\rho_p}{\tau} \tag{23}$$

$$\beta_2 = \frac{D_p}{\tau} \tag{24}$$

where D_p is the solid-phase dispersion coefficient and τ is a relaxation time. Based upon extensions of dusty gas mechanics, the relaxation parameter for dense flows is

$$\tau = \frac{m_b}{6\pi r_p \mu/(1-\phi)^{n-1}} \tag{25}$$

where m_b is the buoyant mass of the particle, namely

$$m_b = \frac{4\pi r_p^3 \left(\rho_p^* - \rho_f^*\right)}{3} \tag{26}$$

The Stokes drag coefficient is divided by the hindered settling function to account for the effects of particle loading.

Thus, for dense flows, the solid-phase linear momentum balance is written as

$$\sum F_p = 0 = \frac{\rho_p}{\tau}\left(v_f - v_p\right) - \frac{D_p}{\tau}\frac{\partial \rho_p}{\partial z} - \rho_p g \tag{27}$$

where one-dimensional axial flow is assumed. If the intrinsic densities of both the fluid and solid are constant and the Richardson-Zaki hindered settling function is valid, then (27) is written as

$$\phi\left(v_f - v_p\right) - \phi v_t (1-\phi)^{n-1} - D_p \frac{\partial \phi}{\partial z} = 0 \tag{28}$$

Now, the continuity equations are utilized in order to obtain a relationship between the fluid velocity, v_f, and the particle velocity, v_p. Adding the liquid-phase and solid-phase continuity equations yields

$$\frac{\partial\left(\epsilon v_f\right)}{\partial z} + \frac{\partial\left(\phi v_p\right)}{\partial z} = 0 \tag{29}$$

Integrating (29) yields

$$\epsilon v_f + \phi v_p = k \tag{30}$$

where k is a constant of integration. It is known when $v_p = 0$ that $v_f = \frac{v_o}{\epsilon}$ thus allowing evaluation of the constant, which yields

$$\epsilon v_f + \phi v_p = v_o \tag{31}$$

Substitution of (31) into (28) yields, upon rearrangement, the expression

$$\phi v_p = \phi v_o - \phi v_t (1-\phi)^n - (1-\phi) D_p \frac{\partial \phi}{\partial z} \tag{32}$$

The relationship (32) is substituted into the solid-phase continuity equation (9) giving

$$\frac{\partial \phi}{\partial t} = -\frac{\partial \left(\phi v_s - (1-\phi) D_p \frac{\partial \phi}{\partial z} \right)}{\partial z} \tag{33}$$

where v_s is defined as

$$v_s \equiv v_o - v_t (1-\phi)^n \tag{34}$$

Equation (33) is a convection-dispersion model which has been derived from two-phase flow theory. This analysis provides a link between two-phase modeling and flux-based modeling, though the two approaches result in different final forms.

Boundary conditions of the Dankwerts type are assumed, namely

$$\phi v_s - (1-\phi) D_p \frac{\partial \phi}{\partial z} = 0 \quad @ \quad z = 0 \tag{35}$$

$$\phi v_s - (1-\phi) D_p \frac{\partial \phi}{\partial z} = \phi \frac{dh}{dt} \quad @ \quad z = h \tag{36}$$

where $\frac{dh}{dt}$ is the bed-height velocity. The system (33)–(36) is a nonlinear PDE, two-point boundary value problem with a non-stationary upper boundary.

A comparison between experimental and simulation results is now presented for two flow configurations. The first considers changes in the fluidization velocity; the second looks at changes in the fluid properties. The simulation results are based upon solution of (33)–(36) for the given flow configuration.

Fluid flow rate change

We consider the following problem. A bed of spherical particles of radius r_p and density ρ_p^* is initially fluidized by a fluid of viscosity μ and density ρ_f^* at a superficial velocity v_1. At steady state under these conditions, the solid fraction ϕ_1 is determined from

$$\frac{v_1}{v_t} = f_1(\phi_1) \tag{37}$$

where the terminal settling velocity of a single particle in the Stokes flow regime is

$$v_t = \frac{2 r_p^2 \left(\rho_p^* - \rho_f^* \right) g}{9 \mu} \tag{38}$$

and the hindered settling function is given by the Richardson-Zaki expression

$$f_1(\phi_1) = (1-\phi_1)^n \tag{39}$$

The steady-state height of the bed under these conditions is determined from a particle mass balance

$$h_1 = \frac{\phi_s h_s}{\phi_1} \tag{40}$$

where ϕ_s and h_s are the solid fraction and height of the settled bed, respectively, before fluidization.

Now, suppose that the fluid entering the bottom of the bed experiences a step change in the fluidization velocity at $t = 0$, so that the bed adjusts to a new steady state given by (37)–(40) but with the subscript "2" (for the new flow conditions) replacing "1". The question to be addressed is how the transient occurs as the bed moves to the new steady state. The two-phase model (33)–(36) is applied in order to address this question.

Experimental system

A STREAMLINE 50 column (5 cm ID) was used in the fluidization studies with STREAMLINE DEAE absorbent (Pharmacia Biotech). The resin has a mean diameter of 200 μm and a mean density of 1.2 g/cm^3. Feedstock was delivered to the column by means of a computer-controlled peristaltic pump (Cole-Parmer). A 20% aqueous ethanol solution with an approximate density of 0.98 g/cm^3 and a viscosity of 0.011 g/cm ·s at 25°C was used to fluidize the resin.

The bed was considered to be in a stable state of fluidization if channeling was observed to be minimal in the lower region of the column. The steady-state bed height was recorded for a range of flow rates. The Richardson-Zaki parameters, n and v_t, were determined from linear regression of log-log plots of v_o versus ϵ. The Stokes settling velocity was also determined using (6) and measured properties. The regressed and theoretical Stokes values differ considerably, with values of 0.17±0.01 cm/s and 0.44±0.13 cm/s, respectively. The regressed value for the Richardson-Zaki exponent is 4.31.

Discussion

Figure (6) shows a comparison of simulations and the experimental data for five experimental runs where the volumetric flow rate was increased from 15 to 20, 25, 35, 45, and 55 mL/min. The two-phase model shows excellent agreement with the experimental data. Step

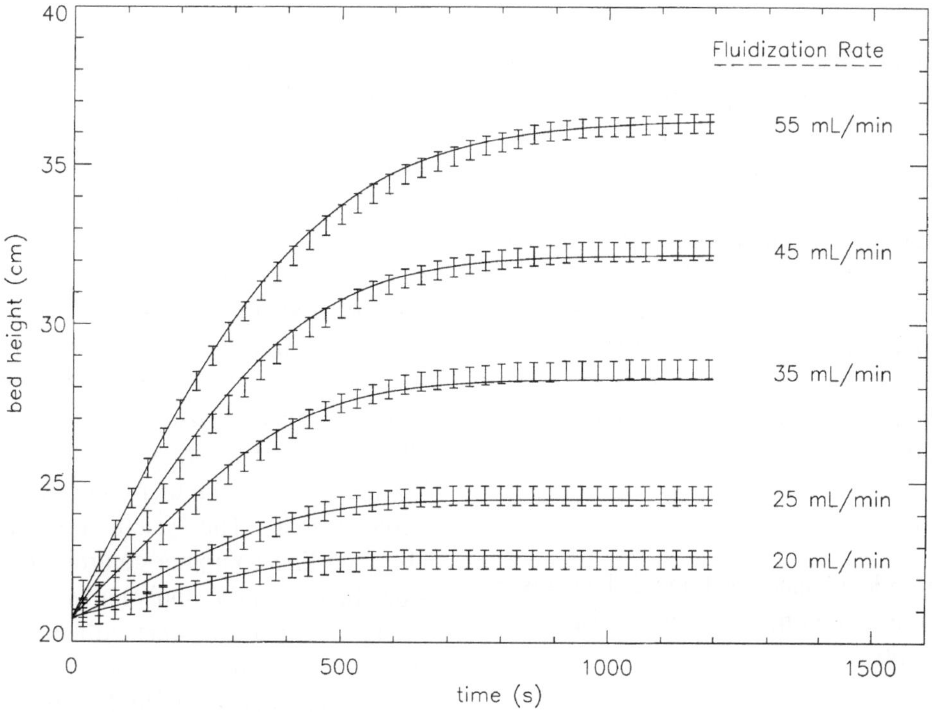

Figure 6. Dynamic response of bed height to a series of step increases in the fluidization velocity. An initial fluidization velocity of 15 mL/min was used. (—) two-phase model and symbols are the experimental data with error bars of ±3 mm.

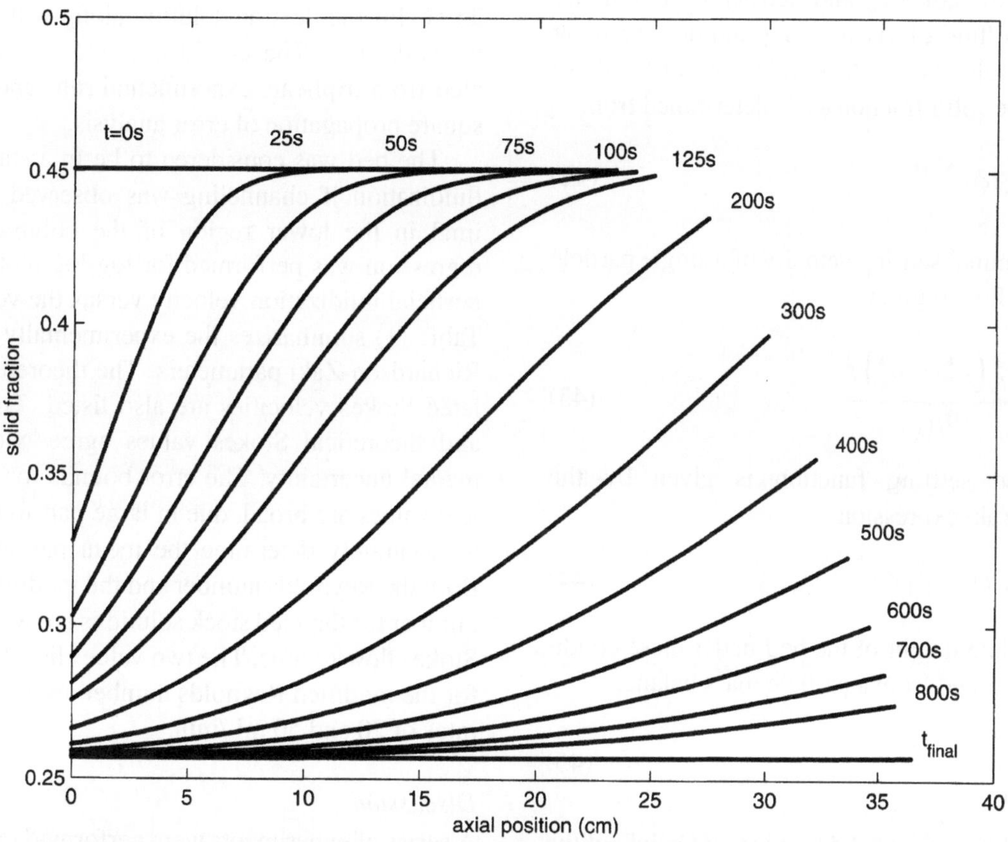

Figure 7. A plot of the solid fraction versus axial position for a step change from water to the dextran solution at 40 mL/min.

decreases in the fluidization velocity were also studied (Thelen and Ramirez, 1998).

Figure (7) shows the axial profile of the solid fraction at various times for a step increase in the fluidization velocity. The results indicate a smooth transition between the initial and final axial solid fraction profiles.

The solid-phase dispersion coefficient was treated as a free parameter and fitted by least-squares regression. Table (2) shows a listing of the solid-phase dispersion coefficients and sensitivity limits. In the porous flow literature, the dispersion coefficient is often modeled by a power law function of the form (Ramirez et al., 1980)

$$D_p = D(v_f)^{\alpha} \tag{41}$$

where α has theoretical bounds of 1 and 2. The power law model was found to fit the solid-phase dispersion coefficient data well.

Fluid property changes

The following situation is considered. A bed of spherical particles of radius r_p and density ρ_p^* is initially fluidized by a fluid of viscosity μ_1 and density ρ_1^* at a superficial velocity v_o. At steady state under these conditions, the solid fraction ϕ_1 is determined from

$$\frac{v_o}{v_{t,1}} = f_1(\phi_1) \tag{42}$$

where the terminal settling velocity of a single particle in the Stokes flow regime is

$$v_{t,1} = \frac{2r_p^2\left(\rho_p^* - \rho_1^*\right)g}{9\mu_1} \tag{43}$$

The hindered settling function is given by the Richardson-Zaki expression

$$f_1(\phi_1) = (1 - \phi_1)^n \tag{44}$$

The steady-state height of the bed under these conditions is determined from a particle mass balance

$$h_1 = \frac{\phi_s h_s}{\phi_1} \tag{45}$$

where ϕ_s and h_s are the solid fraction and height of the settled bed, respectively, before fluidization.

Now, suppose that the fluid entering the bottom of the bed experiences a step change in fluid properties (viscosity and/or density) at $t = 0$, so that the bed adjusts to a new steady state given by (42)–(45) but with the subscript "2" (for the new fluid properties) replacing "1". The question to be addressed is how the transient occurs as the bed moves to the new steady state. The simplified two-phase model (33)–(36) is applied in order to address this question.

Experimental system

A STREAMLINE 50 column (5 cm ID) was used in the fluidization studies with STREAMLINE SP adsorbent. The resin has a mean diameter of 186 μm and a mean density of 1.184 g/cm^3 (Chang and Chase, 1996). Feedstock was delivered to the column by means of a computer-controlled peristaltic pump (Cole-Parmer). The system was configured so as to allow on-line computer-controlled switching of feed stock between two supply tanks.

Two feed stocks were utilized in the experiments in order to determine the bed-expansion characteristics for step changes in fluid properties. In all studies, the bed was initially fluidized with deionized (DI) water then a step change was made to an aqueous dextran solution. Table (1) shows physical properties at 25 °C for the dextran solution along with confidence intervals (CI). The confidence intervals were evaluated from triplicate experimental runs and root-mean square propagation of error analysis.

The bed was considered to be in a stable state of fluidization if channeling was observed to be minimal in the lower region of the column. A linear regression was performed for log-log plots of the superficial fluidization velocity versus the void fraction. Table (1) summarizes the experimentally determined Richardson-Zaki parameters. The theoretically calculated Stokes velocities are also listed. The regressed and theoretical Stokes values agree within experimental uncertainty. The error bounds for the theoretical values are broad, due in large part to the inability to accurately determine the mean particle diameter. Both the Reynolds number and the modified Reynolds number for the feed stock solutions are well within the Stokes flow regime. The two values listed in Table (1) for the modified Reynolds number are for fluidization rates of 20 and 50 mL/min.

Discussion

A series of experiments were performed in order to determine the bed expansion properties for a step change in the fluid properties. In these experiments, the bed was initially fluidized with DI water at a given flow

Table 1. Parameters and data for flow rate and fluid property experiments

feed stock	density g/cm^3	95 % CI g/cm^3	viscosity cp	95 % CI cp	particle Reynolds number	modified Reynolds number
water	0.9972	–	0.890	–	0.82	0.065-0.14
dextran	1.006	0.00007	1.347	0.0109	0.35	0.039-0.082
	n (regressed)	95 % CI	V_t (regressed) cm/s	95 % CI cm/s	V_t (theoretical) cm/s	95 % CI cm/s
water	5.19	5.01-5.37	0.374	0.342-0.408	0.395	±0.073
dextran	4.96	4.73-5.20	0.229	0.207-0.253	0.249	0.047

Table 2. Dispersion coefficient data for flow rate and fluid property experiments.

Feedstock initial/final	velocity initial/final cm/s	dispersion best-fit cm^2/s	coefficient 10 % limits cm^2/s
water	0.0127 to 0.0170	0.08	0.06–0.12
water	0.0127 to 0.0212	0.10	0.07–0.16
water	0.0127 to 0.0297	0.18	0.13–0.23
water	0.0127 to 0.0382	0.20	0.16–0.24
water	0.0127 to 0.0467	0.37	0.34–0.40
water to dextran	0.0170	0.19	0.15–0.25
water to dextran	0.0255	0.27	0.22–0.36
water to dextran	0.0340	0.36	0.30–0.44
water to dextran	0.0425	0.45	0.41–0.47

rate. Upon reaching a state of uniform fluidization, as indicated by a constant degree of expansion, a step change was made to a fluid with different physical properties. After the introduction of the step change, the bed height was recorded as a function of time. Figure (8) shows plots of the bed height versus time for a step change from DI water to the dextran solution. The solid lines are the two-phase model and the symbols are the experimental values, including error bounds of ± 3 mm. In Figure (8), the time delay predicted by the two-phase model is in agreement with the experimental data. The time delay is a result of a compacting zone. The compacted zone takes a finite amount of time to reach the top of the bed, at which point the bed begins to expand. Figure (9) show a plot of the solid fraction versus the axial position for a number of time slices. The model results depict how the compacted zone looks when the dispersion effects are considered. In comparing Figures 7 and 9, it is clearly evident that changes in fluidization velocity do not create a compacted zone within the bed, as was observed for the fluid property change. The compacted zone is the reason for the observed time delay associated with a step change in fluid properties.

The solid-phase dispersion coefficient was treated as a free parameter and fitted by least-squares regression. Table (2) shows a listing of the solid-phase dispersion coefficients as obtained by least-squares fit.

Definition of the control problem

Chang and Chase (1996) have addressed some of the issues concerning the interactions between important adsorption properties and hydrodynamic properties in EBA. In particular, they have investigated the relative adsorption efficiencies for two modes of operation, namely operation at a fixed bed height and operation at a fixed flow rate. Chang and Chase (1996) contend

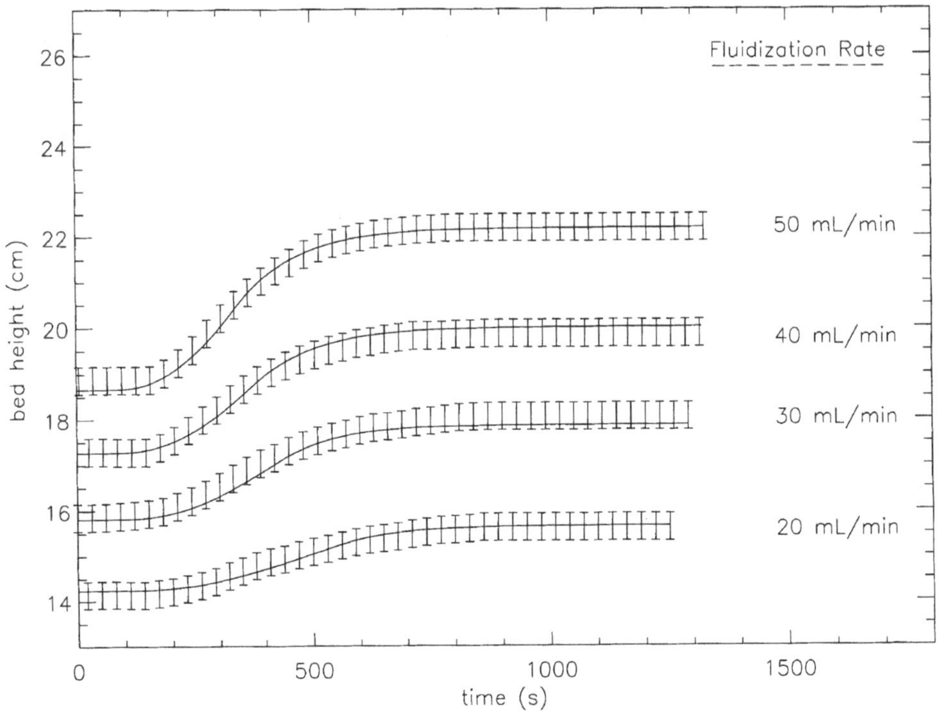

Figure 8. Transient bed-height expansion experiment for a step change in fluid properties. Step change from DI water to the dextran solution. (—) two-phase model and symbols are experimental data with error bars of ±3 mm.

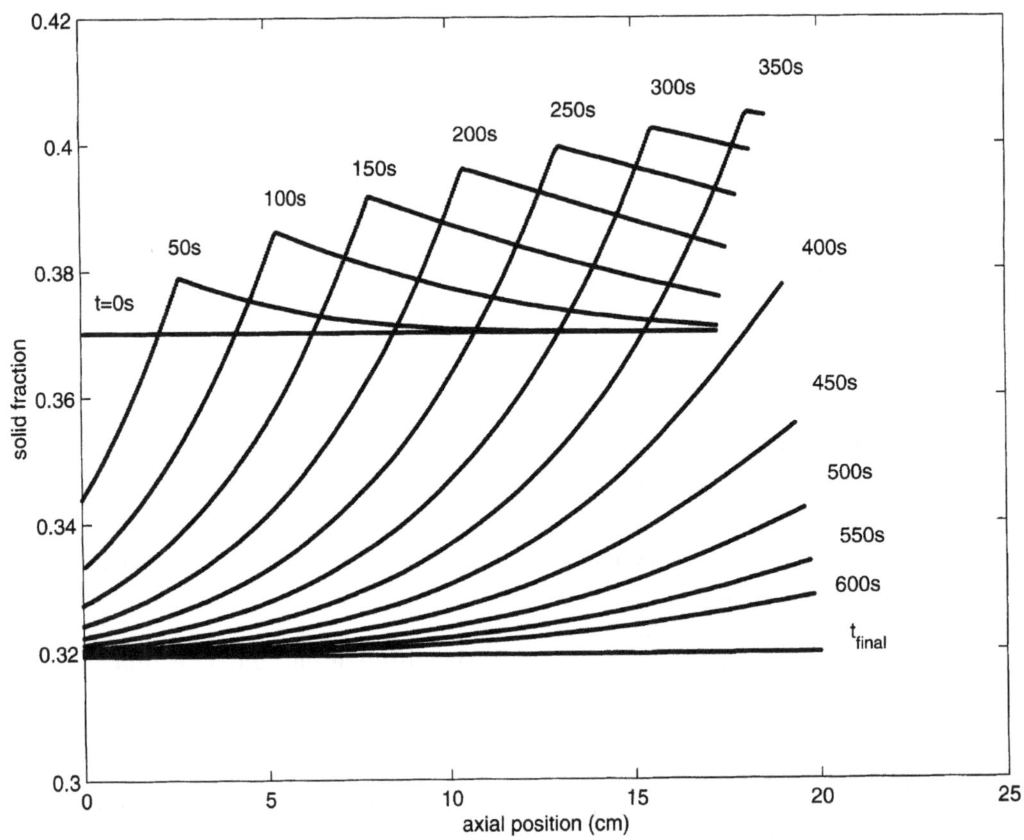

Figure 9. A plot of the solid fraction versus axial position for a step change from water to the dextran solution at 40 mL/min.

that operation at a fixed bed height is most efficient. Specifically, Chang and Chase (1996) argue that operation of the column at a bed height of twice the settled bed height is optimal, where optimal is defined in the sense of maximizing the dynamic binding capacity of the resin.

DeLuca et al. (1996) state that the wide fluctuations in physical properties of biological feedstocks, the viscosity in particular, greatly affect the bed expansion behavior. Furthermore, due to the high optical density of industrial feedstocks, visual inspection and manual control of the degree of expansion is difficult. Practicality is another concern, as the current operating procedure requires the manual placement of a plunger at the upper solid-liquid interface in order to minimize the dead volume in this region. The use of this plunger is not practical at the prepartive scale.

These factors contribute to the difficulties encountered in the operation and control of EBA processes for biological applications. As such, prediction of bed expansion behavior in conjunction with reliable level measurement is of significant importance for accurate monitoring and control of the EBA operation.

The primary control objective is the regulation of the bed at a specified degree of expansion which may be changed during operation by the operator. The control system should provide adequate regulation with load disturbances and changing expansion characteristics due to such effects as temperature swings, changes in physical properties, and protein loading/unloading, the presence of a high loading of cellular debris.

IMC-based PID control

The proportional-integral-derivative (PID) controller is ubiquitous in the process industries. There are numerous reason for this, including their long history of proven operation, the fact that they are well understood by industrial personnel, and the fact that a well-tuned PID controller often meets or exceeds the defined control objectives (Chien and Fruehauf, 1990). However, a poorly tuned controller often fails to meet the defined control objectives, and in many situations PID loop tuning can be difficult, time consuming, and often involves a trial-and-error procedure (Chia and Lefkowitz, 1992). As such, there is motivation to obtain the PID tuning parameters from process models and characteristic of the closed-loop response when such information is available. The internal model control (IMC) concept is introduced as a model-based control strategy.

IMC theory

The IMC framework is a commonly used approach for model-based control (Morari and Zafiriou, 1989). We now look at the IMC design procedure for single-input single-output (SISO) systems. A common approach to model-based control is the formulation of the control problem as an optimization problem. The IMC method is based on the linear quadratic or H_2 optimization function. The H_2 optimization is

$$\min_Q \|e\|_2^2 = \int_0^\infty e^2(t)dt \tag{46}$$

where e is the error signal and the integral square error is minimized for a particular input v. Use of Parseval's theorem allows (46) to be written in the frequency domain

$$\min_Q \|e\|_2^2 = \min_Q \frac{1}{2\pi} \int_{-\infty}^{\infty} |e(i\omega)|^2 d\omega \tag{47}$$

where Q is the controller. If a sensitivity function is defined as

$$e = \epsilon(d - r) \tag{48}$$

where d is a disturbance signal and r is a reference signal, the frequency domain minimization is rewritten as

$$\min_Q \|\epsilon w\|_2^2 = \min_Q \frac{1}{2\pi} \int_{-\infty}^{\infty} |\epsilon(i\omega) w(i\omega)|^2 d\omega \tag{49}$$

and w is defined as the input weighting function. Morari and Zafiriou (1989) have shown that for a SISO system ϵ can be made identically equal to zero by making the controller equal to the plant inverse.

$$Q = \frac{1}{G_m} \tag{50}$$

As such the IMC-based controller is obtained by setting the controller equal to the plant inverse when allowed.

IMC-bbased PID controller design theory

We now show how the IMC design procedure can be used to obtain the PID controller parameters. Figure 10 shows block diagrams for the IMC and classical

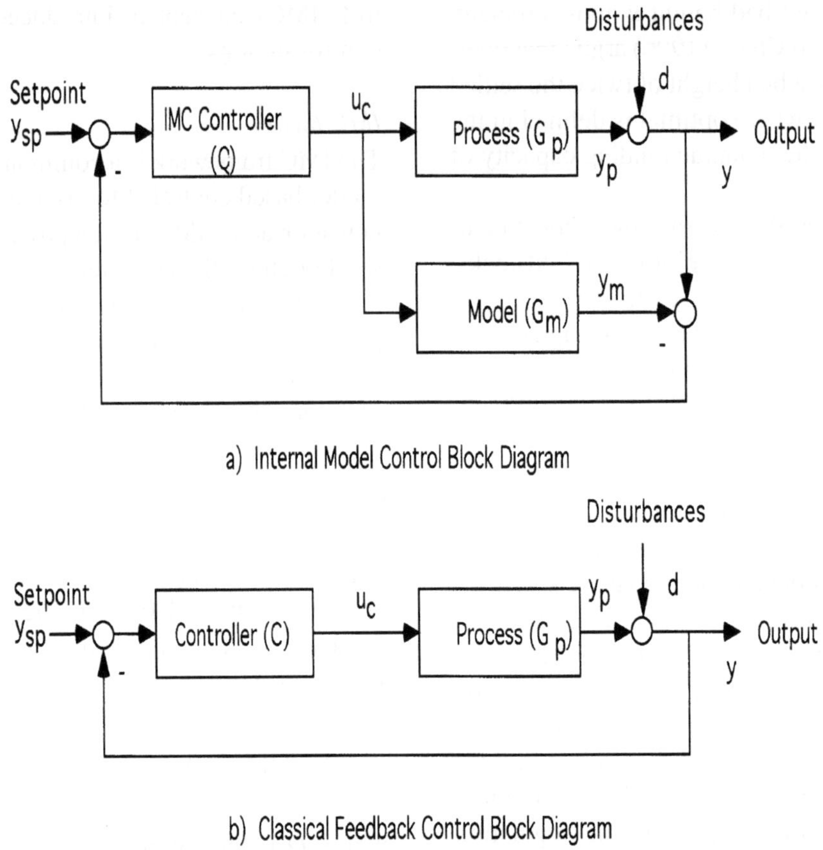

Figure 10. Block diagram representation of the IMC and feedback controllers.

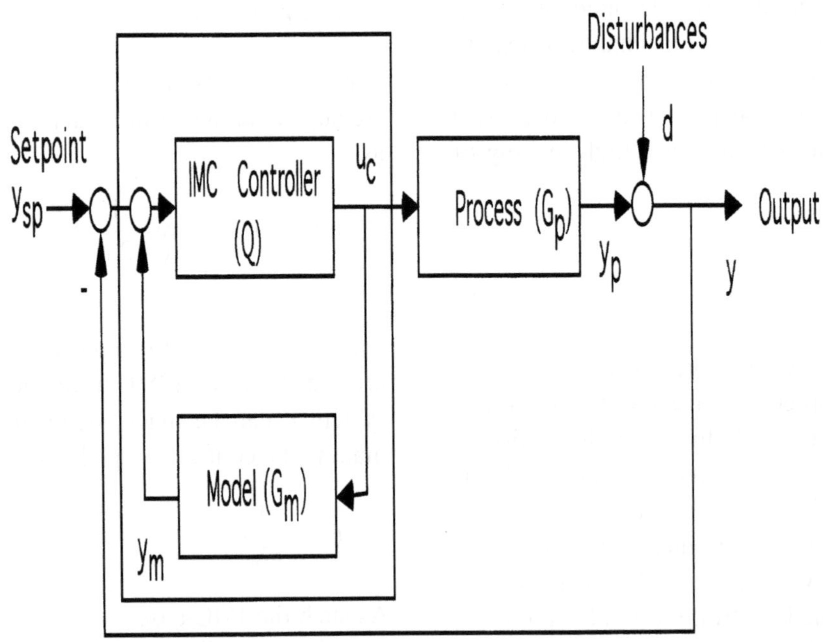

Figure 11. Rearrangement of the IMC block diagram to illustrate the equivalence between the IMC and feedback controllers.

feedback control structures. Figure 11 shows a rearrangement of Figure 10a. From inspection, it is clear that the block diagrams for the IMC and classical feedback controllers are equivalent, where the controllers are related as

$$C(z) = \frac{Q(z)}{1 - G_m(z)Q(z)} \tag{51}$$

As such, the IMC control structure provides a convenient parameterization of the feedback controller. For many process models, the IMC controller can be reduced to a PID controller or a PID controller cascaded with a lag. The IMC-based PID tuning method is discussed is great detail by Rivera et al. (1986) for continuous systems and in Zafiriou and Morari (1985, 1986) for sampled-data systems.

A common approach to the design of digital PID controllers by the IMC-based PID tuning method is to simply discretize the continuous IMC-based PID controller. However, as discussed by Zhu and Saucier (1992), such an approach provides satisfactory performance only when the sampling time is sufficiently small. A more rigorous approach to the IMC-PID tuning method for sampled-data systems is by direct application of (51), which defines the relationship of the feedback controller, $C(z)$, to the IMC controller, $Q(z)$.

For many process models, the IMC controller for the sampled-data system is of too high an order to be implemented by a PID controller. To address this problem, Zhu and Saucier (1992) recommend cascading the PID controller with a lag. The lag-compensated controller is of the form

$$C(z) = D(z)L(z) \tag{52}$$

The PID controller, $D(z)$, is given by the discrete transfer function

$$D(z) = k_c\left(1 + \tau_I \frac{z}{(z-1)} + \tau_D \frac{(z-1)}{z}\right) \tag{53}$$

where k_c is the controller gain, the integral time constant is defined as

$$\tau_I = \frac{T_s}{T_I} \tag{54}$$

and the derivative time constant is defined as

$$\tau_D = \frac{T_D}{T_s} \tag{55}$$

The lag term is defined as (Zhu and Saucier, 1992)

$$L(z) = \frac{1}{1 + A(z)} \tag{56}$$

The compensator $A(z)$ is defined as

$$A(z) = \sum_{i=1}^{m+n-1} a_i z^{-i} \tag{57}$$

where m is the number of sampling intervals corresponding to the process time delay, n is the order of the system, and a_i are the coefficients of the compensator.

Design of IMC-based PID controller for first-order system

Consider now the design of an IMC-based PID controller where the process transfer function is given by the first-order model

$$G(z) = \frac{k_p}{z - \pi_1} \tag{58}$$

The IMC controller is given by

$$Q(z) = \frac{1-\lambda}{z-\lambda}\frac{z-\pi_1}{k_p} \tag{59}$$

Application of equation (51) gives the equivalent feedback controller

$$C(z) = \frac{1-\lambda}{k_p}\frac{z-\pi_1}{z-1} \tag{60}$$

The IMC-based PID controller is given by equating (52) and (60)

$$\frac{1-\lambda}{k_p}\frac{z-\pi_1}{z-1} = k_c\left(1 + \tau_I \frac{z}{(z-1)} + \tau_D \frac{(z-1)}{z}\right) \tag{61}$$

Equation (61) is a polynomial equation; by equating coefficients of like powers of z, the PID tuning parameters can be obtained in terms of the process parameters and the IMC tuning parameter, λ. For the first-order process,

$$k_c = \frac{(1-\lambda)\pi_1}{k_p} \tag{62}$$

$$\tau_I = \frac{(1-\pi_1)}{\pi_1} \tag{63}$$

$$\tau_D = 0 \tag{64}$$

A physical system will have input constraints of the form

$$u^{min} \le u \le u^{max} \tag{65}$$

Input constraints often lead to performance degradation in control systems that employ integral action (Seborg et al., 1989). The performance degradation is due to the effects of integral windup. When control inputs are saturated, as often occurs during set-point transfers, the integral terms tend to build up. Thus even as the output approaches the set-point, the controller will often overshoot the set-point due to the inertia of the integral term. We consider a method to limit integral windup based upon the technique of not integrating the error signal when saturation is active. More general approaches are available to deal with saturation compensation (Campo and Morari, 1990; Kothare et al., 1994; Zafiriou, 1990)

Linearization of infinite-dispersion limit of fundamental model

A simplified model of the process dynamics can be obtained from the infinite-dispersion limit of the two-phase model.

$$\frac{dh_b}{dt} = v_o - v_t\left(1 - \frac{\phi_s h_s}{h_b}\right)^n \tag{66}$$

This equation is nonlinear, so Taylor series linearization is applied to the equation.

$$\frac{dh_b}{dt} = v_o - \beta h_b + \gamma \tag{67}$$

where β and γ are

$$\beta = n v_t\left(1 - \frac{\phi_s h_s}{\bar{h}_b}\right)^{n-1} \frac{\phi_s h_s}{\bar{h}_b^2} \tag{68}$$

$$\gamma = \beta \bar{h}_b \tag{69}$$

The equilibrium operating condition is given by $\bar{h}_b$, $\bar{v}_o$, and $\frac{dh_b}{dt} = 0$, so that

$$0 = \bar{v}_o - \beta \bar{h}_b + \gamma \tag{70}$$

Now (70) is subtracted from (67) in order to write the system in deviation form

$$\frac{dh_b'}{dt} = v_o' - \beta h_b' \tag{71}$$

where the deviation variables are defined as

$$v_o' = v_o - \bar{v}_o \tag{72}$$

$$h_b' = h_b - \bar{h}_b \tag{73}$$

In order to obtain the transfer function representation of (71), the Laplace transform method is applied

$$s h_b'(s) = v_o'(s) - \beta h_b'(s) \tag{74}$$

Rearranging (74), yields the process transfer function

$$G_p(s) = \frac{h_b'(s)}{v_o'(s)} = \frac{\frac{1}{\beta}}{\frac{1}{\beta}s + 1} \tag{75}$$

For implementation on a computer-controlled system, the discrete transfer function is required. The composite system, including process transfer function and zero-order hold device is

$$G(s) = G_p G_{zoh} = \frac{\frac{1}{\beta}}{\frac{1}{\beta}s + 1} \frac{1 - e^{-sT_s}}{s} \tag{76}$$

where the Laplace transfer function of a zero-order hold device is given by (Deshpande and Ash, 1988)

$$G_{zoh}(s) = \frac{1 - e^{-sT_s}}{s} \tag{77}$$

and T_s is the sampling period. The z-transform equivalent of (76) is given by

$$G(z) = \frac{\frac{1}{\beta}\left(1 - e^{-\beta T_s}\right)}{z - e^{-\beta T_s}} = \frac{k_p}{z - \pi_1} \tag{78}$$

Using the regressed values for v_t and n and a steady-state expansion of 52 cm, the value of β is 0.00383 s^{-1}. This value of β and a sampling time of 5 s yields first-order constants of $k_p = 4.95\ s$ and $\pi_1 = 0.981$.

Experimental system

A STREAMLINE 50 column (5 cm ID) was used in the fluidization studies with STREAMLINE DEAE adsorbent (Pharmacia Biotech). STREAMLINE DEAE is a weak anion-exchange resin with the functionalization (-O-CH_2CH_2-$N^+(C_2H_5)_2H$). The resin has an ionic capacity of 0.13 to 0.21 mmol Cl^-/mL adsorbent. The resin has a binding capacity of approximately 40 mg BSA/mL adsorbent at a linear velocity of 300 cm/h, 50 mM Tris-HCl, and a pH of 7.5. The resin has a mean daimeter of 200 μm and a mean density of 1.2 g/cm^3.

The column was filled with resin until a settled bed height of 26.0 *cm* was reached. Deionized water was used to estimate the Richardson-Zaki parameters. A linear regression of the logarithm of the fluidization velocity versus the logarithm of the void fraction yielded a Stokes settling velocity of 0.581 cm/s and a Richardson-Zaki exponent of 5.31.

Dextran T500 (Pharmacia Biotech) was used as a viscosity modifying agent. Bovine serum albumin

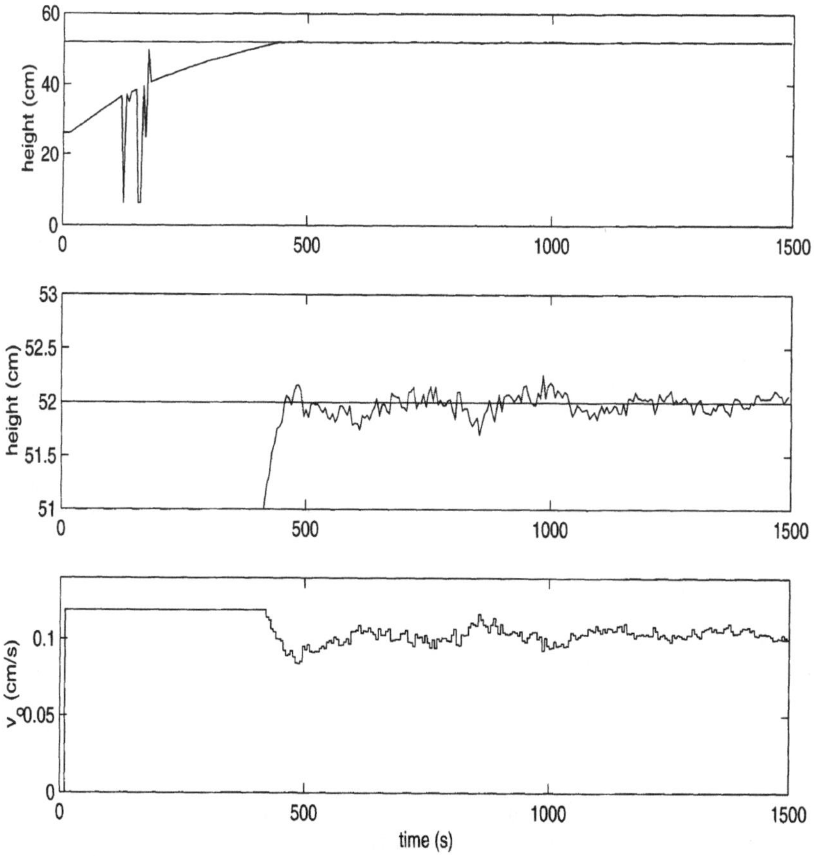

Figure 12. Experimental implementation of IMC-based PID controller tuned using infinite-dispersion model. The column is fluidized with deionized water. Setpoint change from fully settled bed to a $2x$ expansion.

(BSA) was used in the protein uptake studies. *E. coli* were grown in a minimal media for the cell-laden experiments. The cell-culture media consisted of 6 g of tryptone (Difco Laboratories), 3 g of yeast extract (Difco Laboratories), and 8 g NaCl per liter of deionized water.

The acoustic level measurement system consisted of a Panametrics model 5072PR pulser-receiver connected to a 500 kHz immersion style transducer (Panametrics). The hardware consisted of an AT-MIO-16E-1 board (National Instruments) which resided on an ISA expansion slot on a personal computer (Gateway, Pentium 180 Mhz). The ultrasonic signal was preprocessed using a BNC-2080 board. The signal was then sent to the AT-MIO-16E-1 board. A peristaltic pump, controllable via a 4-20 mA signal, was used to deliver fluid to the column. For computer control of the pump, the AT-MIO-16E-1 board generated a 0-5 V signal which was converted to a 4-20 mA signal by a custom-designed voltage to current converter. Labview (National Instruments) was used for development of the software interface for control, signal processing, and data acquisition.

Experimental verification of controllers

In this section, the IMC-based PID controller is implemented on an experimental EBA column. The first experiment considers the setpoint change from the fully-settled bed to a $2x$ expansion. The bed is fluidized with deionized water. The second test considers the setpoint change from $2x$ to $1.9x$ expansion. Again, the bed is fluidized with deionized water. The third test considers the regulation of the bed at a $2x$ expansion when subjected to a load disturbance of a $1.75x$ increase in the fluid viscosity. The bed was initially fluidized with deionized water and a step change was made to an aqueous dextran solution. The fourth test considers the regulation of the bed at a $2x$ expansion when subjected to a load disturbance of a solution with a cellular concentration of approximately 10^8 cells/mL, a protein loading of 2 mg BSA/mL, and a dextran loading such that the final viscosity was $1.75x$ that of the initial fluidization medium. The *E. coli* were grown in the minimal media previously described.

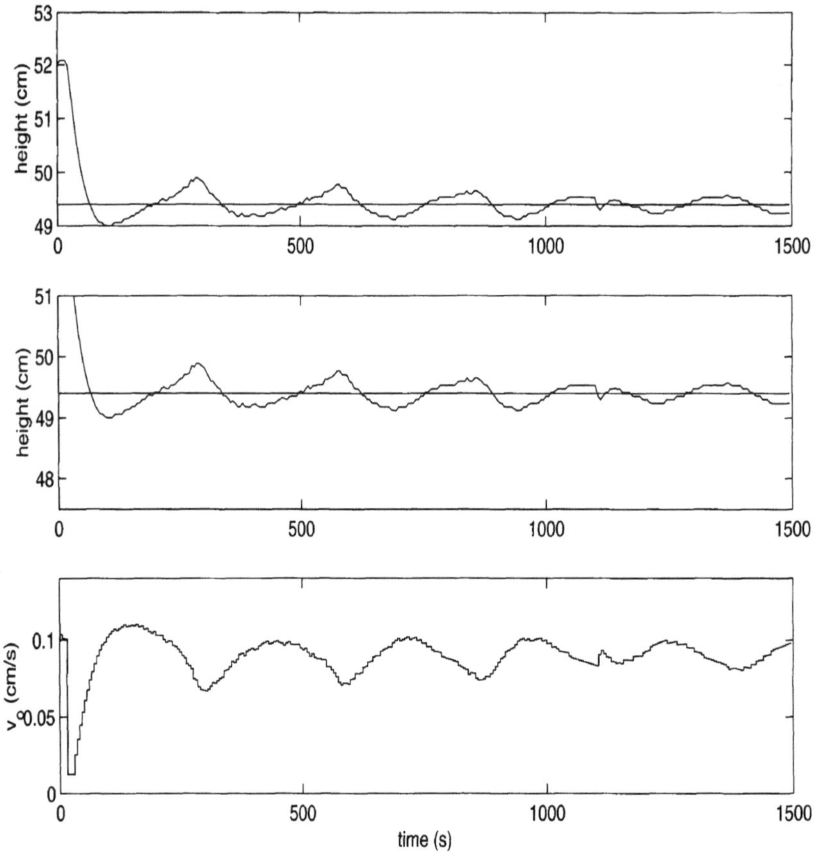

Figure 13. Experimental implementation of IMC-based PID controller tuned using infinite-dispersion model. The column is fluidized with deionized water. Setpoint change from a $2x$ to a $1.9x$ expansion.

PID control based on Infinite-Dispersion Model

The IMC-based PID controller designed using the infinite-dispersion model is used for experimental verification of the controller. A tuning constant of $\lambda = 0.8$ was used in the simulations. This results in the following PID parameters

$$k_c = 0.0396\, s^{-1} \tag{79}$$

$$\tau_I = 0.0194 \tag{80}$$

The controller was significantly de-tuned in order to minimize controller activity due to the noisy bed height measurements.

Figure 12 shows the results for the IMC-based PID controller for a setpoint change from the fully-settled bed at a height of 26 cm to a $2x$ expansion at 52 cm. The bed was fluidized with deionized water throughout the experiment. The setpoint transfer was accomplished with negligible overshoot and very little oscillation. There were a number of measurement faults in the initial stages of the experiment. Due to the fact that the controller was saturated, these faults did not result in a performance reduction.

Figure 13 shows the experimental results for a setpoint change from a $2x$ to a $1.9x$ expansion. The bed is fluidized with deionized water throughout the duration of the experiment. This experimental run resulted in no measurement faults, and the measurements had considerably less noise than typical of most experiments.

Figure 14 shows the experimental results for regulation of that bed at a $2x$ expansion when subjected to a load disturbance. The load disturbance was initiated by making a step change in the fluidizing liquid. A step change was made from deionized water to a aqueous dextran solution with a viscosity $1.75x$ that of water. Initially, the measurements were quite noisy and a number of faults were observed. These faults resulted in a number of controller saturations. However, after about 500 sec, the measurements returned to normal noise levels. The controller was found to perform extremely well throughout the experiment. The bed height was maintained at 52 ± 0.4 cm for the entire experiment.

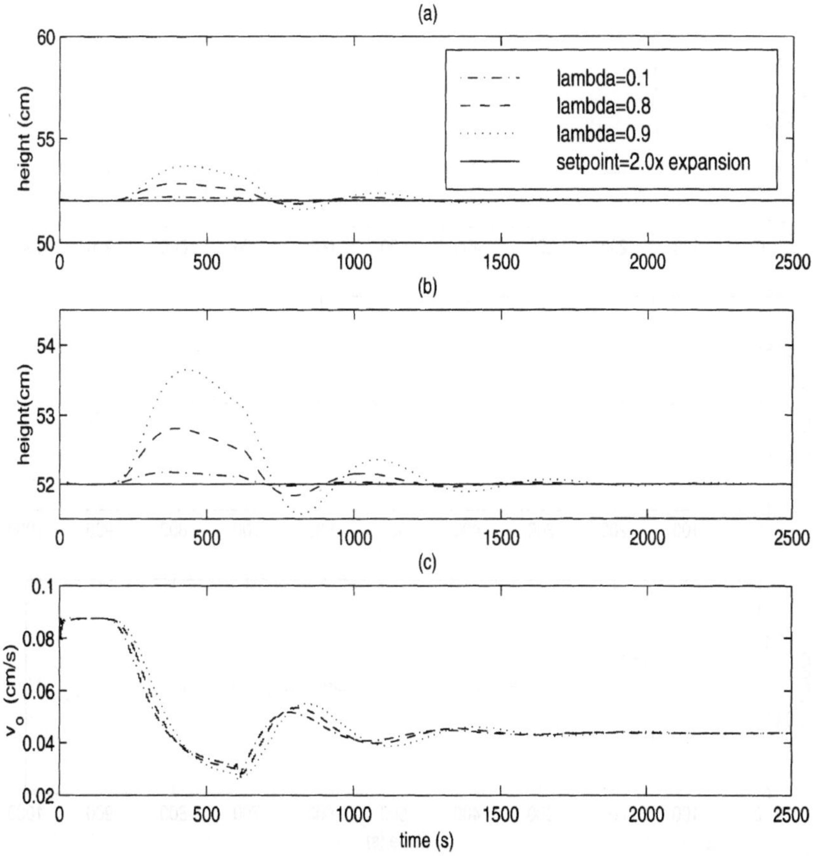

Figure 14. Experimental implementation of IMC-based PID controller tuned using infinite-dispersion model. The column is initially fluidized with deionized water. Regulated at a $2x$ expansion when subjected to a load disturbance of an aqueous dextran solution with viscosity $1.75x$ that of water.

Figure 15 shows the experimental results for regulation of that bed at a $2x$ expansion when subjected to a load disturbance. The load disturbance was initiated by making a step change in the fluidizing liquid. A step change was made from deionized water to a mixture containing cells, BSA, and dextran. The solution viscosity was $1.75x$ that of water. The measurements were quite stable and only a single fault was observed. This fault resulted in near controller saturation. The controller was found to perform extremely well throughout the experiment. The bed height was maintained at 52±0.4 cm for the entire experiment.

Conclusions

We have shown that ultrasonic backscattering can be used to accurately measure the bed height in expanded-bed adsorption columns for both particulate-free and particulate-laden feedstocks. For the particulate-free feedstock, the bed height was measured to a resolution of 0.074 cm, while for the particulate-laden feedstock the bed height was measured to a resolution of 0.15 cm. The cells in the particulate-laden feedstock caused an attenuation of the ultrasonic signal, and hence a decrease in the signal-to-noise ratio. The reduction in the signal-to-noise ratio was the cause of the decrease in the measurement resolution for the particulate-laden feedstock as compared to the particulate-free feedstock.

Two-phase theory has been applied to solid-liquid fluidization in the Stokes flow regime. As the inertial terms can be neglected in the Stokes flow regime, the linear momentum balances are simplified to algebraic constraints. The simplified linear momentum balance considered three forces; namely, drag, buoyant, and diffusive. The particle-phase force balance utilized concepts from dusty gas mechanics. The force balance expression was then combined with the continuity equations, to obtain a governing equation for solid-liquid fluidization in the Stokes regime. This simplified two-phase equation was then applied to an experimental EBA system. Two flows were investigated, namely, step changes in the fluidization velocity and step changes in the fluid properties. Experimental

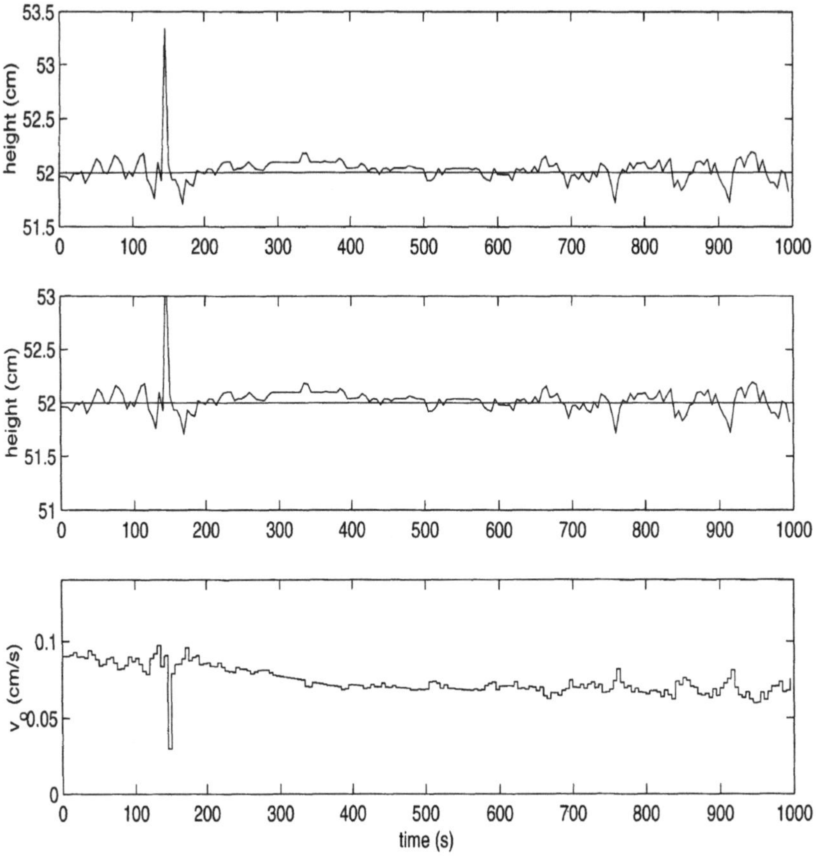

Figure 15. Experimental implementation of IMC-based PID controller tuned using infinite-dispersion model. The column is initially fluidized with deionized water. Regulated at a $2x$ expansion when subjected to a load disturbance of fermentation-like mixture, including cells, protein, and dextran with viscosity $1.75x$ that of water.

data and model simulations were in excellent agreement for both flow configurations. Furthermore, the two-phase model provided an accurate prediction of the observed time delay in bed expansion for step changes in fluid properties, as well as a mechanistic explanation for its observation.

Lastly, the IMC-based PID tuning procedure was applied to the regulation of the bed-expansion of a model EBA system. A simplified first-order model was derived from the infinite-dispersion limit of the two-phase model. The parameters of the simplified first-order model were used to tune the PID controller. Three test cases were studied; namely start-up of the column from the fully-settled bed to 2x expansion, a step change from $2x$ to $1.9x$ expansion, and regulation at $2x$ expansion when subjected to a load disturbance. The IMC-based PID controller was found to perform adequately in all cases.

References

Asif M, Petersen JN, Kaufman EN, Cosgrove JM and Scott TC, (1994) A dynamic model of the hydrodynamics of a liquid fluidized bed. *Ind. Eng. Chem. Res.*, 33: 2151.

Bailey JE and Ollis DF (1986) **Biochemical Engineering Fundamentals**; McGraw-Hill, New York.

Campo PJ, and Morari M (1990) Robust control of processes subject to saturation nonlinearities. *Computers and Chemical Engineering*, 14: 343–358.

Chang YK and Chase HA, (1996) Development of operating conditions for protein purification using expanded bed techniques: The effect of the degree of bed expansion on adsorption performance. Biotechnol. Bioeng., 49: 512.

Chia TL and Lefkowitz I, (1992) Robust PID Tuning Using IMC Technology-Pt. 1, Intech, October: 36.

Chien I-L and Fruehauf PS, (1990) Consider IMC Tuning to Improve Controller Performance. Chem. Eng. Prog., October: 33.

Coulson JM, Richardson JF, Backhurst JR and Harker JH, (1991) Chemical Engineering: Particle Technology and Separation Processes, 4th ed., Pergamon Press, Oxford.

Davis RH, (1995) Velocity of sedimenting particles in suspensions. in: **Sedimentation of Small Particles in a Viscous Fluid**, Tory EM (ed.) Computational Mechanics Publications, South Hampton, U.K.

De Luca L, Hellenbroich D, Titchener-Hooker NJ, Chase HA (1994) A study of the expansion characteristics and transient behaviour of expanded beds of adsorbent particles suitable for bioseparations. Bioseparation, 4: 311–318.

Fan L-T, Schmitz JA and Miller EN, (1963) Dynamics of liquid-solid fluidized bed expansion. AIChE J., 9: 149.

Fay B (1993) Ultrasonic Backscattering: Fundamentals and Applications. **Physical Acoustics: Fundamentals and Applications**. In: Oswald L and Breazeale MA, (ed.); Plenum Press: New York.

Gailliot FP, Gleason C, Wilson JJ and Zwarick J (1990) Fluidized bed adsorption for whole broth extraction. Biotechnol. Prog. 6: 370–375.

Gibilaro LG, Waldram SP, Foscolo PU, (1984) A simple mechanistic description of the unsteady state expansion of liquid fluidized beds. Chem. Eng. Sci. 39: 610.

Hjorth R, Kampe S and Carlsson M. (1995) Analysis of some operating parameters of novel adsorbents for recovery of proteins in expanded beds. Bioseparation, 5: 217–223.

Kothare MV, Campo PJ, Morari M and Nett, CN, (1994) A unified framework for the study of anti-windup designs. Automatica, 30: 1869–1883.

Lee SM (1989) The primary stages of protein recovery. *Biotechnol. Prog.* 11: 103–118.

Marble FE, (1970) Dynamics of dusty gases. Annual Review of Fluid Mechanics 2: 397.

Morari M and Zafiriou E (1989) **Robust Process Control**. Prentice Hall: Englewood Cliffs.

Panametrics (1995) Ultrasonic Transducers for Nondestructive Testing, April.

Peddieson J Jr (1974) On continuum description of solid-fluid suspensions. Dev. Theor. Appl. Mech., 7: 355.

Pharmacia Biotech, (1996) Application Note, STREAMLINE Scale Up, July.

Poncelet D, Naveau H, Nyns E-D and Dochain D, (1990) Transient response of a solid-liquid model biological fluidized bed to a step change in fluid superficial velocity. J. Chem. Tech. Biotechnol., 48: 439.

Ramirez WF, Shuler PJ and Friedman F, (1980) Convection, dispersion, and adsorption of surfactants in porous media. Soc. Pet. Eng. J., December: 430.

Richardson JF and Zaki WN, (1954) Sedimentation and fluidization: Part I. Trans. Instn. Chem. Eng. 32: 35.

Rivera DE, Morari M and Skogestad S, (1986) Internal Model Control. 4. PID Controller Design. Ind. Eng. Chem. Process Des. Dev., 25: 252.

Seborg DE, Edgar TF, Mellichamp, DA, (1989) Process Dynamics and Control. John Wiley and Sons, New York.

Slis PL, Willemse TW and Kramers H (1959) The response of the level of a liquid fluidized bed to a sudden change in the fluidizing velocity. Appl. Sci. Res., 8: 209.

Stokes, GG, (1901) Mathematical and Physical Papers.

Thelen TV, Mairal AP, Thorsen CS and Ramirez WF (1997) Application of Ultrasonic Backscattering for Level Measurement and Process Monitoring of Expanded-Bed Adsorption Columns. Biotechnol. Prog., 13: 681.

Thelen TV and Ramirez, WF, (1997) Bed-height dynamics of expanded beds. Chem. Eng. Sci., 52: 3333.

Thelen, TV and Ramirez WF, (1998) Two-Phase Flow Theory for Solid-Liquid Fluidization in the Stokes Flow Regime. AIChE Journal, To appear.

Truesdell C (1962) Mechanical basis of diffusion. J. Chem. Phys., 37: 2336.

Zafiriou E (1990) Robust Model Predictive Control of Processes with Hard Constraints. Computers and Chemical Engineering, 14: 359–371.

Zafiriou E and Morari M (1985) Digital Controllers for SISO systems: A Review and a New Algorithm. Int. J. Control, 42: 855.

Zafiriou E and Morari M (1986) Design of robust digital controllers and sampling-time selection for SISO Systems, Int. J. Control, 44: 711.

Zhu JJ and Saucier MF, (1992) An IMC-Based Extended PID Controller for Sampled-Data Systems, Automatic Control Conference, 601.

Bioseparation **8:** 33–41, 1999.
© 1999 *Kluwer Academic Publishers.*

Factors affecting dispersion in expanded bed chromatography

Shuichi Yamamoto[1,*], Nagomu Akazaki[1], Oliver Kaltenbrunner[2,†] & Peter Watler[2]
[1]*Department of Chemical Engineering, Yamaguchi University, Tokiwadai, Ube 755-8611, Japan and* [2]*Amgen, Inc., Thousand Oaks, CA, USA (*author for correspondence) e.mail:shu-yama@po.cc.yamaguchi-u.ac.jp Fax: +81-836-35-9933*

Received 9 October 1998; accepted in revised form 2 March 1998

Key words: dispersion, expanded bed, fluidized bed, HETP, residence time distribution

Abstract

Factors affecting the dispersion of solutes in expanded bed chromatography were experimentally investigated to characterize the behavior in small columns. Pulse response curves were measured with a vitamin B^{12} tracer, and HETP (height equivalent to a theoretical plate) values were calculated from peak variance and retention time. Approximately 15 min were required to attain a stable steady state expanded bed height with constant HETP values. HETP values ranged from 0.8 to 1.6 cm and did not change appreciably with the degree of expansion (1.5–3.5 fold), column diameter (1.6 and 2.6 cm), column temperature (293–308 K) or settled bed height (ca. 4–11 cm). A very small column (1.6 cm diam. and 4.2 cm-settled bed height) was successfully expanded and axial mixing measured could be useful for conducting scale down experiments.

Introduction

Chromatographic separation using a fluidized bed with very low mixing is desirable for the capture, clarification and concentration of proteins (Chase 1994; Pharmacia, 1997). Low axial mixing fluidized bed systems, or expanded beds (EB), are claimed to give the same adsorption performance as packed beds (Chase, 1994; Pharmacia, 1997).

However, in contrast to the large number of investigations on zone spreading in packed bed liquid chromatography (LC), the fundamental mechanism in expanded beds has not yet been thoroughly investigated. For example, the axial mixing as well as the mass transfer between the mobile and stationary phases is still unclear.

For a homogeneously packed bed, the contribution of axial dispersion in the interstitial spaces to the HETP is typically 2 – 5 particle diameters (d_p) and is not sensitive to the mobile phase linear velocity (Yamamoto et al., 1988). Such dispersion is primarily due to variable flow path distances in the interstitial space. Although several papers have been published on the axial mixing in expanded beds (Batt et al., 1995; Thommes et al., 1995, Karau et al., 1997), it has only been measured under limited experimental conditions. Even the relationships between axial mixing and degree of expansion are not well established.

To minimize the use of expensive feed and reagents during optimization and characterization it is often desirable to scale down the expanded bed system. Column diameter and settled bed height are key variables often considered for such column scaling. Other key variables which affect axial mixing in expanded beds are temperature, average bead diameter, particle size distribution and bead density.

This study employed columns of relatively small diameter (1.6 and 2.6 cm) and Streamline media which was developed especially for use in expanded beds. In this system a dead space between the top of the expanded bed and the flow distributor was maintained to avoid occluding the frit. To measure the degree of axial mixing in packed and expanded beds, residence time distribution curves were generated with a vitamin B^{12} tracer. HETP values were then calculated from the residence time distribution (RTD) curves. HETP values

† Part of this work was done while O.K. was at Yamaguchi University

were used to assess the effect of superficial velocity and dead space on axial mixing and peak broadening. In addition, the time needed to attain a stable, steady state expanded bed height was measured.

Materials and methods

Stationary phase

Media consisted of cross linked agarose ion exchange beads, Streamline-SP (particle size = 100–300 μm, density = 1200 kg/m^3), Amersham-Pharmacia Biotech, Uppsala, Sweden. Streamline-AC base matrix (without protein A ligand, particle size = 80–165 μm, density = 1300 kg/m^3) and colored Streamline base matrix were generous gifts of Amersham-Pharmacia Biotech. Colored Streamline was prepared as follows: The base matrix was expanded with water at 5 cm/min. Beads which rose to the top (approximately 50% of the total) were withdrawn and set aside. The remaining bottom beads were removed and colored. The colored and uncolored beads were then combined. This mixture of beads was used to qualitatively assess circulation patterns in the expanded bed. To study axial disperion in fixed beds, non-porous glass beads (average particle diameter d_p = 5330 μm) and plastic (polystyrene) beads (d_p = 1480 μm) were used.

Columns

Column tubes were jacketed XK16 (1.6 cm ID) and XK26 (2.6 cm ID) (Amersham-Pharmacia Biotech, Uppsala, Sweden). The end-piece was modified with a conical expansion from inlet tube to the frit. The length of the conical expansion was 1.2 cm for XK16, and 2.0 cm for XK26. System temperature was maintained constant within ±0.5 °C.

Residence time distribution curves

Residence time distribution curves were generated using a solution of 1 mg/mL Vitamin B^{12} as the tracer. This colored solution enabled qualitative observation of tracer zone movement in the column. Injection volume (V_F) for the XK16 was 0.2 mL and 1 mL for the XK26. Acetone was also used as a tracer for comparison. No significant difference was found between the vitamin B^{12} and acetone data. The mobile phase was in most cases Milli-Q quality water. A column packed with relatively small beads (d_p=40–60 μm) was used to verify that the injection pulse was symmetrical. Any minor injection pulse distortion is expected to be negligible since zone spreading with larger beads is much greater.

Instrumentation

A Pharmacia P-50 pump and a six-way injection valve (V-7, Pharmacia) were used for delivering the mobile phase and injecting the sample pulse. The UV detector (Pharmacia UV-1) was connected to a personal computer (PC) via an A-D converter (C1141, Contec, Osaka, Japan). A BASIC program was written for logging the UV detector output via the A-D converter to the PC.

Results

Bed expansion

Figure 1 shows a typical bed expansion behavior. In most experiments the expanded bed height stabilized to a constant height after 10–15 min. At this point, the axial mixing in the EB is quite stable and the measured HETP values are constant (Figure 1). In addition, colored Streamline media was used to visually confirm proper bed expansion. When the bed was expanded to a stable height, the beads separated into distinct red and white zones (bottom and top sections respectively). This qualitative test confirmed the absence of macro-mixing and gross axial circulation.

Figures 2a and 2b shows the dependence of bed expansion on superficial velocity (u_0) for various sedimented bed heights (Z_0) at two different column diameters. The degree of expansion (Z/Z_0) increases linearly with superficial velocity and is independent of the initial sedimented bed height. In addition, the slope of the regression line in Figures 2a and 2b is the same indicating that degree of expansion is independent of column diameter.

Calculation of HETP from residence time distribution curves

The mixing in the unit operation system such as reactor and separator can be assessed by residence time distribution (RTD) curve measurements (Levenspiel, 1972; Swaine and Daugulis, 1988). Either pulse or step input response curves can be used for the RTD analysis. In this study, the pulse response curve method was used as it is more sensitive and is suited

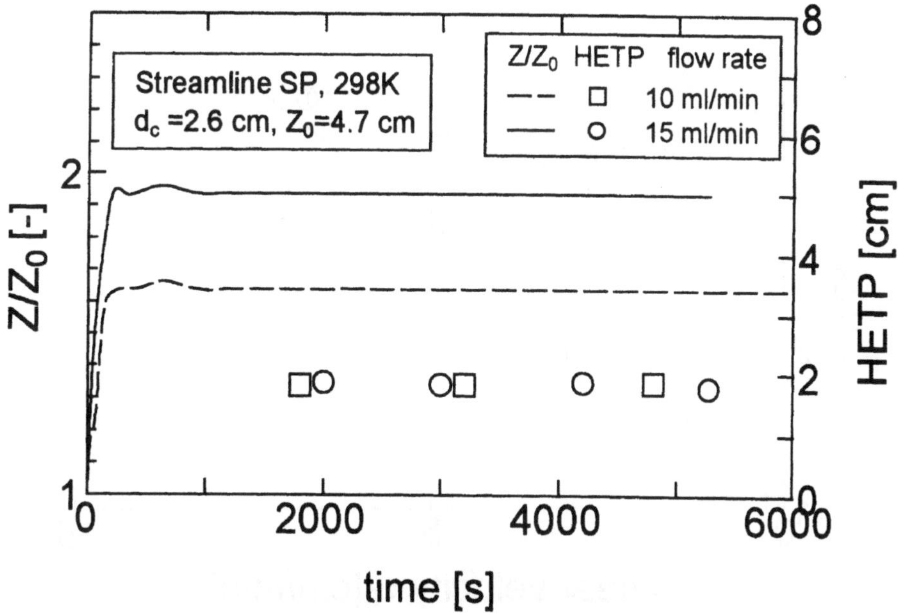

Figure 1. Degree of expansion Z/Z_0 and HETP as a function of time. Z_0 = sedimented bed height = 4.7 cm, Z = expanded bed height, volumetric flow rate F_v = 15 mL/min, column diameter d_c = 2.6 cm, 298 K, media = Streamline SP.

for the visual observation of the movement of the zone in the bed. The first absolute moment of the response curve μ_1', which implies the peak retention time, is given by

$$\mu_1' = \int_0^\infty Ct\,dt / \int_0^\infty C dt = \int_0^\infty C dt/\mu_0, \quad (1)$$

where $C=C(t)$ is the solute concentration at the bed exit (elution curve or chromatogram), t is the time from the start of the pulse injection and μ_0 is the zero-th moment (area of the curve). The normalized second central moment μ_2, which implies the peak variance, is given by

$$\mu_2 = \int_0^\infty C(t-\mu_1')^2 dt/\mu_0. \quad (2)$$

When the response curve is symmetrical and Gaussian, $\mu_1' = t_R$ and $\mu_2 = \sigma^2$ where t_R is the peak retention time and σ^2 is the peak variance. The measured moments (suffix obs) were corrected for the contribution due to the sample injection $t_F = V_F/F_v$ (V_F = injection volume and F_v = volumetric flow rate) by

$$\mu_1' = \mu_{1,\mathrm{obs}}' - (t_F)/2; \quad (3)$$

$$\mu_2 = \mu_{2,\mathrm{obs}} - (t_F)^2/12. \quad (4)$$

The HETP is calculated from the moments by

$$\mathrm{HETP} = Z/N = Z\mu_2/(\mu_1')^2, \quad (5)$$

where N = plate number and Z = bed height. There are several dimensionless groups describing the axial mixing, which can be related to N. Peclet number based on the bed height, Pe,z (which is also called Bodenstein number Bo) is defined as Pe,z = Zu/D_L = $2N$ where D_L = axial dispersion coefficient and u = interstitial velocity (Thömmes et al., 1995). Peclet number based on the particle diameter Pe,p = ud_p/D_L is very often employed for describing HETP of packed bed chromatography (Yamamoto et al., 1988). The number of tranfer units for axial dispersion, (NTUd) is given by NTUd = Pe,z = $2N$ (Vermeulen et al., 1984; Arnold et al., 1985).

Figure 3 shows typical RTD pulse response curves for the EB and the fixed bed. The plate numbers obtained by Equations (1)–(5) were similar to those calculated by assuming a Gaussian peak shape [i.e., $N = 8(t_R/w)^2$]. The response curves calculated from the plate model (Yamamoto et al., 1988) using these N values were in good agreement with the experimental curves (Figure 3).

Dead space

In conventional packed bed chromatography, there is no significant dead space between the bed and the outlet frit. However, a certain dead volume is needed to maintain a stable EB and to avoid accumulation of particles on the top frit. In order to investigate the effect of this extra column volume on zone spreading, the variance (σ^2) was measured using a packed glass bead column with dead spaces varying from 0

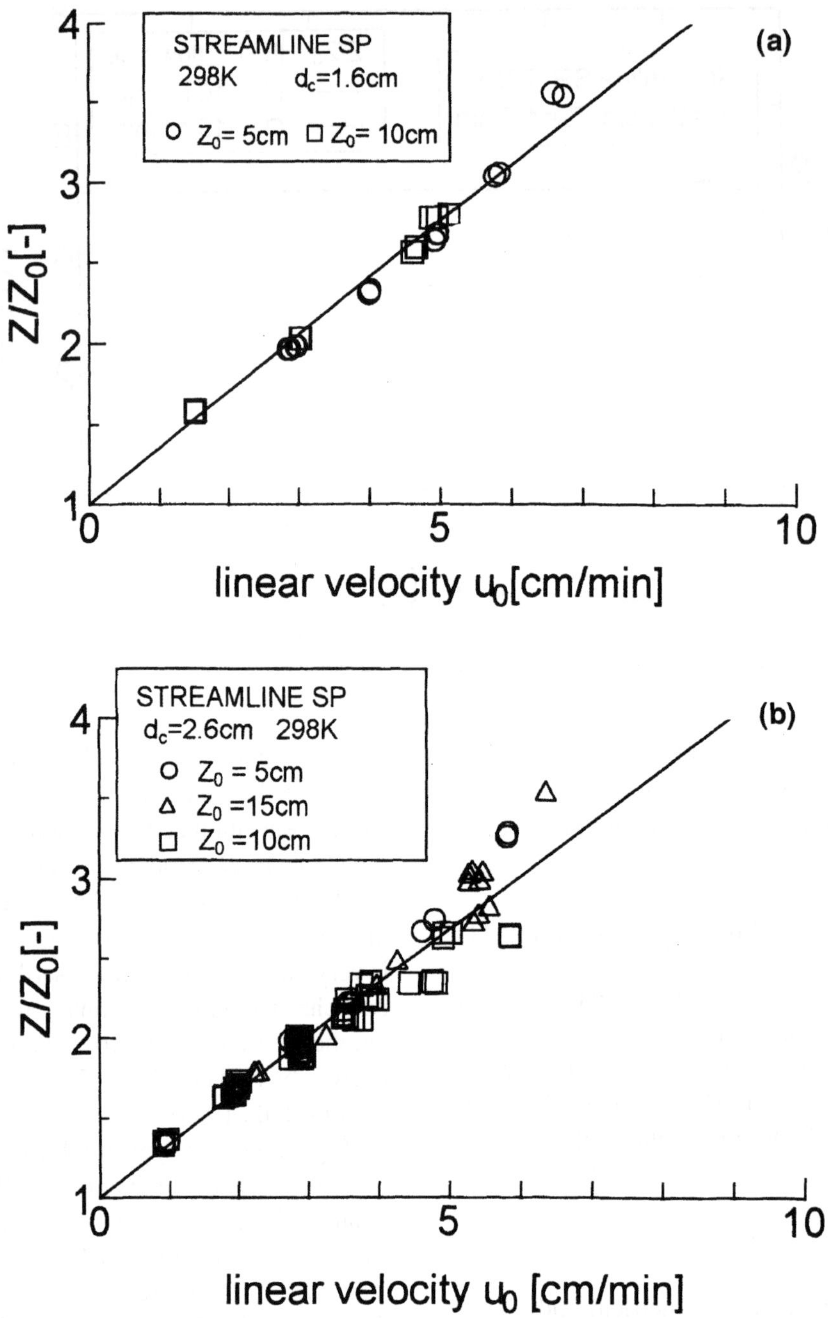

Figure 2. Degree of expansion Z/Z_0 as a function of linear velocity $u_0 = F_V/A_C$. A_C = column cross sectional area. F_V = volumetric flow rate.

to 2.3 cm. The overall zone spreading is assumed to be a sum of the zone spreading from the extra column space and from axial dispersion and mass transfer effects. As shown in Figure 4, for a constant packed bed height, the variance increases with increasing dead space. However, even when the length of the dead space is 2 cm, the contribution to the overall variance is only 1–4 mL^2. This is small compared with the overall variance of 50–150 mL^2. In this study, the distance between the top of the EB and the support net of the outlet end-piece was maintained 0.5 – 1.5 cm. The extra column broadening effect was found to be negligible in the present system.

Plate height – linear velocity relationships

The dependence of HETP and hence axial mixing on superficial velocity was quite weak as shown in Figures 5–8 although for larger values of Z_0, Karau et al. (1997) reported a moderate increase in axial mixing with increasing velocity. Although temperature affects bed expansion (Pharmacia, 1997), the HETP

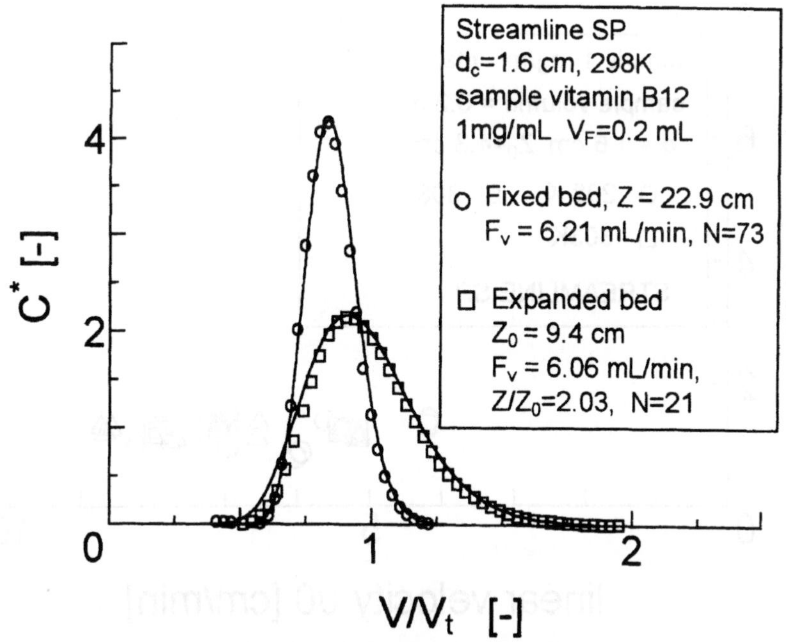

Figure 3. Typical pulse response curves. 298 K. Streamline SP. □ = EB (d_c = 1.6 cm, Z_0 = 9.4 cm, Z = 19.1 cm, u_0= 3.0 cm/min); ○ = fixed bed (FB) (d_c = 1.6 cm, $Z_0 = Z$ = 23 cm, u_0 = 3.1 cm/min). Sample = Vitamin B^{12}. The solid curves are the calculated curves by the plate model with N = 21 for the EB and N = 73 for the FB (Yamamoto et al., 1988). $C^* = C(t)V_t/C_0V_F$ where $C_0V_F = \mu_0$ (μ_0 = zeroth moment of the curve). V_t = total (fixed or expanded) bed volume.

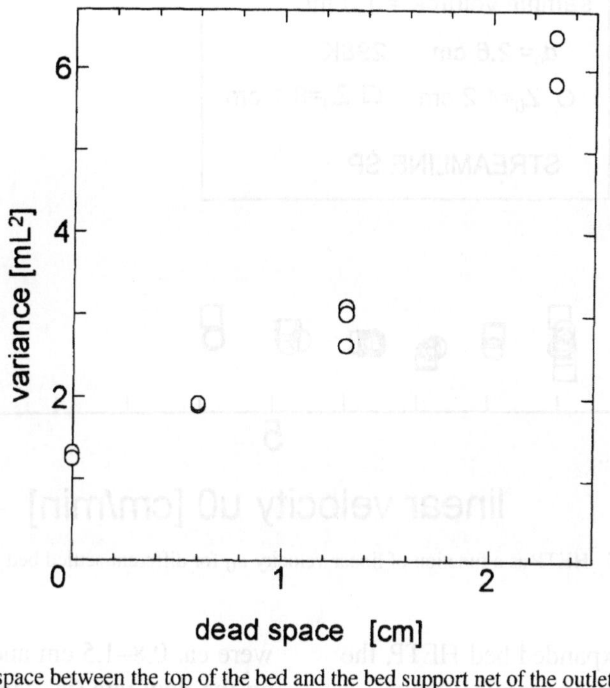

Figure 4. Dispersion due to the dead space between the top of the bed and the bed support net of the outlet end-piece. 298 K, packed glass bead column (1.6 cm diam.,10 cm height). As the cross sectional area is 2 cm^2, 2 mL implies 1 cm between the top of the bed and the bed support net.

was not sensitive to temperature (Figure 5), settled bed height (Figure 6), column diameter (Figure 7) or bead diameter (Figure 8).

The distribution coefficient of vitamin B^{12} is ca. 0.7, which implies diffusive mass transfer limitations within the media bead. These mass transfer limitations were measured using a packed bed of Streamline media and caused HETP values to increase with the linear mobile phase velocity from 0.05 to 0.3 cm (Figure 9). When these HETP due to diffusive mass transfer is

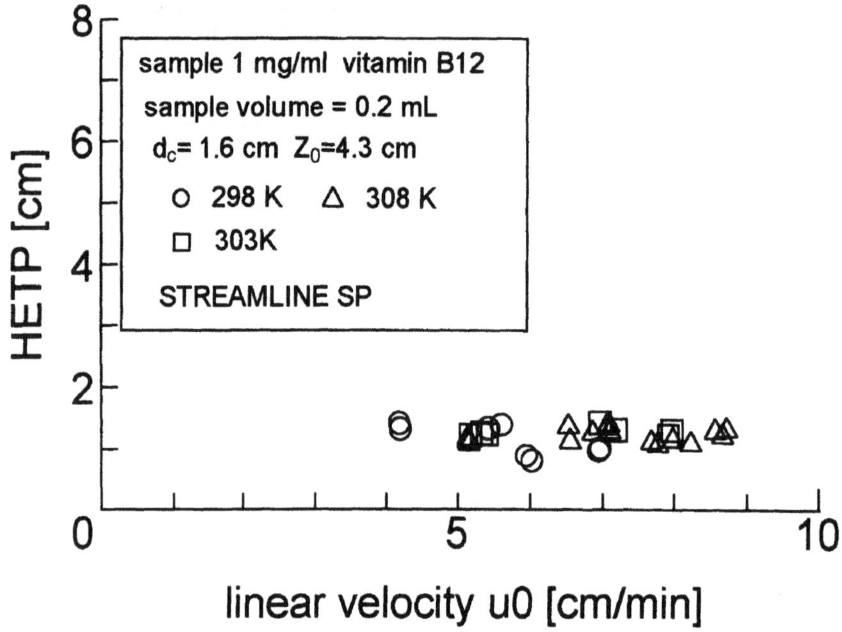

Figure 5. HETP as a function of linear velocity u_0 for different temperatures.

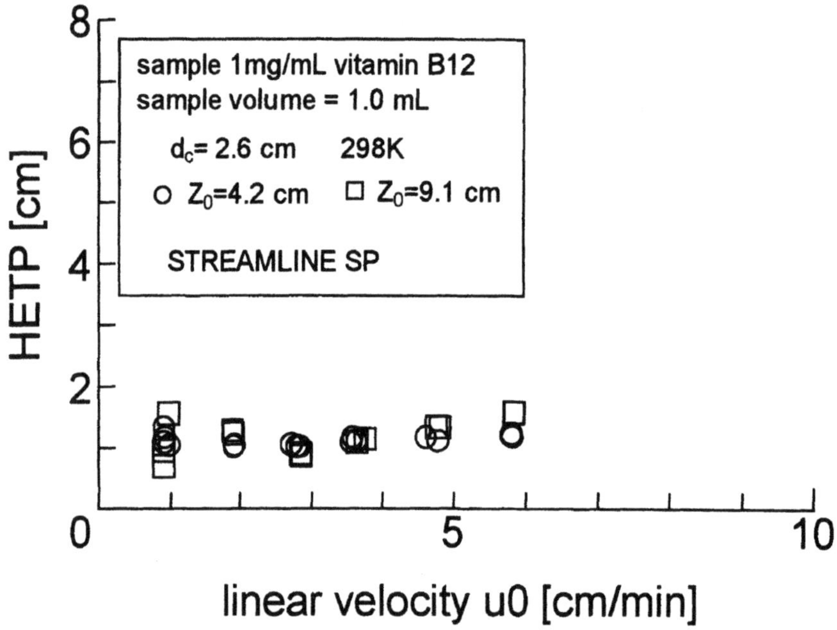

Figure 6. HETP as a function of linear velocity u_0 for different settled bed heights.

subtracted from the overall expanded bed HETP, the HETP at high velocities becomes somewhat lower.

Discussion

HETP values calculated from the expanded bed literature data, are in the range of ca. 1.0–2.0 cm (Batt et al., 1995; Thömmes et al., 1995; Chang & Chase, 1996; Karau et al., 1997). The HETP values in this study were ca. 0.8–1.5 cm and did not significantly depend on the flow rate (ie. degree of expansion), settled bed height, column diameter or the temperature. Krishnaswamy et al. (1978) proposed a correlation for axial dispersion in liquid fluidized bed systems. Although the particles were much larger in the systems they examined (d_p = 0.5–6 mm), the HETP values for our system were calculated by their correlation equation. The calculated HETP values range between 0.7 and 2 cm for Z/Z_0 = 2–3 and Z_0 = 10–20 cm.

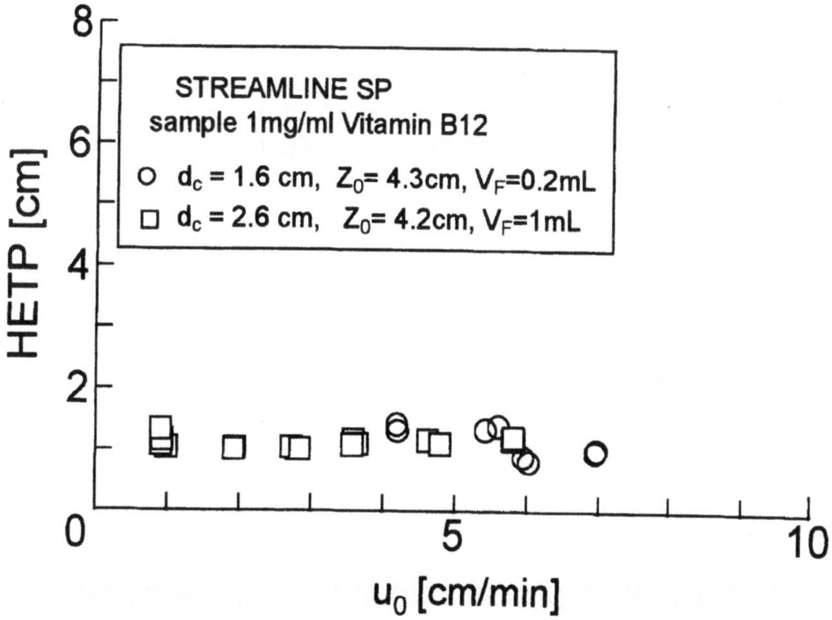

Figure 7. Figure 7 HETP as a function of linear velocity u_0 for different column diameters.

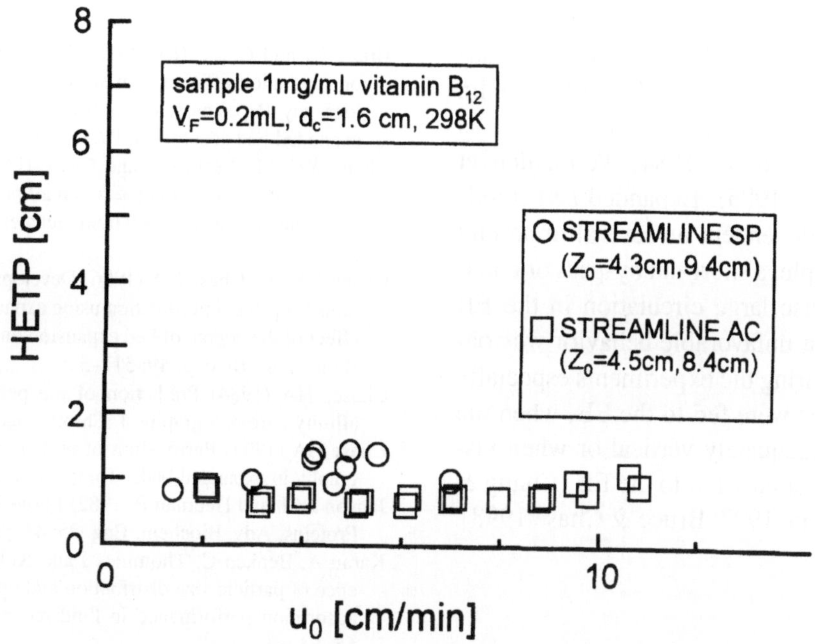

Figure 8. HETP as a function of linear velocity u_0 for different bead sizes.

If the EB is operated at rather high flow rates to obtain a long expanded bed, e.g., 60 cm, the plate number will be ca. 30 even when HETP is rather high, i.e., 2 cm. As the pressure drop is known to be a limiting factor in the scale-up of packed bed chromatography (Janson and Hedman, 1982), large diameter, short bed height columns are usually used for packed bed process chromatography. However, pressure drop is not a restriction with expanded beds and consequently, beds longer than 100 cm can be used for process chromatography.

Scale-down of the EB columns is needed for adsorption capacity measurements in order to avoid using large quantities of expensive proteins and reagents. The axial mixing of the smallest EB column used in this study (d_c = 1.6 cm, Z_0 = 4 cm) was shown to be similar to that of a larger one (d_c = 2.6 cm, Z_0 = 10 cm). Thömmes et al. (1995) reported that the axial mixing decreases with an increase of the height to diameter ratio (Z/d_c).

Although this paper presented stable and favorable axial mixing data, the overall performance of

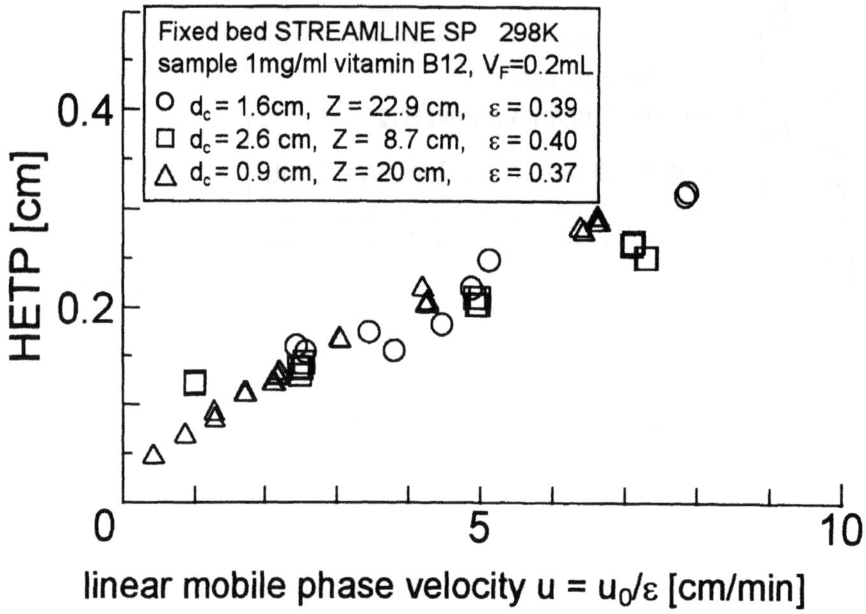

Figure 9. HETP as a function of linear mobile phase velocity *u* for packed beds.

the EB must be assessed by the total zone spreading such as that from the mass transfer in the media and from the adsorption-desorption rates as well as that from the mixing (Chase, 1984; Vermeulen et al., 1984; Arnold et al., 1985). Expanded bed stability is also very much dependent on several important precautions. For example, a short-cut by-pass or channeling which may cause large circulation in the EB must be avoided. Such unfavorable behavior was occasionally observed during the experiments especially when small air bubbles were fed to the EB, when the EB column was not adequately vertical or when viscous solutions were suddenly fed to the EB (Chang & Chase, 1996; Pharmacia, 1997; Bruce & Chase 1998).

Acknowledgment

This work was supported by a Grant-in Aid for scientific research (A1, No.08305031) from the Ministry of Education, Science, Sports and Culture, Japan.

References

Arnold FH, Blanch HW and Wilke CR (1985) Analysis of affinity separations I: Predicting the performance of affinity adsorbers, Chem. Eng. J. 30: B9–B23

Batt BC, Yabannarvar VM and Singh V (1995) Expanded bed adsorption process for protein recovery from whole mammalian cell culture broth. Bioseparation 5: 41–52.

Bruce LJ and Chase HA (1998) The effect of column alignment on separation efficiency in scaled down expanded bed chromatography. Paper presented at the 2nd international conference on expanded bed adsorption, P7.3.

Chang YK, McCreath GE and Chase HA (1995) Development of an expanded bed technique for an affinity purification of G6DPH from unclarified yeast cell homogenates. Biotechnol. Bioeng. 48: 355–366.

Chang YK and Chase HA (1996) Development of operating conditions for protein purification using expanded bed techniques: The effect of the degree of bed expansion on adsorption performance. Biotechnol. Bioeng. 49: 512–526.

Chase, HA (1984) Prediction of the performance of preparative affinity chromatography. J. Chromatogr. 297: 179–202.

Chase HA (1994) Purification of proteins by adsorption chromatography in expanded beds. Trends Biotech. 12: 296–303.

Janson J-Ch and Hedman P (1982) Large Scale Chromatography of Proteins. Adv. Biochem. Eng. 25: 43–99.

Karau A, Benken C, Thömmes J and Kula M-R (1997) The influence of particle size distribution and operating conditions on the adsorption performance in fluidized beds. Biotechnol. Bioeng. 55: 54–64.

Krishnaswamy PR, Ganapathy R and Shemilt LW (1978) Correlating parameters for axial dispersion in liquid fluidized systems, Can. J. Chem. Eng. 56: 550–553.

Levenspiel O (1972) Chemical Reaction Engineering. Wiley, New York.

Pharmacia Biotech (1997) Expanded bed adsorption – principles and methods. Pharmacia Biotech, Uppsala.

Swaine DE and Daugulis AJ (1988) Review of liquid mixing in packed bed biological reactors. Biotechnol. Prog. 4: 134–148.

Thömmes J, Halfar M, Lenz S and Kula M-R (1995) Purification of monoclonal antibodies from whole hybridoma fermentation broth by fluidized bed adsorption. Biotechnol. Bioeng. 45: 205–211.

Thömmes J, Halfar M, Lenz S and Kula M-R (1995) Hydrodynamics and performance in fluidized bed adsorption. Biotechnol. Bioeng. 48: 367–374.

Vermeulen T, LeVan MD, Hiester N.K and Klein G (1984) Adsorption and Ion Exchange. In Perry RH, Green DW and Maloney JO (eds), Perry's Chemical Engineer's Handbook, 6th ed. (pp. 16-1–16-48). McGraw-Hill, New York.

Yamamoto S, Nakanishi K, Matsuno R and Kamikubo K (1979) Operational conditions for media chromatography – prediction of elution curves. Agric. Biol. Chem. 43: 2499–2504.

Yamamoto S, Nakanishi K and Matsuno R (1988) Ion Exchange Chromatography of Proteins. Marcel Dekker, New York.

Address for correspondence: Shuichi Yamamoto, Department of Chemical Engineering, Yamaguchi University, Tokiwadai, Ube 755-8611, Japan. Tel.: +81-836-35-9419; Fax: +81-836-35-9933; E-mail:shu-yama@po.cc.yamaguchi-u.ac.jp

Bioseparation **8:** 43–51, 1999.
© 1999 Kluwer Academic Publishers.

Physical and biochemical characterization of a simple intermediate between fluidized and expanded bed contactors

John Chi-Wei Lan, Grant E Hamilton & Andrew Lyddiatt
Biochemical Recovery Group, Centre for Bioprocess Engineering, School of Chemical Engineering, University of Birmingham, Edgbaston, Birmingham, B15 2TT, UK

Key words: expanded bed, fluidized bed, adsorption kinetics, particulate feedstock, protein adsorption

Abstract

Physical and biochemical comparison has been made of the performance of a simple fluidised bed contactor and a commercial expanded bed contactor, characterised by identical dimensions, and operated at various settled bed heights with two anion exchange adsorbents. The contactors were tested with various feedstocks comprising bovine albumin in the absence and presence of 20 g dry cell weight biomass litre^{-1}. Earlier classification of the simple contactor as a single-stage, well mixed fluidised bed is reviewed. The relative merits of STREAMLINE DEAE and DEAE Spherodex LS as fluidisable, anion exchange adsorbents are discussed.

Abbreviations: EB – Expanded Bed; BRG – Biochemical Recovery Group; EBH – expanded bed height; SBH – settled bed height; DCW – Dry cell weight.

Introduction

Fluidised bed adsorption has been the subject of intensive study in our laboratory since 1986. Early recognition was recorded of the benefits of the application of relatively dense adsorbent materials, such as ion exchange or immunoaffinity derivatives of kielseguhr-agarose composites, in fluidised bed contactors for the direct recovery of protein products from particulate feedstocks (Wells and Lyddiatt, 1987; Lyddiatt, 1990). These latter included whole fermentation broths, microbial cell disruptates and other biological extracts. The unsophisticated nature of early commercial adsorbents and experimental prototypes designed and assembled in our laboratories (Wells and Lyddiatt, 1987; Gibson and Lyddiatt, 1993; Gilchrist and Lyddiatt, 1994) invited the adoption of a simple design of fluidised bed contactor based upon glass tubing of varied lengths (10–500 cm) and diameters (2 to 5 cm) fitted with a variety of metal and plastic basal meshes (50–90 μm) to act as the fluid distributor. It was assumed that such contactors might most efficiently operate if fed from a reservoir of feedstock with a recirculating stream of product-rich, particulate feedtstock. In such circumstances, it was concluded by ourselves (Wells and Lyddiatt, 1987; Morton and Lyddiatt, 1992) and others (Chase, 1994) that the contactor performance was best interpreted as a well-mixed fluidised bed which approached an adsorption equilibrium with the initial product concentration (Co) in the feedstock reservoir. The adoption of the application of feedstock recirculation to the establishment of processes of direct product sequestration, wherein a fluidised bed contactor was operated as the closed, external loop to a productive fermenter (Morton and Lyddiatt, 1994), further reinforced this view.

In the period since 1986, the work of Chase and others has established the technique of expanded bed adsorption. Here the fluidisation of a population of particles, characterised by a range of densities and particle diameters, creates a stable, stratified bed wherein single-pass flows of feedstock achieve fractional adsorption of protein solutes in a manner that rivals fixed bed chromatography (reviewed in Chase, 1994; Hjorth *et al.*, 1995; Thoemmes, 1997). Such technology has been commercialised with customised contactor designs and adsorbent materials, the cost of which reflects the extensive development programmes required for their establishment.

Such developments have invited a re-examination of the simply designed fluidised bed contactor (referred herein as the BRG contactor) in respect of suitability for single-pass, adsorptive fractionation of protein products contained in particulate feedstocks. We report here various comparative tests of the physical and biochemical performance of the BRG contactor (2.5 cm dia) and an equivalent commercial expanded bed (EB) contactor operated with two types of anion exchange adsorbent. One such adsorbent (STREAMLINE DEAE) was custom designed for expanded bed operations. The second (DEAE Spherodex LS) was a high performance fixed bed adsorbent which, by chance of enhanced density (1.35 kg l^{-1}), proved suitable for fluidised bed operations (Morton and Lyddiatt, 1992).

Materials and methods

STREAMLINE DEAE and DEAE Spherodex LS were obtained from Amersham Pharmacia Biotech (Sweden) and Biosepra (France) having a particle diameter range of 100 to 300 μm and density of 1.2 and 1.35 kg l^{-1} respectively. Adsorbents were equilibrated in 10 mM Tris-HCl, pH 7.5 (buffer A). Bovine serum albumin (BSA) (MW= 67 KD) was obtained from Sigma (Poole, U.K.) as a standard adsorbate. Blocks of Bakers' Yeast were supplied from British Fermentation Products (BFP). BSA was dissolved at 3 mg ml^{-1} in buffer A with or without the addition of 20 g DCW $litre^{-1}$ of yeast cells. Protein concentrations were measured by absorption spectrophtometry at 280 nm for cell-free experiments and by Coomassie blue analysis of supernatants from centrifuged samples in particulate experiments (Bradford, 1976). Adsorption and desorption experiments were carried out using a STREAMLINE 25 column (Amersham Pharmacia Biotech, Sweden) and a BRG column (University of Birmingham) driven by a common pump and valve switching system. Both contactors had identical dimensions but the BRG device was equipped with a simple basal distributor comprising a 98 μm mesh. The relationship of expanded bed height (H) in response to increasing flow velocity was determined in terms of the degree of bed expansion (H/Ho%), where Ho represents the settled bed height (SBH). The theoretical plate number (N) determinations in triplicate were conducted at a fixed linear flow rate (200 cm hr^{-1}) according to the Amersham Pharmacia Biotech STREAMLINE 25 operation manual using a step-input method. The axial dispersion coefficient (D_{axl}) and Bodenstein number (Bo), which express the state of liquid dispersion and fluidization behavior, were calculated by

$$D_{axl} = \frac{U \times H \times \sigma_\theta^2}{2 \times \varepsilon^2} \qquad (1)$$

$$B_O = \frac{U \times H}{D_{axl}} \qquad (2)$$

where U is the superficial flow velocity, σ_θ is the dimensionless variance and ϵ is bed voidage. Equilibrium adsorption isotherms were determined according to Chase (1984) and Mao and Hearn (1996). The biochemical experiments using 3 mg ml^{-1} BSA in the absence or presence of biomass were operated in a single pass process in both contactors at a constant superficial flow velocity of 200 cm hr^{-1}. The adsorbent was loaded with feedstock until C/Co approached 1 (where C and Co describe the protein concentration in the bed effluent and inlet respectively), whereupon unbound protein and cell debris was washed out with buffer A. Following flow-reversal, the adsorbed protein was eluted with 0.45 M NaCl in buffer A in fixed bed mode.

Results and Discussions

Bed expansion

The degree of bed expansion contributes to adsorption efficiency as a composite function of liquid distribution, liquid properties and particle and process conditions (Hjorth *et al.*, 1995). Bed expansion for STREAMLINE DEAE adsorbent was considerably greater than for DEAE Spherodex LS (see Figure 1) in identical flow conditions despite a similar particle size range between 100 and 300 μm. This can be attributed to the difference in density between the two particles. The material expansion in the BRG contactor might be expected to exhibit an unstable fluidization behavior because of channeling generated by the simple distributor. This was not visibly apparent in the present study where SBH values of 15.8, 23.1 and 29.8 cm were examined. In support of this, De Luca et al. (1994) concluded that distributor design is not critical for flow distribution provided that a sufficient SBH is available to dissipate jet streams generated by the flow distributor.

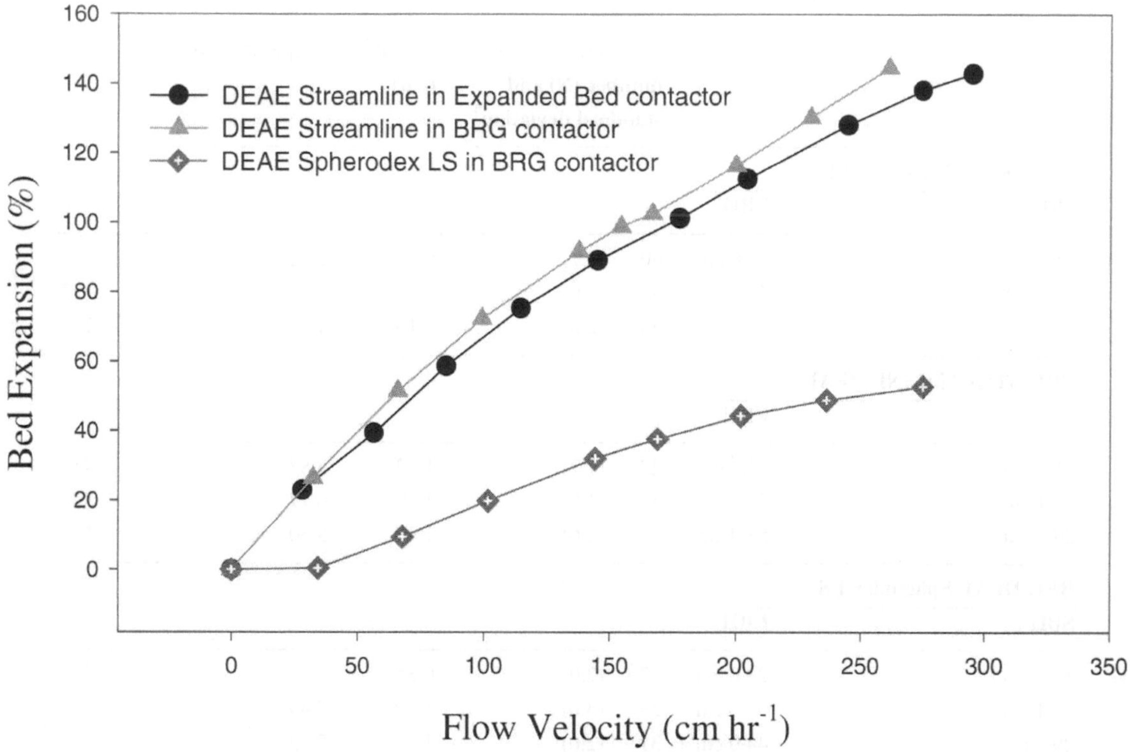

Figure 1. Comparison of bed expansion behaviour of STREAMLINE DEAE and DEAE Spherodex LS in Expanded Bed and BRG contactors. The bed expansion tests were carried out at SBH=23.1 cm at room temperature in buffer A.

Summary of physical characteristics

The theoretical plate number (N), axial dispersion coefficient (D_{axl}), HETP and Bodenstein numbers (Bo) as estimated in both contactors operated at various SBH are summarized in Table 1. Bodenstein number (Bo) is a dimensionless variable which represents the axial dispersion and fluidisation behaviour in the passage of materials through the contactor. Comparison of the two contactors confirmed that the EB contactor exhibited superior physical properties i.e. diminished mixing and stable fluidization throughout the bed. However, results also indicated that the BRG contactor was characterized by less mixing than might be expected from such a simple distributor design. The increase of apparent plate and Bodenstein numbers with increased SBH in the BRG contactor, loaded with either STREAMLINE DEAE or DEAE Spherodex LS, may be attributed to a dampening effect upon the liquid jets entering the bed. Table 1 indicates that STREAMLINE DEAE operated in the EB contactor at SBH = 23.1 cm had the greatest plate number although the reason for this remains unclear. All Bodenstein numbers and theoretical plate numbers (N; recorded with standard deviation in Table 1) are derived from data collected from triplicate experiments conducted with common conditions of fluid flow and bed expansion. Contactors were routinely assembled in a vertical alignment by means of spirit level and plumbline. Deviation from increasing values of Bo may be influenced by non-verticality of small-scale contactors (H.A. Chase, pers. comm.). However, the most important observations from the physical studies recorded in Table 1 are the comparative values and trends of theoretical plate numbers (N) recorded for the two contactors operated at selected settled bed heights.

Equilibrium kinetic studies

The Langumuir isotherm relationship between the equilibrium concentration of a solute in the solid phase and liquid phase (q^* and c^*) can be described by

$$q^* = \frac{q_m \times c^*}{K_d + c^*} \tag{3}$$

where q_m is the maximum adsorption capacity and K_d is the dissociation constant of the adsorption interaction (Chase, 1984). BSA binding isotherms in buffer A and a plot of q^* against c^* yield estimates of the rate constants and dissociation constant of

Table 1. A Comparison of the physical characteristics of adsorbent solid phase

		Theoretical plate number (N) and standard deviation[1]	HETP[2] (cm)	$D_{axl} \times 10^{-6}(m^2s^{-1})^4$	Bo[4]
EB STREAMLINE DEAE					
SBH[3]	**EBH[3]**				
15.8 cm	30.6 cm	60 (1.7)	0.51	2.70	86
23.1 cm	49.1 cm	91 (2.6)	0.54	3.09	127
29.8 cm	59.9 cm	53 (2.0)	1.13	4.59	72
BRG STREAMLINE DEAE					
SBH	**EBH**				
15.8 cm	32.9 cm	27 (1.7)	1.22	5.87	41
23.1 cm	50.5 cm	38 (1.0)	1.33	6.77	53
29.8 cm	58.8 cm	42 (2.6)	1.40	8.30	57
BRG DEAE Spherodex LS					
SBH	**EBH**				
15.8 cm	23.0 cm	27 (2.0)	0.85	4.25	29
23.1 cm	33.6 cm	28 (3.6)	1.20	7.49	31
29.8 cm	44.0 cm	31 (2.6)	1.42	5.23	36

[1]Theoretical plate number was determined at 200 cm hr^{-1} and is the mean value of three separate determinations. Standard deviation was calculated according to statistical methods.
[2]HETP (H/N) values were related to EBH/SBH at 200 cm hr^{-1}
[3]SBH and EBH were defined as settled bed height and expanded bed height respectively.
[4]D_{axl} and Bo were defined and calculated as described in the text.

the two adsorbent and feedstock types which are displayed in Table 2. A comparison indicated that DEAE Spherodex LS yielded a higher estimated maximum adsorption capacity (q_m) and a lower dissociation constant (K_d) for BSA adsorption. The presence of yeast cells (20 g DCW $litre^{-1}$) resulted in a reduction of the estimated maximum adsorption capacity for both STREAMLINE DEAE and DEAE Spherodex LS. The trend is important here rather than the absolute values of q_m which depend upon the stoichiometric assumptions of the Langmuir isotherm in the equilibrium batch binding experiments.

Comparative performance of STREAMLINE DEAE

The breakthrough curves of non-particulate and particulate feedstocks applied to EB and BRG contactors are depicted in Figures 2 and 3, whilst the biochemical performance is summarized in Table 3. The EB contactor yielded steeper breakthrough curves than the BRG contactor when fed with non-particulate feedstock, which was indicative of more efficient adsorption. However, the breakthrough performance of the BRG contactor approached that of the EB contactor when operated at the highest SBH (29.8cm). The performance with clarified feedstock confirmed the physical measurements in Table 1. An SBH value of 23.1 cm in the EB contactor yielded the highest dynamic and apparent saturation capacities (51 and 98 mg ml^{-1} respectively). The breakthrough curves in both contactors commonly only *approach* saturation (C/Co=1) during the time course of experiments with cell-free feedstock and this may be explained by channelling phenomena within the bed. The effect was less pronounced in the EB contactor. In the presence of biomass, where C exceeded Co = 3 mg ml^{-1} (see Figures 2,4), this was attributed to the lytic release of intracellular protein and other solutes. The phenomenon proved difficult to predict or control in extended experimentation (6 to 8 hours) with large volumes of concentrated broth (6 litres feedstock at 20 g DCW $litre^{-1}$). Inspection of Table 3 reveals that estimates of apparent saturation capacity commonly exceed the values recorded for q_m in Table 2. However, extended adsorption of protein in single pass fluidised bed operations from feedstock containing in-

Table 2. The kinetic investigation of protein adsorption.

Adsorbent	Cells	K_d(M)	K_d(mg ml^{-1})	q_m(mg ml^{-1})
STREAMLINE DEAE	(–)	$5.07*10^{-6}$	0.34	70
	(+)	$7.04*10^{-6}$	0.48	61
DEAE Spherodex LS	(–)	$2.99*10^{-7}$	0.02	88
	(+)	$5.22*10^{-7}$	0.04	73

(–): Measured without added biomass.
(+): Estimated in the presence of added biomass (20g DCW litre^{-1} yeast cells).

creasing concentration of yeast pigments, nucleic acid and protein in addition to the original constant challenge of BSA and intact yeast (3 g and 20 g DCW litre^{-1} respectively) will be extremely heterogeneous process. It is very likely that the high values for saturation capacities reflect complex and wash-resistant ion exchange and hydrophobic interactions between ill-defined system components in addition to *bona fide* ion exchange adsorption of BSA. The original concept of a simple binary experimental system (i.e. BSA plus yeast biomass) is thus shown to be flawed for the experiments performed here, but closely reflects representative practical operations.

In contrast to cell free studies, the breakthrough curves of protein challenge for adsorption from particulate feedstocks were markedly steeper in both EB and BRG contactors. It appeared to indicate that the presence of cells reduced mixing and channelling in both contactors, and thereby achieved a greater efficiency of protein binding. Plate number was shown to increase for STREAMLINE DEAE operated with SBH equal to 5 cm in a BRG contactor (i.d. = 2.2 cm) in the presence of cells (data not shown). The method of determination adopted (STREAMLINE 25 user manual, Amersham Pharmacia Biotech) was inappropriate for greater values of SBH because of the degradative influence of exposure of yeast cells to acetone encountered over extended timescales (> 1 h).

The maximum dynamic capacity at C/Co = 0.1 in the presence of cells was obtained for STREAMLINE DEAE operated at SBH = 23.1 cm and was equivalent to 70 mg BSA per ml settled adsorbent. A similar value (64 mg ml^{-1}) was obtained in the BRG contactor at SBH = 29.8 cm (see Table 3). It is interesting to note that these performances coincided with maximum plate numbers estimated for EB and BRG contactors in the absence of cells (compare Tables 1 and 3).

Comparative performance of DEAE spherodex LS

Figure 4 records the breakthrough curves of protein capture on DEAE Spherodex LS in the absence and presence of cells in the BRG contactor *alone*. Apparent saturation capacities of 120 and 106 mg ml^{-1} were respectively achieved for the adsorption from clarified and unclarified feedstocks at SBH of 29.8 cm. Both values were greater than those achieved for STREAMLINE DEAE in the BRG contactor operated at the same settled bed height (see Table 3). This was in line with the equilibrium adsorption data presented in Table 2. However, in contrast to STREAMLINE DEAE, the dynamic capacity of DEAE Spherodex LS at C/Co = 0.1 was significantly reduced in operations with particulate feedstocks. However, with one exception at SBH = 29.8 cm, the efficiency of protein recovery (relative to the total challenge) was consistently higher (>78%) for DEAE Spherodex LS loaded to effective saturation in the absence or presence of cells.

It should be emphasised that recovery efficiency is shown in Table 3 in respect of apparent saturation capacity in order to further illustrate the quality of breakthrough behaviour depicted in Figures 2-4 (high and low values respectively for steep and shallow breakthrough curves). No recommendation for practical process operation is intended since, as noted in the Conclusions, washing and protein desorption would be conventionally initiated at C/Co = 0.1. Although not undertaken here, examination of Figures 2–4 and the dynamic capacities in Table 3 suggest that such recoveries would be improved for STREAMLINE DEAE when operated with cell suspensions. This would not be true for DEAE Spherodex LS and invites the conclusion that physical behaviour such as channelling of STREAMLINE DEAE in both contactors restricts efficient protein capture in the absence of cells, whilst damping of perturbances in the presence of biomass improves performance in that circum-

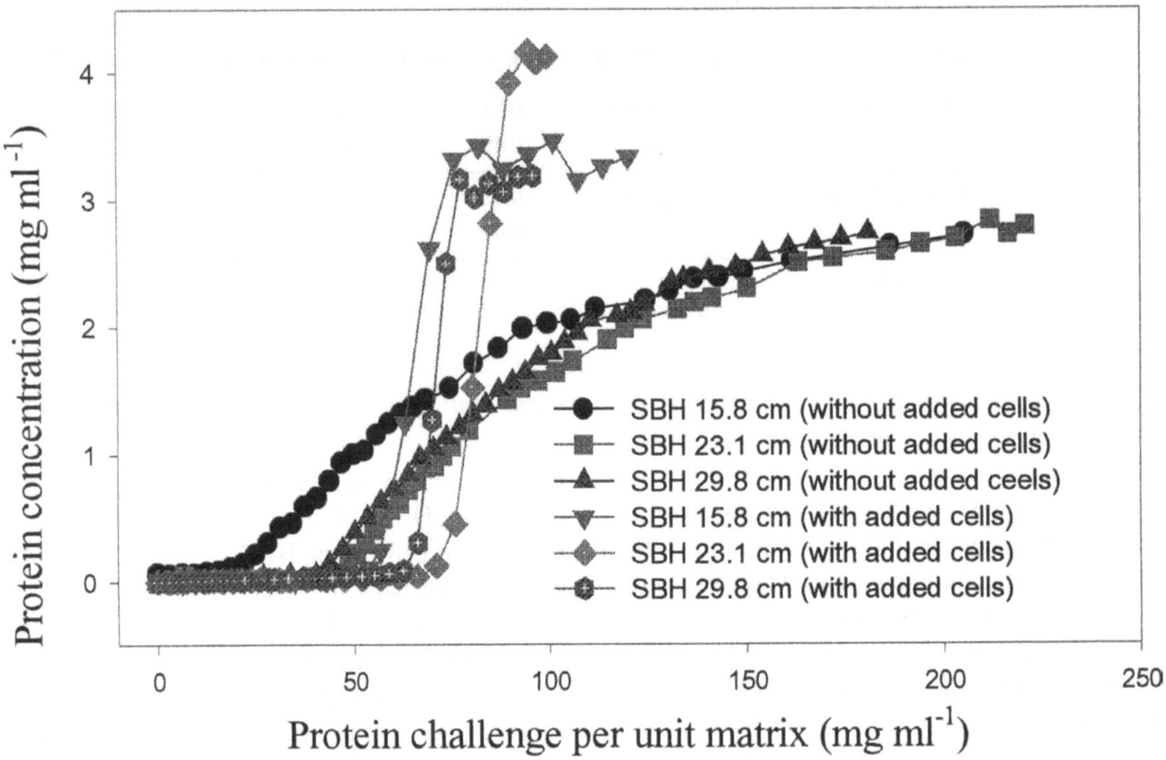

Figure 2. The protein adsorption to STREAMLINE DEAE in the absence and presence of biomass in the Expanded Bed contactor. The experiments were operated at 200 cm hr^{-1} with various SBH at room temperature. 3 mg ml^{-1} BSA (Co) dissolved with or without 20g DCW $litre^{-1}$ yeast cells in buffer A was bound to STREAMLINE DEAE in a single pass system.

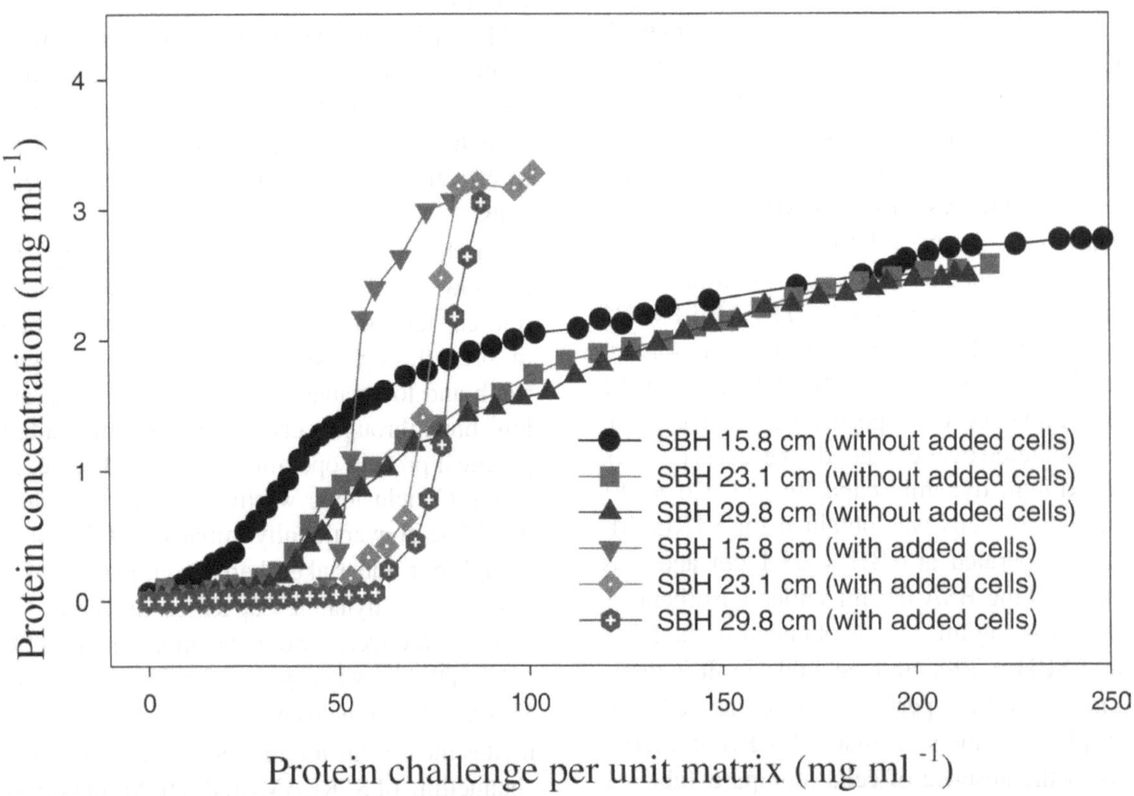

Figure 3. The protein adsorption to STREAMLINE DEAE in the absence and presence of biomass in the BRG contactor. The experiments were operated at 200 cm hr^{-1} with various SBH at room temperature. 3 mg ml^{-1} BSA (Co) dissolved with or without 20g DCW $litre^{-1}$ yeast cells in buffer A was bound to STREAMLINE DEAE in a single pass system.

Table 3. A Comparative Summary of Biochemical Performance

	Apparent saturation capacity (mg ml^{-1})[1]		Dynamic capacity at C/Co=0.1 (mg ml^{-1})[2]		Recovery efficiency of protein challenge (%)[3]	
STREAMLINE DEAE EB SBH	(–)[4]	(+)	(–)	(+)	(–)	(+)
15.8 cm	73	60	27	57	33	52
23.1 cm	98	64	51	70	45	76
29.8 cm	84	61	48	66	37	69
STREAMLINE DEAE BRG SBH	(–)	(+)	(–)	(+)	(–)	(+)
15.8 cm	82	66	15	56	32	78
23.1 cm	88	72	35	58	43	68
29.8 cm	90	70	37	64	39	80
DEAE Spherodex LS BRG SBH	(–)	(+)	(–)	(+)	(–)	(+)
15.8 cm	108	100	90	65	84	79
23.1 cm	114	102	101	75	87	82
29.8 cm	120	106	110	77	90	82

1. Apparent saturation capacity was described as adsorbed protein (mg) per settled adsorbent volume (ml) and was determined by mass balance following loading and washing of adsorbate.
2. Dynamic capacity was estimated from a mass balance of feedstock input and breakthrough outlet of protein up to the point of C/Co = 0.1.
3. Recovery efficiency of protein challenge was defined as the percentage ratio of total eluted protein to total protein feeding challenge.
4. (–): Protein adsorption in the absence of yeast cells (+): Protein adsorption in the presence of yeast cells

stance. For DEAE Spherodex LS, physical performance is good in both cases, but adsorption capacity is incrementally reduced by the presence of complex anionic cell material. It is presently not known if differences in bed expansion have an influence upon the breakthrough curves recorded for STREAMLINE DEAE and DEAE Spherodex LS.

Conclusion

The bed expansion characteristics observed for the two adsorbents (STREAMLINE DEAE and DEAE Spherodex LS) in the BRG contactor did not vary significantly with settled bed height (15.8 to 29.8 cm) and were closely similar to performances observed earlier in the EB contactor (G Hamilton and A Lyddiatt; unpub). Comparative estimates of theoretical plate number, axial dispersion and Bodenstein number (Table 1) for the EB and BRG contactors indicated that the view of the latter as a simple, well-mixed fluidised bed was an oversimplification for the range of SBH values studied. Batch binding studies of BSA adsorption to STREAMLINE DEAE and DEAE Spherodex LS in the absence and presence of biomass (20 g DCW yeast litre^{-1}) suggested that the latter has a higher affinity and capacity for albumin under the experimental conditions (10 mM Tris-HCl, pH 7.5 as solvent system).

Breakthrough analyses of albumin binding to STREAMLINE DEAE in the absence of cells indicated that the performance of EB and BRG contactors approached equality at the greatest values of SBH (23.1 and 29.8 cm), but the breakthrough curves proved rather shallow. Similar experiments conducted with particulate feedstocks (20 g DCW yeast litre^{-1}) yielded steeper breakthrough curves for STREAMLINE DEAE operated in both EB and BRG contactors (refer to Figures 2 and 3). Here the presence of biomass served to *increase* the dynamic capacity (at C/C0 = 0.1) in both contactors to values which were closely similar for all values of SBH. This is an important observation, since 10% product breakthrough is the maximum likely tolerance in an efficient adsorption operation. It was concluded that the presence of cells as complex anions dampened the mixing and

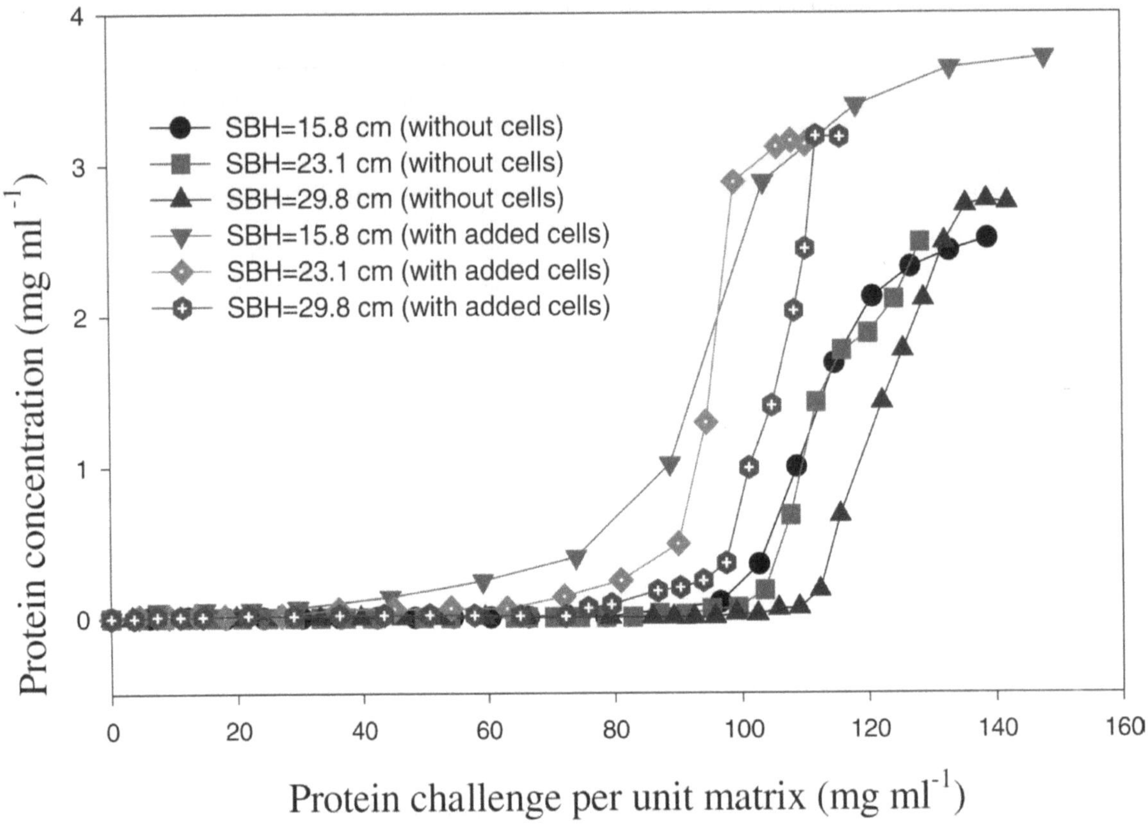

Figure 4. Protein adsorption of DEAE Spherodex LS in the absence and presence of biomass in the BRG contactor. The experiments were operated at 200 cm hr^{-1} with various SBH at room temperature. 3 mg ml^{-1} BSA (Co) dissolved with or without 20g DCW $litre^{-1}$ yeast cells in buffer A was bound to DEAE Spherodex LS in a single pass system.

channelling phenomena which contributed to performances observed with cell-free feedstocks (Figures 2 and 3). That this might be a phenomenon specific for STREAMLINE DEAE was indicated by study of albumin binding to DEAE Spherodex LS in the absence and presence of cells in the BRG contactor (see Figure 4) and the EB contactor at SBH equal to 13 cm (G Hamilton and A Lyddiatt; unpub). Here breakthrough curves were similar, with the presence of biomass serving only to reduce the dynamic and saturation capacities of the adsorbent (see Figure 4 and Table 3). Negatively charged cells would be expected to compete with albumin at pH 7.5 for association with positively charged DEAE-groups on the adsorbent. However, both dynamic and effective saturation capacities for albumin on DEAE Spherodex LS in the absence and presence of cells were the greatest seen in this study, and were complemented by the percentage recoveries of protein achieved in fixed bed elution in 0.45 M NaCl in buffer A.

We conclude that the simple, economic design of the BRG contactor, comprising a 2.5 cm glass tube fitted with a 98 μm basal mesh sheet as flow distributor, compares favourably in physical and biochemical performance with commercial EB contactors. Performance is closely equivalent with albumin binding to STREAMLINE DEAE in the presence of 20 g DCW yeast $litre^{-1}$. The impact of biomass upon the dynamic capacity and shape of breakthrough curve for STREAMLINE DEAE deserves further study.

Acknowledgement

JC-WL gratefully acknowledges the support for his postgraduate studentship from the School of Chemical Engineering, University of Birmingham. GEH was funded by the Birmingham BBSRC Rolling Programme in Biochemical Engineering.

References

Bradford M M. (1976) A rapid and sensitive method for the quantitation of microgram quantities of protein utilising the principle of protein-dye binding. Anal Biotech 72: 249–254

Chase HA (1984) Prediction of the performance of preparative affinity chromatography. J Chromatography. 297: 179–202

Chase HA (1994) Purification of proteins by adsorption chromatography in expanded beds. Trends in Biotechnology 12: 296–303

Gibson NB and Lyddiatt A (1993) Cellulose composites in liquid fluidised bed adsorption and recovery of protein. In: Kennedy JF, Phillips GO and Williams PA (eds.) Cellulosics – materials for selective separations and other technologies (pp. 55–62) Ellis-Horwood, UK.

Gilchrist GR, Burns MT and Lyddiatt A (1994) Solid phases for protein adsorption in liquid fluidised bed. In: Pyle DL (ed.) Separations for Biotechnology 3 (pp. 329–335) Royal Society of Chemistry, Cambridge.

Hjorth R, Kampe S and Carlsson M (1995) Analysis of some operating parameters of novel adsorbents for recovery of protein in expanded beds. Bioseparation 5:217–223

Luca L. De, Hellenbroich D, Titchener-Hooker NJ and Chase HA (1994) A study of the expansion characteristics and transient behaviour of expanded beds of adsorbent particles suitable for bioseparations. Bioseparation 4: 311–318

Lyddiatt A (1990) Immunoaffinity adsorption: Practical application in laboratory research and production. In: Zola H (ed.) Focus on Immunological Methods (pp. 181–203) CRC Press, USA.

Mao QM & Hearn MTW (1996) Optimization of affinity and ion-exchange chromatographic processes for the purification of proteins. Biotechnology and Bioengineering. 52: 204–222

Morton PH & Lyddiatt A (1992) Direct recovery of protein products from whole fermentation broths. In: Slater MJ (ed.) Ion Exchanger Advances (pp. 237–244) Elsevier Applied Science, Netherlands.

Morton PH and Lyddiatt A (1994) Direct integration of protein recovery from productive fermentations. In: Pyle DL (ed.) Separations for Biotechnology 3 (pp. 329–335) Royal Society of Chemistry, Cambridge.

STREAMLINE^{TM} 25 Column User Manual, Amersham Pharmacia Biotech, SE-751 84 Uppsala, Sweden

Thoemmes J (1997) Fluidised bed adsorption as a primary recovery step in protein purification. Adv Biochemical Engineering. 58: 185–230

Wells CM and Lyddiatt A (1987) Liquid fluidised bed adsorption in biochemical recevery from biological suspensions. In: Verrall MJ and Hudson MJ (eds.) Separations for Biotechnology (pp. 436–445) Ellis-Horwood, Chichester.

Address for correspondence: A. Lyddiatt, Biochemical Recovery Group, Centre for Bioprocess Engineering, School of Chemical Engineering, University of Birmingham, Edgbaston, Birmingham, B15 2TT, UK.

Bioseparation **8:** 53–67, 1999.
© 1999 *Kluwer Academic Publishers.*

On-line monitoring of the purification of GST-(His)$_6$ from an unclarified *Escherichia coli* homogenate within an immobilised metal affinity expanded bed

R.H. Clemmitt, L.J. Bruce & H.A. Chase
Department of Chemical Engineering, University of Cambridge, Pembroke Street, Cambridge, CB2 3RA, United Kingdom

Received 10 September 1998; accepted in revised form 20 March 1999

Key words: expanded bed adsorption, glutathione s-transferase, immobilised metal affinity chromatography, on-line monitoring, polyhistidine tails

Abstract

The use of a rapid chromatographic assay to monitor the level of a specific protein during its downstream processing by expanded bed adsorption is described. An expanded bed column (5 cm diameter) has been modified to allow the abstraction of liquid samples at various heights along the bed, in an automated, semi-continuous manner throughout the separation. The withdrawn samples were filtered in-line and the level of the target protein assayed by a rapid on-line chromatographic method. Using this technique it was possible to monitor the development of adsorbate profiles during the loading, washing and elution phases of the application of an unclarified feedstock. The potential of the technique is demonstrated using the separation of histidine tagged glutathione s-transferase (GST-(His)$_6$) from an unclarified *Escherichia coli* homogenate using an expanded bed of Ni^{2+} loaded STREAMLINE Chelating™. The level of GST-(His)$_6$ in the abstracted homogenate samples was measured using Zn^{2+} loaded NTA-silica as an affinity chromatographic sensor. The approach described demonstrates potential for the on-line monitoring and control of expanded bed separations and for providing a greater understanding of adsorption/desorption and hydrodynamic processes occurring within the bed.

Abbreviations: CDNB – 1-chloro-2,4-dinitrobenzene; dH^{2}O – distilled water; EBA – expanded bed adsorption; EDTA – ethylenediaminetetraacetic acid; FPLC – fast protein liquid chromatography; GST-(His)$_6$ – histidine tagged glutathione s-transferase; IDA – iminodiacetic acid; IPTG – isopropyl β,D-thiogalactopyranoside; K_d – dissociation constant; LBamp – Luria broth (25 g l^{-1}) with ampicillin (100 μg ml^{-1}); NTA – nitrilotriacetic acid; q_m – equilibrium binding capacity

Introduction

Expanded bed adsorption enables proteins to be recovered directly from particulate containing feedstocks, thereby combining clarification, concentration and initial purification in a single operation (Chase, 1994; Hjorth, 1997; Thömmes, 1997). The higher voidage fraction within the bed enables the free passage of suspended solids, whilst the target protein is adsorbed to the stationary phase. The loose, fluidised nature of the expanded bed also allows the withdrawal of liquid samples for the on-line monitoring of developing adsorbate profiles. Such sampling may be performed without seriously affecting the separation efficiency, as determined by the minimal effect on the liquid dispersion within the bed or the resulting breakthrough curves (Bruce and Chase, 1998).

Traditionally, the adsorption process is followed by assaying the levels of the key components in the effluent stream from the bed. However, control strategies based on effluent concentrations can result in the wastage of the adsorbate contained within the voids of column. On-line monitoring of the levels of key components at various heights within the bed allows for tighter control of the separation and greater process efficiency. During the batchwise operation of adsorption processes such as expanded beds, the information may be used to control the duration of the adsorption, washing and elution stages. The adsorption phase should be continued until the level of the target protein just begins to break through the column. Further loading would result in uneconomic loss of product, less loading would result in under utilisation of the adsorbent and low productivity. The duration of the wash stage should be such that cell debris and other contaminants have been reduced to an acceptable level, whilst minimising buffer consumption and product loss. During elution, effluent monitoring can be used to ensure that only those fractions containing the majority of the product are pooled, whilst minimising the collection of contaminating molecules.

Various types of on-line assay have been used for monitoring the level of proteins during bioprocesses, such as flow injection analysis for enzyme activity (Richardson et al., 1996a), enzyme-linked immunosorption analysis (Nilsson et al., 1994) or optical biosensors (Holwill et al., 1996). Chromatographic assay systems may also be used to yield the same information using either non-specific adsorbents such as ion exchangers, (Chase et al., 1998) or more specific affinity adsorbents. Non-specific media may be used to resolve most protein mixtures adequately, however the chromatograms produced can be difficult to analyse in a truly on-line manner. Also, the sensitivity of such devices is governed by the dissociation constant of the interaction, the higher values associated with complex feeds such as cell homogenates make adsorption inefficient at low concentrations. The use of highly selective affinity media for chromatographic monitoring has a number of advantages. The selectivity of the interaction makes peak identification easy as the eluted protein is well 'resolved' from other proteins. The lower dissociation constants make adsorption efficient at low concentrations. And unlike non-specific adsorbents where sample size must be small compared to the bed volume, in order to achieve the required resolution, larger volumes may be applied greatly increasing the sensitivity of the assay. Factors important in the design of a successful chromatographic assay include the choice of interaction type (with appropriate adsorption behaviour in terms of the equilibrium binding capacity, q_m and the dissociation constant, K_d, and suitable elution schemes), matrix, bed volume, flowrate, sample volume and a consideration of the fouling properties of the ligand and matrix (Chase, 1986).

Our particular interest is in the simplification of schemes for the purification of proteins. By combining expanded bed methodology with an affinity fusion approach, one obtains a simple protein production and facilitated purification process. The fusion of a sequence of six histidines to a protein confers a higher affinity than any host cell protein, enabling efficient purification using metal loaded STREAMLINE Chelating™ directly from unclarified cell homogenates. Recent work in this laboratory, has demonstrated that STREAMLINE Chelating™ can be used to purify hexa-histidine tagged glutathione s-transferase (GST-$(His)_6$) to electrophoretic purity at a yield of 85% directly from unclarified *E. coli* homogenates (Clemmitt and Chase, unpub.). The adsorbate, GST-$(His)_6$, would be a good target for chromatographic assay using its affinity for either immobilised metals or glutathione. This separation was scaled up 25× from the 1 cm dia. column used originally to a 5 cm dia. column and formed the basis of a model system for the development of this approach to on-line monitoring.

In this paper, we demonstrate the withdrawal of liquid samples of an unclarified *E. coli* homogenate from various points within the expanded bed, and the subsequent on-line monitoring of the front of adsorbate (GST-$(His)_6$) during the loading, washing and elution stages of the separation process. The approach described here could be developed to provide on-line information for the control and optimisation of expanded bed adsorption or for providing information on the adsorption/desorption mechanisms within the bed, to aid in design and understanding of expanded bed separation processes.

Materials and methods

All experiments were performed at room temperature unless otherwise noted.

Materials

STREAMLINE Chelating™ (100–300 μm, 1.2 g ml^{-1}, 34.1 μmol Cu^{2+} ml^{-1} gel) used in this work was a gift from Pharmacia Biotech (Uppsala, Sweden). This has the metal chelating ligand iminodiacetic acid (IDA) attached to the agarose-quartz composite STREAMLINE matrix. NTA-silica (16–24 μm, 1 g) was purchased from Qiagen Ltd. (West

Sussex, UK). *E. coli* DH5α, containing the plasmid pGEX-5X-1 encoding the IPTG inducible expression of GST-$(His)_6$, was a gift from Dr Jonas Lidholm of Pharmacia & Upjohn Diagnostics (Uppsala, Sweden). Polypropylene glycol (average MW 1025) was purchased from BDH (Poole, United Kingdom). All other chemicals used were obtained from Sigma (Poole, United Kingdom). The 0.45 μm membrane filters (0.5 ml void volume, 13 mm dia.) were purchased from Whatman (Whatman, UK). The GST Detection Module used for the assay of GST activity in clarified bacterial sonicates was a gift from Pharmacia Biotech (Uppsala, Sweden).

Methods

Fermentation

E. coli DH5α pGEX-5X-1 was supplied in the form of an agar stab, and the bacteria were transferred to 30 ml LBamp broth (25g l^{-1} LB broth base, 100 μg ml^{-1} ampicillin) and grown overnight. Glycerol was then added to 20%, and the culture stored as 1 ml aliquots at −20 °C. When required, one of these cultures was used to inoculate 30 ml LBamp in a 250 ml flask, this was then left for 6–7 hours at 30 °C, 200 rpm in an LH orbital incubator (Inceltech (UK), Berkshire). This sub-culture was then transferred to 1 l of LBamp in a 2 l flask which was left overnight at 30 °C, 200 rpm. This was used to inoculate the 5 l fermentor.

All fermentations were performed in an LH 2000 fermentor (Inceltech (UK), Berkshire) with a working volume of 5 l. The defined growth medium was a modified M 9 media, with 25 g l^{-1} glycerol as the carbon source. The pH was maintained at 6.8 by addition of 33% (v v^{-1}) ammonia or 2 M nitric acid. The fermentation temperature was controlled at 30 °C, whilst O_2 was continuously sparged at a rate of 4 l min^{-1}. The agitation speed was 800 rpm throughout the fermentation. Foaming was controlled manually by the addition of polypropylene glycol 1025. After approximately 9 h fermentation, and at an OD 600 of 10–12, IPTG was added to 1 mM (final concentration). Then, after further fermentation of 5–6 h the bacteria were harvested by centrifugation at 10000 rpm, 4 °C for 15 min (JA-10, Beckman) and stored as pellets at −20 °C until use. Three fermentations were performed to generate the material for the expanded bed.

Homogenisation

The frozen pellets were thawed on ice, and then resuspended in 100 mM sodium phosphate, 0.5 M NaCl, 20 mM imidazole pH 7.5 to a tenth the volume at the end of fermentation (~1500 ml). The cell pellet was thoroughly dispersed using a cell suspender (Ultra-Turrax T 18/10, Sartorious Instruments, Surrey, UK), operated at 1/3 speed, for six 20 second pulses, with cooling from an ice bath. Endonuclease was then added (Sigma E-1014, activity 419 U μl^{-1}, 60 μl) at a level of 8.33 μl per litre final volume (after the dilution following disruption).

The cells were then disrupted using 6 passes through a continuous flow ultrasonicator (Lucas Dawe Ultrasonics, UK) at 52 ml min^{-1} and a power setting of 8, with cooling from a water jacket supplied with a 50% ethylene glycol/water mix. The level of released GST-$(His)_6$ was then measured and the homogenate diluted using the suspension buffer to the desired final concentration. Finally the pH of the resulting preparation was adjusted to 7.5, using 1 M NaOH.

Expanded bed adsorption

Expanded bed adsorption was carried out in a modified column based on a 5 cm dia. STREAMLINE 50 column (Pharmacia Biotech, Uppsala, Sweden). The modified column had a custom designed and built glass tube (length 100 cm), with sampling ports at heights of 5, 15 and 30 cm, in the place of the supplied tube. Each port consisted of a short, sintered glass tube inserted into 0.5 cm diameter holes in the wall of the column and held in place using silicone sealant, the sinter terminating flush with the inner wall surface (Figure 1).

The adsorbent used in the column was STREAMLINE Chelating™. The column was filled with 300 ml of adsorbent, giving a settled bed height of 15 cm. Prior to homogenate loading, the column was equilibrated, in expanded mode at 48 ml min^{-1} (147 cm h^{-1}) with distilled water and then charged with Ni^{2+} ions using 2 column volumes of 50 mM $NiSO_4$ in distilled water. Non- and loosely bound metal ions were washed from the bed using 5 column volumes of 100 mM sodium acetate, 0.5 M NaCl pH 4.0. The bed of adsorbent was then equilibrated with loading buffer (100 mM sodium phosphate, 0.5 M NaCl, 20 mM imidazole pH 7.5) at 48 ml min^{-1} (147 cm h^{-1}) so as to achieve twice settled bed height (30 cm). After approximately 1 h, when the bed was stable, the feedstock was applied at the same flowrate and then progressively decreasing flowrates to maintain a constant degree of bed expansion (Chang and Chase, 1996). Following loading, the bed was washed with

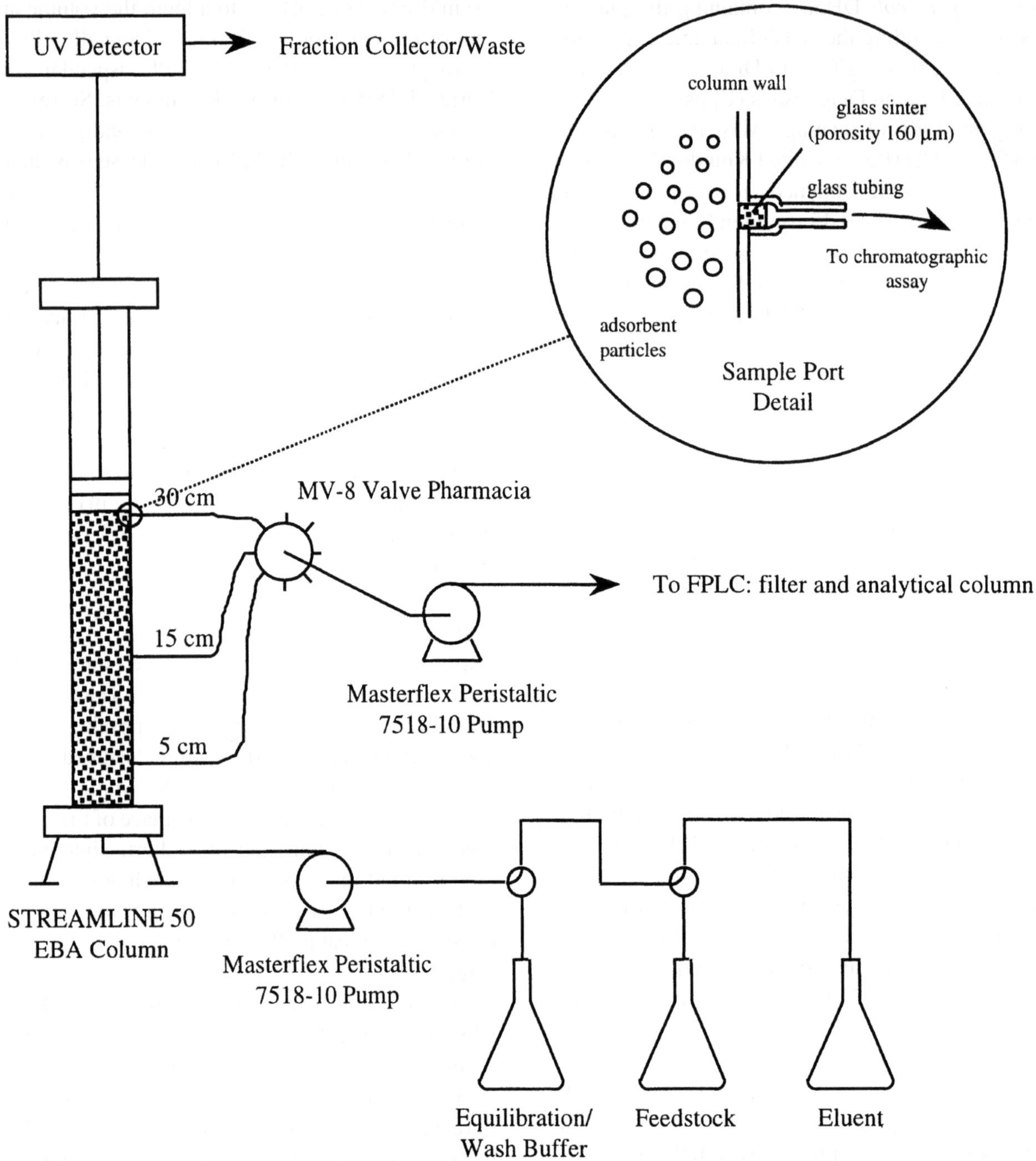

Figure 1. Modified expanded bed adsorption column and sample extraction system (with sampling port detail).

loading buffer, and then eluted using 100 mM sodium phosphate, 0.5 M NaCl, 500 mM imidazole pH 7.5, again in expanded mode. The bed was regenerated with 50 mM EDTA in distilled water to remove strongly bound proteins together with the remaining Ni^{2+} ions adsorbed by the IDA groups.

On-line monitoring assay

Samples of the liquid phase were withdrawn from the bed at heights of 5 cm (16.7%), 15 cm (50%) and 30 cm (100%), in an automated, semi-continuous fashion using a peristaltic pump, at 1.3 ml min^{-1}, as in Figure 1. The samples withdrawn from the bed were assayed using rapid immobilised metal affinity chromatography on a 1 ml bed of Zn^{2+} loaded NTA Silica contained within an HR5/5 column (Pharmacia Biotech, Uppsala, Sweden). The chromatographic assay was automated using the FPLC chromatography system (Pharmacia Biotech, Uppsala, Sweden) containing P-500 piston pumps, an LCC-500 controller,

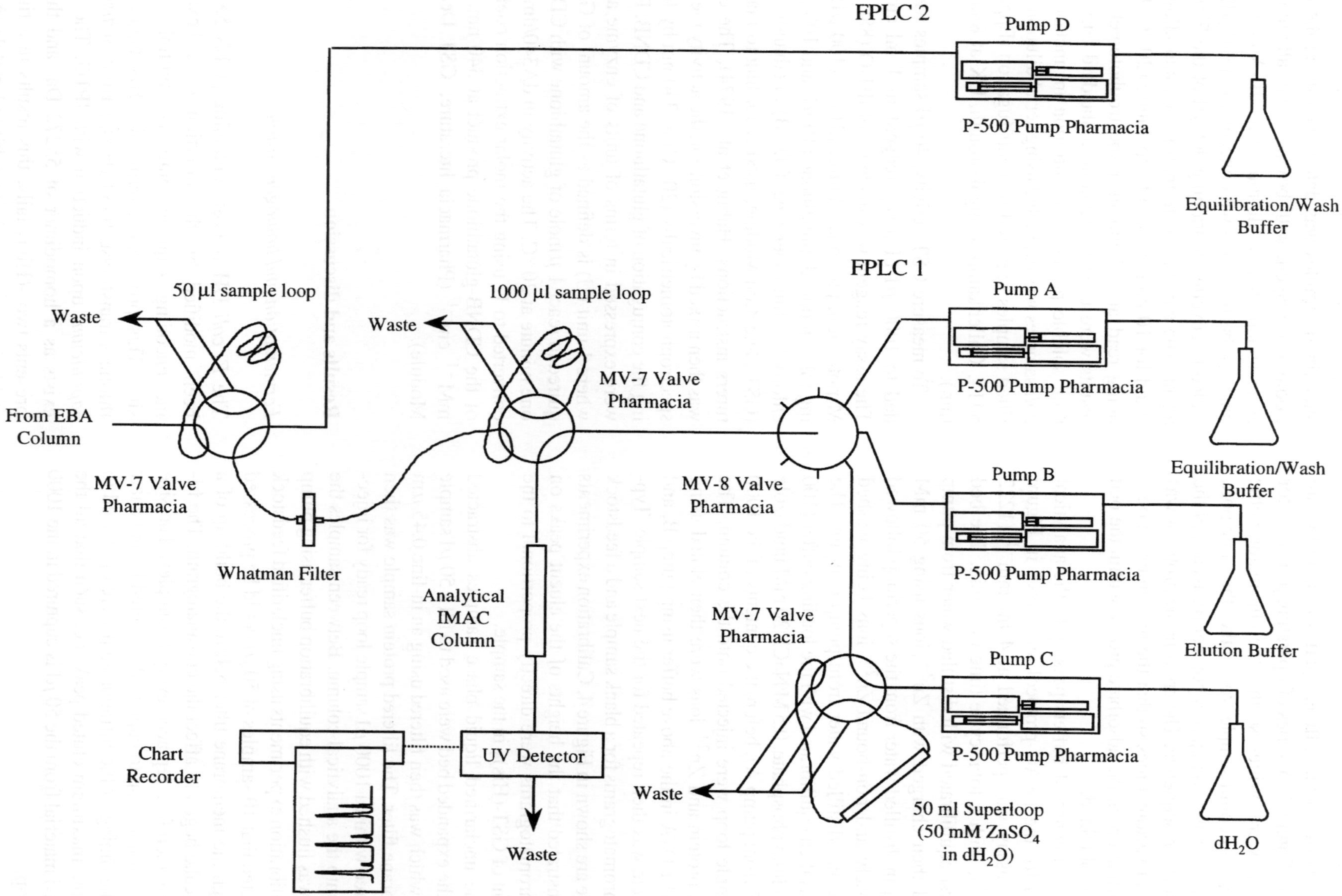

Figure 2. Schematic of FPLC arrangement for filtering and subsequent rapid chromatographic monitoring. The concentration of GST-(His)$_6$ in the withdrawn samples were evaluated from the elution peak of the chromatogram during the separation.

MV-7 and MV-8 solenoid valves, and a UV-1 monitor to measure the absorbance at 280 nm. The interconnections between the different parts of the FPLC are shown in Figure 2. A series of operating protocols for the FPLC were written so that the chromatographic assay system could sample from either the same port, alternately from two adjacent ports or from each of the three ports in sequence. The appropriate protocol was chosen at various times so that the maximum amount of information on the adsorbate profile within the bed could be obtained.

The sequence of buffers passing to the analytical column for the IMAC protocol are shown in Figure 3. The assay was performed at 4 ml min^{-1} (1222 cm h^{-1}), with a total cycle time of 4 min. The bed was first equilibrated with distilled water from pump C, and then charged with Zn^{2+} ions using 50 mM $ZnSO_4$ in distilled water from the superloop followed by loosely and non-bound Zn^{2+} ions being washed away using distilled water from pump C again. The bed was then equilibrated with the loading buffer (100 mM sodium phosphate, 0.5 M NaCl, 20 mM imidazole pH 7.5 from pump A), before the contents of the 1000 μl sample loop were injected onto the column. The bound protein and Zn^{2+} ions were then eluted using 50 mM EDTA in the above buffer using pump B, and the cycle was then repeated for the next sample. Typical chromatograms for a blank sample and a feedstock sample are shown in Figure 4. Calibration experiments demonstrated that the heights of the eluent peaks on the chromatograms were directly proportional to the amount of GST-$(His)_6$ in the sample.

The unclarified liquid phase samples abstracted from the expanded bed were used to fill a 50 μl sample loop, which was then filtered using an in-line 0.45 μm membrane filter. The filtered protein sample was then captured within a 1000 μl sample loop ready for injection onto the analytical column. Between samples the filter was flushed with equilibration buffer using pump D. Calibration experiments using unclarified feedstock indicated that 10 samples of 50 μl could be processed through the membrane filter before the build up of a filter cake began to affect the chromatogram. The filter was therefore replaced every 6 samples. The size of the sample loop to capture the filtered homogenate and the timing of the filtration step were optimised to give the maximum eluted peak, i.e. such that all the filtered material from the 50 μl is captured in the 1000 μl loop.

Protein determination and enzyme assays

The total protein concentration in the clarified and unclarified samples was determined using the Pierce Coomassie reagent with bovine serum albumin used as a standard. Serial dilutions of the BSA in the equilibration buffer of known concentration (40 μl) were added to cuvettes containing 960 μl of the Pierce reagent. The cuvettes were then mixed and allowed to stand for 10 min at room temperature (20 °C) before being read at 595 nm in a spectrophotometer. The results were used to generate a standard curve. The total soluble protein concentration in unclarified *E. coli* samples was evaluated using serial dilution of clarified samples prepared by centrifugation (Eppendorf 5414, Gallenkamp, Loughborough, UK at 8800 *g*, 8 min).

To measure GST activity, 40 μl samples were added to 1000 μl of assay reagent in 1 ml cuvettes. The assay reagent was made up of dH_2O (880 μl), 1 M potassium phosphate buffer pH 6.5 (100 μl), 30.8 mg ml^{-1} reduced glutathione (10 μl) and 100 mM 1-chloro-2,4-dinitrobenzene (CDNB) in ethanol (10 μl) (GST Detection Module, used according to manufacturers instructions, Habig et al., 1974). The cuvette was then mixed by inversion and the activity measured spectrophotometrically (20 °C) at 340 nm by following the conjugation of glutathione and CDNB. Results were expressed in terms of units of enzyme activity where 1 unit (U) is defined as the amount of GST required to react 1 μmole of glutathione with CDNB in one minute at 20 °C. The activity in dA340/dmin was converted to U using the molar extinction coefficient of the CDNB-glutathione product at 340 nm, ε=9.6 mM^{-1} cm^{-1} (Pharmacia literature, GST Detection Module).

Results and discussion

Fermentation and homogenisation

The *E. coli* DH5α host containing pGEX-5X-1 had been modified by the insertion of a hexahistidine encoding sequence into the multiple cloning site. This produces C-terminally His-tagged glutathione s-transferase (GST-$(His)_6$) from *Schistosoma japonicum*, upon induction with IPTG. The protein exists as a homodimer of 55272 Da, and therefore presents two -$(His)_6$ tails, this results in a tight and stable interaction with immobilised metals (Nieba et al., 1997).

(i) Detail of the material passing onto the chromatographic assay column through the course of the 4 minute assay

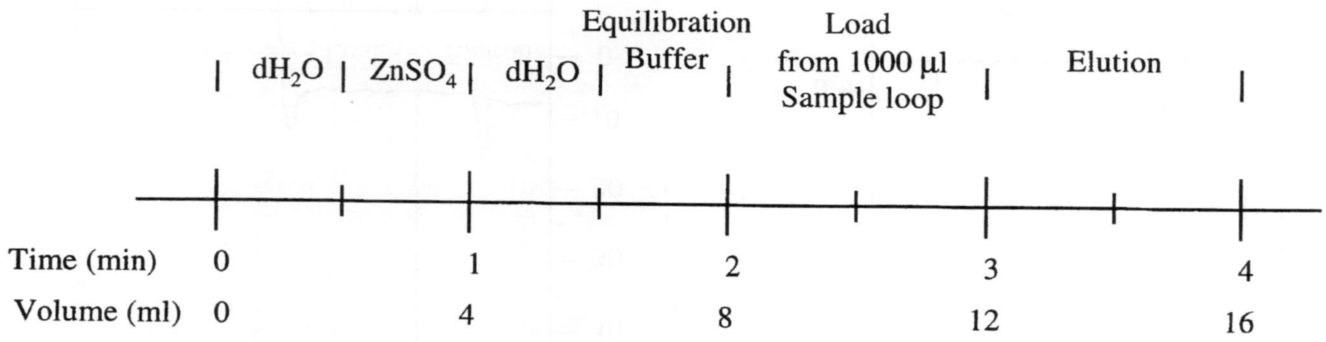

(ii) Detail of the sampling point, sample filtration and loading point

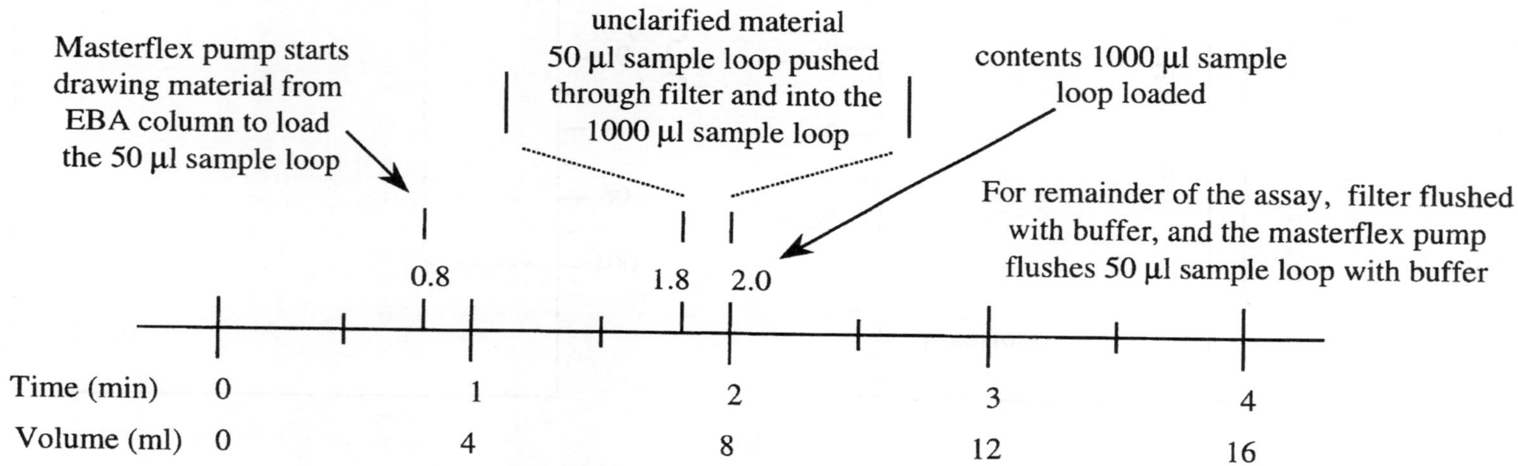

Figure 3. Sequence of buffers passing to the analytical NTA silica column, (i) analytical column preparation, sample loading and elution and (ii) timing of sample port selection, sample filtration and synchronisation with loading point of analytical column.

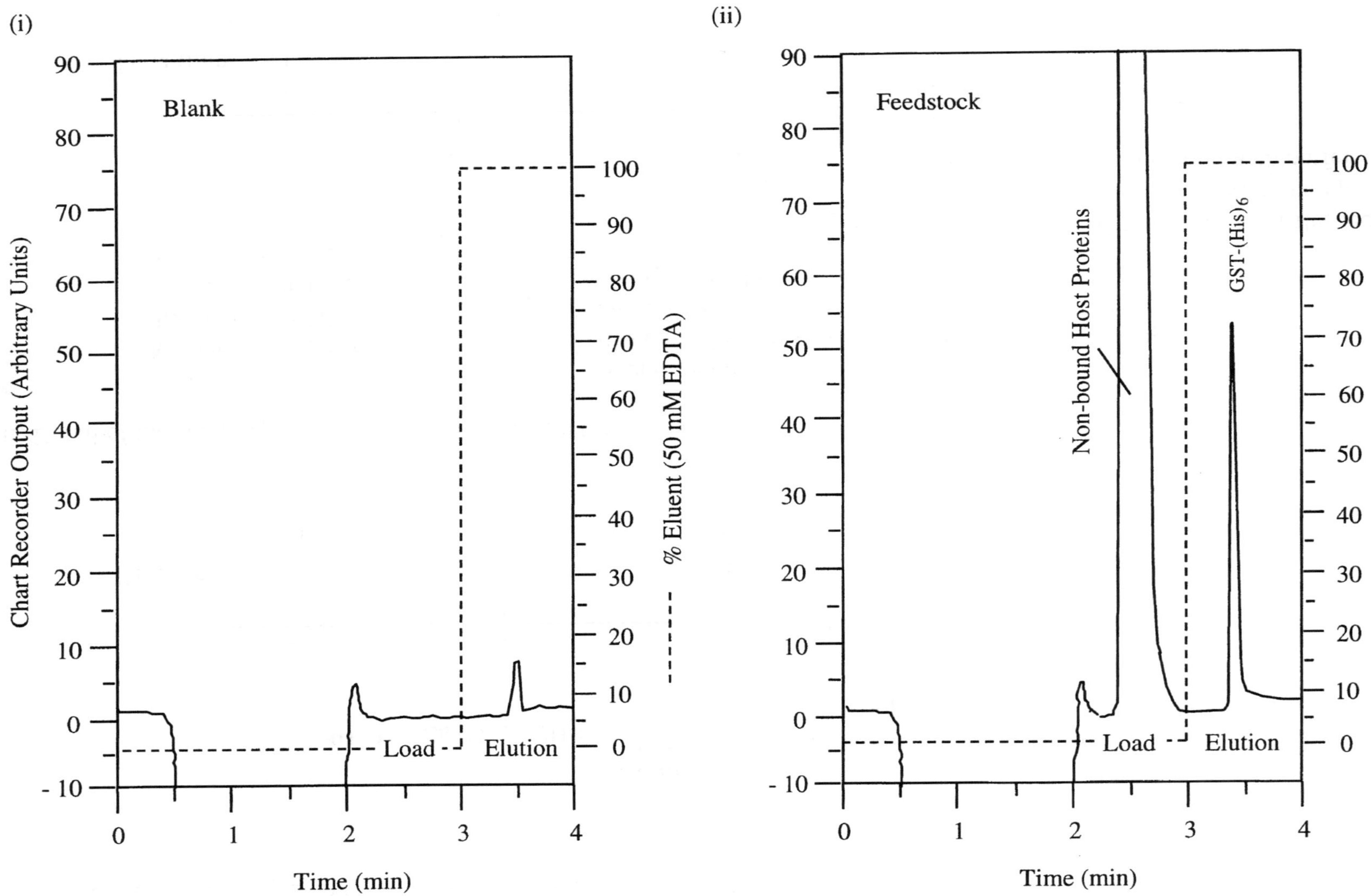

Figure 4. Schematic diagram of the separation of homogenate components using rapid chromatographic assay (i) control assay – running buffer injected, (ii) *E. coli* homogenate feedstock injected at concentration used to load the expanded bed.

The large scale fermentations were designed such that, following induction with IPTG, the level of soluble GST-(His)$_6$ activity had reached a plateau of 12.4 U ml^{-1} ($\pm$ 0.6 U ml^{-1}) at the end of the fermentation. The level of expression at harvest was estimated as 30% total protein, and the dry weight of *E. coli* was 8 g l^{-1}.

The resuspended *E. coli* cell paste was disrupted using the continuous flow ultrasonicator in a multiple pass fashion, until the level of GST-(His)$_6$ had reached a maximum. The amount of endonuclease added to the cell suspension and the disruption as a whole were not optimised as the aim of this experimentation was simply to develop the on-line monitoring concept and not to optimise the large scale separation. The resulting homogenate was then diluted to 7 l, to give a GST-(His)$_6$ activity of 23 U ml^{-1}, total protein of 7.0 mg ml^{-1}, turbidity of 3.5, and a dry weight of 15 g l^{-1}.

Chromatographic assay

The GST-(His)$_6$ direct extraction was selected for development of the chromatographic on-line monitoring principle because the purification target, GST-(His)$_6$ may be detected using its affinity for either immobilised glutathione (Smith and Johnson, 1988) or immobilised metal ions (Nieba et al., 1997; Clemmitt and Chase, unpub.). Desirable characteristics in the chosen adsorbent are a small particle size, for rapid kinetics and high resolution, and a mechanically stable matrix for application at high flowrates. The chosen ligand should have high capacity, tight binding and high selectivity for the target. As glutathione is only available commercially immobilised on Sepharose-4B, which is highly compressible at high flowrates and associated pressure drops, we decided to develop an immobilised metal affinity chromatographic assay. Ni^{2+} loaded NTA-silica has been used successfully for the detection and purification of histidine tagged proteins. The small 16–24 μm dia., macroporous particles have a recommended maximum flowrate of 3000 cm h^{-1} and maximum pressure of 3000 psi, and the initial intention was to use the adsorbent without recharging the Ni^{2+} for the whole monitoring experiment. However, although the adsorbent initially had a high total binding capacity for GST-(His)$_6$ of 340 U ml^{-1} (at 1 ml min^{-1}, 306 cm h^{-1}: calculated from eluted peak in a 1 ml HR5/5 column), following further feedstock applications to 100% breakthrough, the capacity was found to deteriorate and the bed changed from a light green to a brown colour. This was probably due to the slow reduction and leaching of Ni^{2+}. We therefore decided to recharge the metal ion between individual assays.

The protein retained on chelating adsorbents can be eluted using either a step increase to ~500 mM imidazole, a decrease in pH to ~4.0 or by using EDTA to elute the bound metal ions. Whilst imidazole is used to elute the GST-(His)$_6$ from the expanded bed, the presence of a high level of imidazole in the elution stream disguised the smaller eluted protein peak from the analytical column (500 mM imidazole has an A280 nm ~0.5). Analysis of this stream would require some form of on-line gel filtration, which whilst possible would result in lengthly assay times. It was hoped that the assay results could be simply interpreted in terms of on-line GST-(His)$_6$ levels being equivalent to eluted peak height. A pH step down to 3.0 resulted in no discernible GST-(His)$_6$ peak, probably meaning that the bound protein was precipitated during elution. A 50 mM EDTA solution buffered in the equilibration buffer was used to compare the baseline perturbation when stripping the 1 ml bed of loaded Cu^{2+}, Ni^{2+} and Zn^{2+}. The complex eluted from the Zn^{2+} bed resulted in the smallest peak in the absorbance at 280 nm during EDTA elution (Figure 4 shows a typical baseline trace for Zn^{2+} elution). We therefore chose to use Zn^{2+} to load the NTA-silica, with subsequent elution using buffered EDTA, which made recharging the metal ion between runs a strict necessity.

The assay protocol was therefore developed as described in the methods section and Figure 3. The volumes applied during the bed preparation, metal loading, washing and pre-equilibration steps were set arbitrarily, whilst the loading step was set at 4 ml in order to ensure that the tail of non-bound protein was completely 'resolved' from the eluted peak. The elution step with EDTA was also set at 4 ml to ensure that all of the elution peak is captured on the chart recorder. It would be possible, through careful optimisation, to reduce the length of the assay cycle by reducing the length of the initial bed preparation steps and possibly by increasing the operating flowrate. However as the 4 min cycle time is short compared with the time scale over which the levels of adsorbate are changing during this purification by EBA, further optimisation was not necessary. To monitor a less selective separation in which the available adsorption capacity is low and exhausted rapidly within a few column volumes then a faster assay technique would be required.

The inclusion of a replaceable, in-line filter was needed as the highly turbid *E. coli* homogenate contains a large amount of suspended cell debris which would quickly foul and block the voids of an analytical column. Although in this example this meant that manual intervention was required (to replace the filter periodically), truly automated solutions to this problem are becoming available (Richardson et al., 1996b). The volume of the filter and associated tubing were minimised to reduce dispersion of the 50 μl sample. The filter was not placed directly in-line with the column as during its replacement air bubbles could get into the bed which would adversely affect the chromatographic separation, and possibly mean replacement or repacking of the column.

Expanded bed adsorption

The unclarified *E.coli* homogenate (4600 ml, 7 mg ml^{-1} total protein, 23 U ml^{-1} GST-$(His)_6$) was initially pumped onto the expanded bed at 48 ml min^{-1} (147 cm h^{-1}), however the flowrate was reduced during the run to 13 ml min^{-1} (40 cm h^{-1}) due to the high viscosity of the feedstock. The results of the purification are presented in terms of the chromatogram in Figure 5 and the purification table in Table 1, as measured using the off-line spectrophotometric assays for turbidity, total protein and GST-$(His)_6$ on samples collected during the run. During loading, the chromatogram shows the rapid and sharp breakthrough of particulates with the void volume of the column, followed by the equally sharp breakthrough of general non-bound protein. The level of total protein rapidly plateaus out at around 70% of the inlet value reflecting the fact that the GST-$(His)_6$ represents approximately 30% of the intracellular protein in the *E. coli* at the point of harvest. The GST-$(His)_6$ is completely extracted from the homogenate, until after around 140 min its sharp breakthrough begins. As the aim of the experiment was to fully test the on-line monitoring system, loading was continued beyond the 5% threshold normally employed to prevent adsorbate wastage to about 40% total breakthrough, so that the maximum amount of information could be derived from the 3 ports. Table 1 demonstrates that despite the sub-optimal load, the sharp breakthrough meant that just 5.9% of the GST-$(His)_6$ was lost in the flowthrough. The bed was then washed of solids and non-bound protein in preparation for the elution of bound material. The bed was eluted in expanded mode, a technique gaining in popularity due to the potential simplification of column design (Hjorth, 1998), and of practical necessity here to allow on-line sampling during the elution phase. The step change in buffer to that containing 500 mM imidazole eluted the bound GST-$(His)_6$ in a sharp peak at a yield of 85.4% and a purification fold of 3.0, the apparently low value reflecting the high percentage of the target protein in the feedstock (theoretical maximum purification fold of 3.33).

The above data describe the situation as measured and followed exclusively at the bed exit. The rapid on-line chromatographic assay was used to further characterise this separation during each of the adsorption, load and wash stages.

On-line monitoring and control of the adsorption

The use of this particular unclarified bacterial feedstock, which is typical of many potential expanded bed applications, fully challenged the monitoring system under development. The chromatographic assay performed well, requiring manual intervention only for refilling the Zn^{2+} reservoir and filter changing.

Figure 6 shows the adsorbate profiles as a function of time at ports 1, 2 and 3 during each of the adsorption, wash and elution phases of the above expanded bed separation. The level of GST-$(His)_6$ in each abstracted sample was evaluated using the chromatographic assay system. Before and after performing assays of the EBA procedure, replicate blank samples (running buffer) and homogenate feedstock samples were assayed, the level of GST-$(His)_6$ was calculated from the chart recorder peak height (corrected for the blank peak height), and the data plotted in terms of dimensionless concentration. No significant change in the assay chromatograms for either the blank or feedstock were observed over the course of the run. During the load, the level of GST-$(His)_6$ was seen to rise more rapidly at lower ports. The breakthrough of GST-$(His)_6$ was very shallow at port 1, just 5 cm or 16.7% of the way up the bed, this is due to the presence of a mixed zone just above the distributor. This observation confirms other workers observations that a certain minimum length of column is required for effective adsorption performance (Hjorth, 1995). The breakthrough curve becomes sharper at higher ports, until at the bed exit the level of GST-$(His)_6$ rises in a similar fashion to the off-line data, measured at the bed exit. During the washing stage, the GST-$(His)_6$ took longer to clear from the bottom port than from the top two ports. This could be due to the reversible nature of the equilibrium adsorption process, and since more material is bound to the bed at the base, proportion-

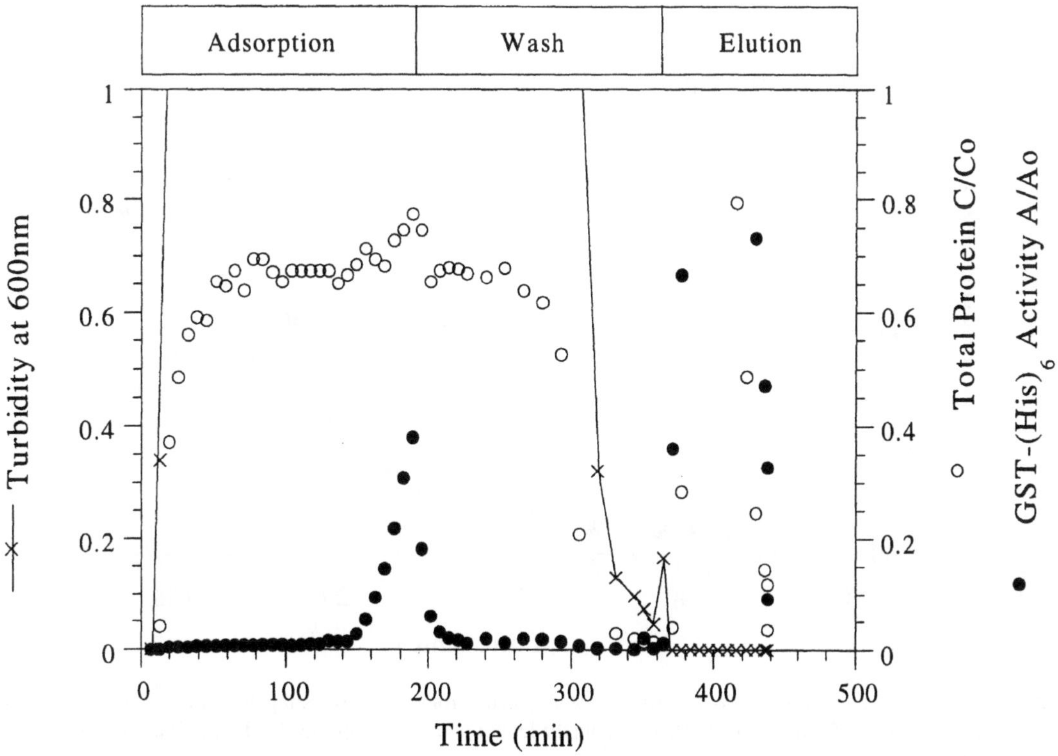

Figure 5. Off-line data: purification of GST-(His)$_6$ from an *E. coli* homogenate using an expanded bed of Ni^{2+} loaded STREAMLINE Chelating. Unclarified *E. coli* homogenate (4600 ml, 7 mg ml^{-1} total protein, 23 U ml^{-1} GST-(His)$_6$) was pumped onto the expanded bed (settled height 15 cm) at a flowrate of 48 ml min^{-1} (147 cm h^{-1}) and then progressively decreasing flowrates to maintain the same degree of bed expansion (approximately 2×). The bed was washed and then eluted in expanded mode using 500 mM imidazole in 100 mM sodium phosphate, 0.5 M NaCl pH 7.5. Regeneration was performed with 50 mM EDTA in dH_2O. Fractions were collected throughout the separation and assayed for turbidity at 600 nm, total protein and GST-(His)$_6$ activity.

Table 1. Purification table for the expanded bed adsorption of GST-(His)$_6$ using Ni^{2+} STREAMLINE Chelating. Unclarified *E. coli* homogenate (4600 ml, 7 mg ml^{-1} total protein, 23 U ml^{-1} GST-(His)$_6$) was pumped onto the expanded bed (settled height 15 cm) at a flowrate of 48 ml min^{-1} (147 cm h^{-1}) and then progressively decreasing flowrates to maintain the same degree of bed expansion (approximately 2×). The bed was then washed and then eluted in expanded mode using 500 mM imidazole in 100 mM sodium phosphate, 0.5 M NaCl pH 7.5. Regeneration was performed with 50 mM EDTA in dH_2O. Fractions were collected throughout the separation and assayed for turbidity at 600 nm, total protein and GST-(His)$_6$ activity

Purification stage	Volume (ml)	Total protein (mg)	Total activity (U)	Yield (%)	Specific activity (U mg^{-1})	Purification factor
Unclarified homogenate	4598	32370	105800	(100)	3.27	(1.0)
Flowthrough	4598	18220	6260	5.9	0.34	–
Wash	1814	4650	490	0.5	0.11	–
Elution (500 mM imidazole)	1505	9220	90410	85.5	9.81	3.0
Regeneration (50 mM EDTA)	1044	260	2210	2.1	8.41	–
Recovery (%)		100	93.9			

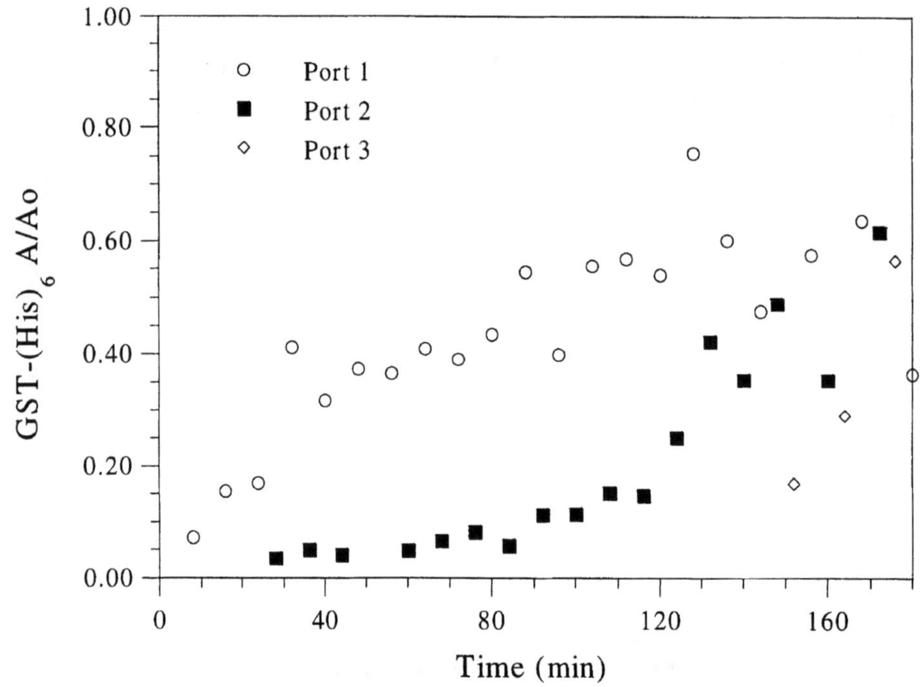

Figure 6. In-bed concentration profiles for GST-(His)$_6$ during expanded bed chromatography. Samples were withdrawn for rapid chromatographic assay from ports 1, 2 and 3, at 5 cm, 15 cm and 30 cm along the length of the bed respectively, during (i) adsorption/loading phase, (ii) washing phase and (iii) elution phase.

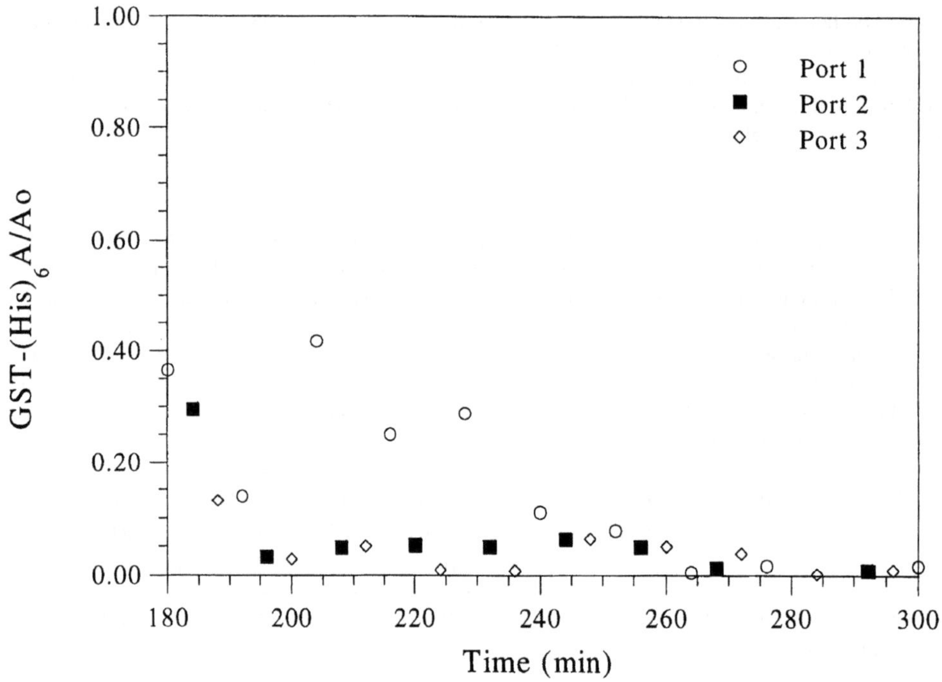

Figure 6.

ately more material may come unbound at the base of the bed. During the expanded elution stage, sampling was performed initially cycling between ports but for the later stages preferentially at the top port. Although the data is limited, the peak of eluted GST-(His)$_6$ can be followed as it passes up the column past each port, growing as more material joins the eluate front. The 4 min cycle time of the chromatographic assay limits the flow of information during the step change to eluent as the majority of the bound material is desorbed

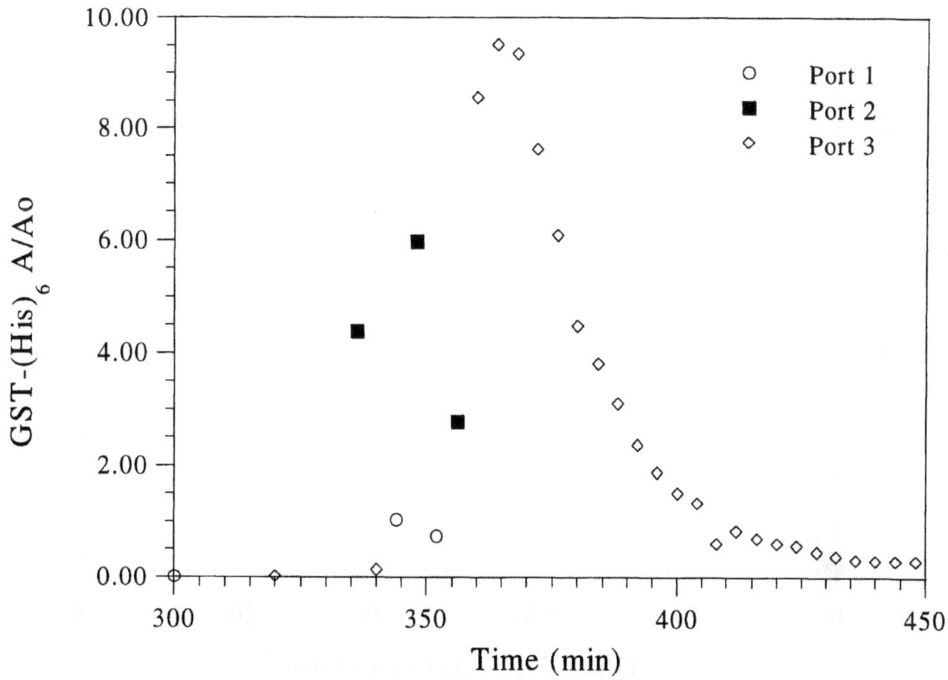

Figure 6.

within a single column volume. However, other results with a purification of lysosyme from cows milk and a faster 1 min chromatographic assay (Chase et al., 1999) demonstrate how this monitoring principle can also be used effectively in the elution phase.

Following the experiment, the fractions collected at the bed exit during the run were also assayed using the chromatographic assay in an off-line fashion for comparison with the spectrophotometric assay data that has already been shown in Figure 5. The correlation plot between the enzyme activity assay and the chart recorder peak height (corrected for blank peak), is shown in Figure 7. The associated plot of GST-$(His)_6$ level in the bed exit versus time measured using the two methods is also shown. The correlation plot demonstrates that the chromatographic assay is an effective method for determining GST-$(His)_6$ concentration for the whole separation, including the elution stage in which far higher levels of enzyme are present. The highest activity measured shows a depressed response from the chromatographic assay, and probably represents the peak rising into the non-linear range of the detector. The overlay plot shows how the chromatographic assay, if applied to the top of the bed/port 3 around the point of breakthrough could be used to tightly control the duration of the load. As soon as GST-$(His)_6$ is detected at port 3, monitoring could be switched to that port only. The absolute tightness of the control would be related to the flow of information from the bioassay, here the 4 min cycle time meaning that a certain tolerance would have to be accepted in the % breakthrough at which the switch to buffer was made.

In this paper we developed a selective and rapid chromatographic assay for measuring the concentration of a protein in unclarified samples abstracted from within an expanded bed. This was employed to monitor the level of that protein through the adsorption, washing and elution phases of its purification by expanded bed adsorption. The resulting system has potential for the monitoring and control of these separations, and could be simply adapted to other on-line assay methods (Richardson et al., 1996a; Nilsson et al., 1994; Holwill et al., 1996). The advantages of this approach compared to those based on laborious and sometimes lengthly assay procedures, whereby the results of a purification may not be known for hours or in some cases up to a day later, are clear.

Conclusions

A chromatographic assay based on a selective metal loaded adsorbent has been developed, together with in-line filtering for measuring the concentration of GST-$(His)_6$ in unclarified cell homogenates. On-line monitoring using the rapid chromatographic assay on samples withdrawn from various heights up the bed

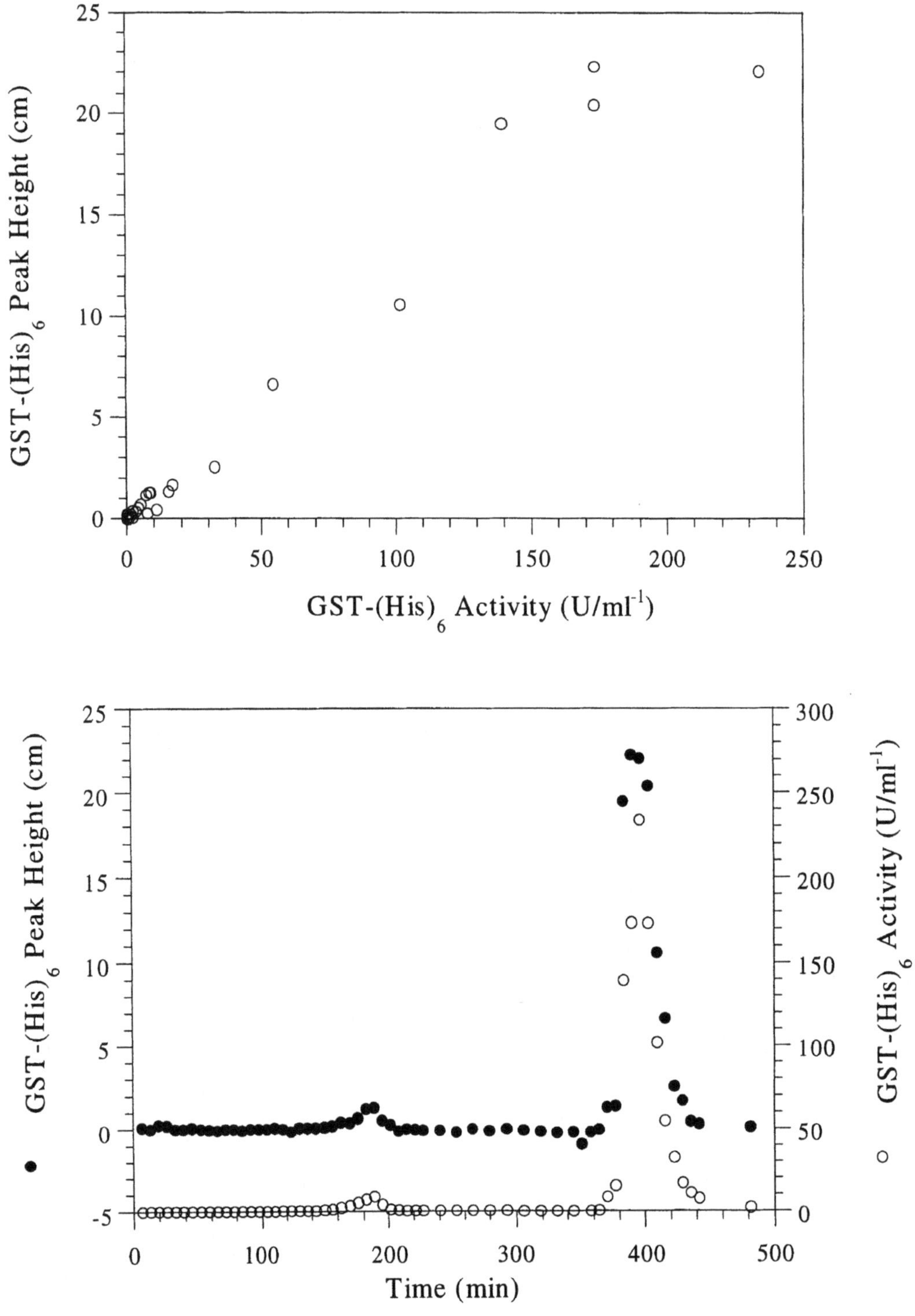

Figure 7. Comparison of the GST-(His)$_6$ activity as measured by rapid chromatographic assay and by spectrophotometric CDNB assay (off-line), using (i) a correlation plot and (ii) an overlay plot of the contents of the expanded bed exit fractions.

was used to follow the whole separation of GST-(His)$_6$ within the adsorption column. The potential of this technique for tighter control and optimisation of expanded bed separation procedures and also for obtaining additional insight into the hydrodynamic and adsorption/desorption processes occurring within the bed was demonstrated. GST-(His)$_6$ was recovered from an unclarified *E. coli* homogenate using Ni^{2+} loaded STREAMLINE Chelating™ at an 85% yield and a purification fold of 3.0.

Acknowledgements

The authors would like to thank the BBSRC for the provision of a studentship for R.H. Clemmitt and also Pharmacia Biotech for the provision of chromatographic equipment and materials. We would also like to thank Dr Jonas Lidholm of Pharmacia & Upjohn Diagnostics for a sample of the *E. coli* DH5α containing PGEX-5X-1 expressing GST-$(His)_6$.

References

Arnold F H (1991) Metal-affinity separations: a new dimension in protein processing. Bio/Technology, 9: 151–156.

Bruce L J and Chase H A (1999) Evaluation of the effect of in-bed sampling on expanded bed adsorption. Bioseparation 8: 53–67.

Chang Y K and Chase H A (1996) Development of operating conditions for protein purification using expanded bed techniques: The effect of the degree of bed expansion on adsorption performance. Biotech. Bioeng. 49: 512–526.

Chase H A, Nash D C and Bruce L J (1999) On-line monitoring of breakthrough curves within an expanded bed adsorber. Bioprocess Eng. 20: 223–229.

Chase H A (1994) Purification of proteins by adsorption chromatography in expanded beds. Trends Biotechnol. 12: 296–303.

Chase H A (1986) Rapid chromatographic monitoring of bioprocesses. Biosensors 2: 269–286.

Habig W H, Pabst M J and Jakoby W B (1974) Glutathione S-Transferases. The first enzymatic step in mercapturic acid formation. J. Biol. Chem. 249: 7130–7139.

Hjorth R (1998) Expanded bed adsorption: Elution in expanded mode. Poster presentation at Second International Conference on Expanded Bed Adsorption, Napa Valley, California, 21–23 June.

Hjorth R (1997) Expanded bed adsorption in industrial bioprocessing: Recent developments. Trends Biotechnol. 15: 230–235.

Hjorth R (1995) Analysis of some operating parameters of novel adsorbents for recovery of proteins in expanded beds. Bioseparation 5: 217–223.

Hochuli E, Bannwarth W, Doebeli H, Gentz R and Stueber D (1988) Genetic approach to facilitate purification of recombinant proteins with novel metal chelate adsorbent. Bio/Technology 6: 1321–1325.

Holwill I, Gill A, Harrison J, Hoare M and Lowe P A (1996) Rapid analysis of biosensor data using initial rate determination and its application to bioprocess monitoring. Process Control and Quality 8: 133–145.

Nieba L, Nieba-Axemann S E, Persson A, Hamalainen M, Edebratt F, Hansson A, Lidholm J, Magnusson K, Karlsson A F and Pluckthun A (1997) BIACORE analysis of histidine-tagged proteins using a chelating NTA sensor chip. Anal. Biochem. 252: 217–228.

Nilsson M, Vijayakumar A R, Holst O, Schornack C, Hakanson H and Mattiasson B (1994) Online monitoring of product concentration by flow-elisa in an integrated fermentation and purification process. J. Ferment. Bioeng. 78: 356–360.

Porath J, Carlsson J, Olsson I and Belfrage G (1975) Metal chelate affinity chromatography: a new approach to protein fractionation. Nature 258: 598–599.

Richardson P, Ravenhall R, Flanagan M T, Holwill I, Molloy J, Hoare M and Dunnill P (1996a) Monitoring and optimisation of fractional protein precipitation by flow-injection analysis. Process Control and Quality 8: 91–101.

Richardson P, Molloy J, Ravenhall R, Holwill I, Hoare M and Dunnill P (1996b) High speed centrifugal separator for rapid on-line sample clarification in biotechnology. J. Biotechnol. 49: 111–118.

Smith D B and Johnson K S (1988) Single step purification of polypeptides expressed in *Escherichia coli* as fusions with glutathione S-transferase. Gene 67: 31–40.

Sulkowski E (1985) Purification of proteins by IMAC. Trends Biotech. 3: 1–7.

Thömmes J (1997) Fluidized bed adsorption as a primary recovery step in protein purification. Adv. Biochem. Eng. Biotech. 58: 185–230.

Address for correspondence: H. A. Chase, Department of Chemical Engineering, University of Cambridge, Pembroke Street, Cambridge, CB2 3BA, United Kingdom. Tel.: 44-(1223)-334777; fax: 44-(1223)-334796; E-mail: hac1000@cam.ac.uk

Bioseparation **8:** 69–75, 1999.
© 1999 *Kluwer Academic Publishers.*

The effect of column verticality on separation efficiency in expanded bed adsorption

L.J. Bruce, S. Ghose[†] & H.A. Chase
Department of Chemical Engineering, University of Cambridge, Cambridge, UK. [†]*Present address: Department of Chemical Engineering and Applied Chemistry, Aston University, Birmingham, UK*

Received 10 September 1998; accepted in revised form 1 November 1998

Key words: column alignment, column verticality, expanded bed adsorption, fluidized bed adsorption

Abstract

The effect of column verticality on liquid dispersion and separation efficiency in expanded bed adsorption columns was investigated using 1 and 5 cm diameter columns. Column misalignment of only 0.15° resulted in the reduction of the Bodenstein number from 140 to 50 for the 1 cm dia. column and from 75 to 45 for the 5 cm dia. column. This degree of misalignment was not detectable by visual assessment of adsorbent particle movement within the column. Depending on the relative importance of transport limitations, kinetic limitations and dispersion to any specific separation, this increase in dispersion with column alignment can significantly affect separation efficiency. Pure protein breakthrough profiles resulting from the application of bovine serum albumin onto STREAMLINE Q XL demonstrated that, at 10% breakthrough, 7.8% more protein could be applied to a vertical 1 cm dia. column compared to the same column misaligned by 0.15°. When an unclarified yeast homogenate was applied to a 1 cm dia. vertical column packed with STREAMLINE DEAE, 10% breakthrough of glucose-6-phosphate dehydrogenase (G6PDH) corresponded to a load 55% greater compared to the same column aligned 0.185° off-vertical. The G6PDH breakthrough curves for vertical and 0.15° off-vertical runs performed using a 5 cm column were essentially indistinguishable.

Introduction

Most development work on expanded bed adsorption (EBA) has been carried out to date using columns with diameters of 5 cm or larger; however, the high cost of many biological products make optimization studies and validation work at this scale prohibitive. Chromatographic processes are typically and most easily scaled by keeping the bed height and linear flow rates constant and increasing only the column diameter. As consumable requirements are proportional to column volume, the requirements for a 1 cm dia. column are 25 times lower than for a 5 cm dia. column. An understanding of the parameters affecting the scale-down of expanded bed adsorption is required, and here the effect of column verticality on liquid dispersion and separation efficiency on 1 and 5 cm dia. columns is investigated.

EBA is related to traditional chromatographic purification processes where the feed solution is loaded on a bed of particles packed in a column. In EBA, however, the particles are not tightly packed. Instead, the flow rate up through a bed of adsorbent particles is carefully controlled to expand the bed, but not to the extent that the adsorbent particles are elutriated from the top of the column. As the inter-particle voidage is increased and the process can tolerate particulate material, such as cells or cell debris, in the feed stream, the need for separate clarification and concentration steps is eliminated.

Movement of adsorbent particles results in a reduction in column efficiency. This dispersive effect is decreased by employing adsorbent particles that will classify when fluidized. With classification, particles are sorted by size, such that the larger/denser particles are found at the bottom of the bed and the smaller/lighter particles are found at the top, and the movement of the fluidized particles is thus reduced. Convective particle movement, or circulation, will still occur in expanded beds due to non-uniformity of liquid flow across the cross sectional area of the column. When a column is obviously not vertical, particle motion up one side of the column and down the other side is observed. Some publications describing EBA have noted the importance of column alignment; however, the relationships between column alignment, measures of mixing within columns and separation efficiency have not been quantified.

Johansson (1994) found that when particles (dia. 200 μm, density 1.150 g ml^{-1}) were fluidized at 300 cm h^{-1} in a 5 cm dia. EBA column and the column was approximately 1° off-vertical, plate number decreased by 55%. Van der Meer et al. (1984) investigated the effect of angles from 0.3° – 3° off-vertical (particle dia. 500 – 700 μm, density 1.200 – 1.400 g ml^{-1}, 1100 cm h^{-1} liquid linear velocities and column diameters ranging from 2 - 6 cm) and found that visual assessment of particle movement in a fluidized bed is a more sensitive indication for column alignment than a simple spirit level. They still observed significant mixing when the off-vertical angles were as small as 0.3°. Hudson et al. (1996) also investigated the effect of column inclination in a liquid fluidized bed not intended for EBA. Using 3 cm dia. particles, liquid velocities of 18000 – 32400 cm h^{-1} and a 8 cm dia. column, the angle off-vertical was varied from 1° – 10°. Particles were observed to move rapidly upward near the high column wall and were recirculated downward near the lower column wall. At higher liquid velocities, the solid circulation pattern was stronger and the effect of inclination on bed expansion was more pronounced. For a vertical column, bed voidage was found to be greatest in the middle of the bed, indicating that more liquid tended to flow up the center of the bed. For an inclined column, bed voidage was greatest near the high wall, indicating that more liquid flows up the high side of the column.

This paper describes an investigation of the effect of column verticality on liquid dispersion (measured using residence time distributions) and separation efficiency (evaluated using breakthrough curves) for 1 cm and 5 cm dia. columns.

Materials and methods

EBA systems based on 1 cm and 5 cm diameter columns

The small scale EBA experiments were carried out on a 1 cm dia., 60 cm long glass column constructed by Soham Scientific (Soham, U.K.). The distributor was made from a sintered glass plate (160 – 220 μm nominal pore size). The larger scale experiments were performed using the STREAMLINE 50 system (Pharmacia Biotech, Uppsala, Sweden). The perforated plate and screen distributor supplied by Pharmacia with the 5 cm dia., 100 cm long column was used for these experiments. A rugged column stand was constructed to hold both the 1 cm and 5 cm dia. columns using V-blocks. Fine threads on the adjustable feet of the column stand allowed for very small adjustments to alignment. A Pro 3600 electronic protractor (Smarttool Technologies, San Jose, CA, USA) was attached to the stand and was used to measure the angle of inclination to an accuracy of 0.02°.

Residence time distributions (RTDs)

RTDs were used to assess the liquid dispersion occurring in the 1 cm and 5 cm dia. columns. The adsorbent, with a settled bed height of ~20 cm, was fluidized with 50 mm NaH_2PO_4 at 260 cm h^{-1} (unless otherwise stated). Before experimentation the expanded bed was allowed to stabilize for at least one hour. Prior to measurement of the residence time distribution, a tracer of 20% v v^{-1} acetone was loaded in the sample loop (0.8 ml and 21 ml for the 1 cm and 5 cm dia. columns respectively, 2.5% of the expanded column volume) and the top adapter was moved to within a few mm of the top surface of the expanded bed. The UV absorbance of the column effluent was measured, and the entire UV absorbance peak was recorded. Moment analysis of the recorded outlet concentration of tracer was used to characterize the resulting distribution as described by Levenspiel (1972). The Bodenstein number (*Bo*) is a dimensionless term employed to relate convective transport of liquid to dispersion in EBA columns:

$$B_o = \frac{uL}{D_{axl}}$$

where u is the linear flowrate, L is the length of the fluidized bed and D_{axl} is the axial dispersion coefficient.

Column alignment

Before the measurement of RTDs or adsorbate breakthrough profiles, the EBA columns were vertically aligned using one of two methods. The more widely used, qualitative, visual method was one technique used to align the 1 cm dia. column. First, the adsorbent particles were fluidized and allowed to stabilize. An obviously misaligned EBA column has adsorbent particles streaming up one side of the column and down the other side. The column was adjusted until the movement of particles was judged to be minimized. Alternatively, the columns were aligned using the digital protractor attached to the column stand. This method was used for all runs performed on the 5 cm dia. column and for most of the runs on the 1 cm dia. column. The column could be set vertical or at a desired angle off vertical.

Bovine serum albumin (BSA) load

To test the effect of column alignment on adsorbate breakthrough, a single protein solution was applied to a 1 cm dia. EBA column. A bed of STREAMLINE Q XL with a settled height of 20 cm was expanded using 50 mm phosphate buffer (pH 7.5) at 260 cm h^{-1}. BSA at ~4 g l^{-1} (>99% pure) in the above phosphate buffer was then applied to the column in either the vertical or a 0.15° off-vertical position. Two runs were performed under each set of conditions. The UV absorbance of the column effluent was monitored and a standard curve was prepared to relate UV absorbance to BSA concentration.

Yeast homogenate load

As a further test of the effect of column alignment on adsorbate breakthrough, an unclarified yeast homogenate was applied to an expanded bed of STREAMLINE DEAE and the breakthrough of glucose-6-phosphate dehydrogenase (G6PDH) was monitored. Yeast was homogenized as described by Bruce and Chase (1998). The yeast homogenate was diluted with 50 mM sodium phosphate to ~20 g solids (dry) l^{-1}. The G6PDH concentration in the feedstock was ~0.9 U ml^{-1} and the pH was 6.0.

The settled bed heights were 20 cm and the adsorbent was fluidized with 50 mM phosphate buffer (pH 6) at 180 cm h^{-1} for the 1 cm dia. column and 220 cm h^{-1} for the 5 cm dia. column. Unclarified yeast homogenate was applied to both the 1 cm and 5 cm dia. columns, and the level of glucose-6-phosphate dehydrogenase (G6PDH) in the collected fractions of the column effluent was determined as described by McCreath et al (1994). One off-vertical and then one vertical run were performed using each of the columns.

Results and discussion

Comparison of column alignment methods

An assessment was made of the variability associated with the two methods of column alignment for the 1 cm dia. column: visual alignment and alignment using a digital protractor. A set of RTDs were measured as described in the Materials and methods section, except the column was first deliberately misaligned, then re-aligned using one of the two methods. Six RTDs were measured using the visual alignment method and five more were measured after the column had been aligned using the digital protractor. Table 1 summarizes the runs performed.

When the column is aligned vertically using the visual method, the standard deviation of the Bodenstein number was roughly eight times larger than when the digital protractor was used. Apparently the 1 cm dia. column is sensitive to small deviations from vertical alignment. It was also observed that misalignment of 0.15° was not detectable by visual assessment of adsorbent movement within the column. Particle movement within the column was clearly visible when the column was greater than 0.25° off-vertical.

The effect of column alignment on liquid dispersion

The effect of column alignment on liquid dispersion was quantified by determining the residence time distributions for a matrix of different column angles for both the 1 and 5 cm dia. columns. The column alignment was varied along two axes (both 'right-left' and 'back-front') in order to get the various axis positions. Five RTDs were performed at each of 16 different axis positions for the 1 cm dia. column, and at each of 13 different axis positions for the 5 cm dia. column, ranging up to a maximum angle off-vertical of 0.4°. The Bodenstein number (Bo) was employed to relate convective transport of liquid to dispersion in the EBA columns. Bo decreases as the amount of mixing in the system increases. Bo was plotted against the two axes in Figure 1. The response surfaces shown are best fit curves, generated using linear regression. They

Table 1. The effect of the column alignment method on mixing. 6 RTD measurements were made using the visual alignment method and 5 RTD measurements were made when the digital protractor was used to align the column optimally vertical

Alignment method	Flow rate (cm h^{-1}	Settled bed height (cm)	Adsorbent	Average *Bo*	Standard deviations of *Bo*
Visual alignment	275	~	STREAMLINE SP	68	19
Use of digital protractor	260	~	STREAMLINE DEAE	117	2.3

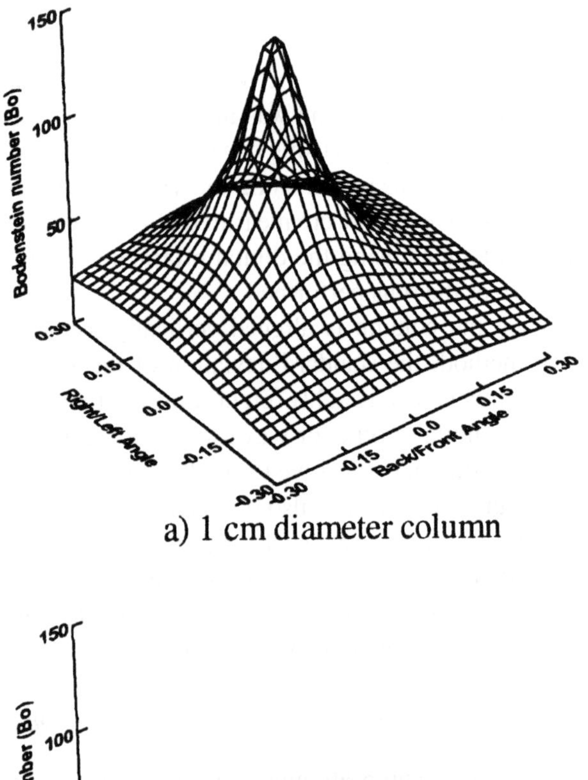

a) 1 cm diameter column

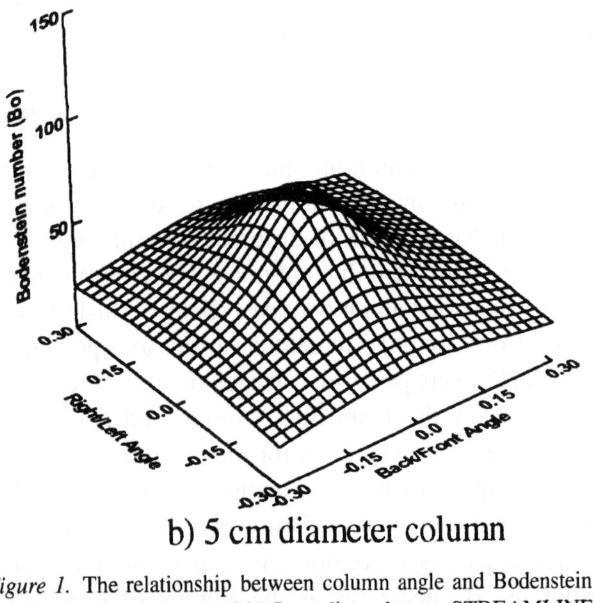

b) 5 cm diameter column

Figure 1. The relationship between column angle and Bodenstein number for the a) 1 cm and b) 5 cm dia. columns. STREAMLINE DEAE was the adsorbent used for both columns, the settled bed height was ~20 cm, the linear flow rate was 260 cm h^{-1}.

are not perfectly centered because the alignment measurement was taken on the column stand not on the columns themselves.

The plots show that the maximum Bo value achieved for the EBA system employing the 1 cm dia. column is higher than for the EBA system for the 5 cm diameter column, indicating that more mixing occurs in the larger system. This may be a result of stirring effects caused by liquid jets issuing from the holes in the perforated plate distributor used with the 5 cm dia. column, although preliminary investigation suggests there is little difference between the flow distribution provided by perforated plates and sintered glass. Further work is needed to determine why there appears to be less mixing in the 1 cm dia. column.

The response surfaces also show that the 1 cm dia. column is more sensitive to vertical alignment. Bo for the 1 cm dia. column dropped from 140 to 50 compared to a reduction from 75 to 45 for the 5 cm dia. column when both columns were misaligned by 0.15°.

In a stable fluidized bed, the upward drag forces of particle-fluid interaction balance the downward gravitational force on the particle. When a column is tilted off vertical, the gravitational force is vertically down, while drag force from the fluid is parallel with the column walls. There will be a resultant force that causes the fluidized particles to move towards the low column wall. In off-vertical fluidized beds, particle motion up the high side and down the low side of the column results. Due to the induced solids circulation patterns, bed voidage is greater near the high side of the column, thus the liquid flow rate across the cross sectional area of the column is not uniform. When the column is vertical there is much less particle dispersion and associated liquid dispersion.

When the column is inclined, each particle in a classified expanded bed would likely circulate in a confined zone. Particles of a given size would move

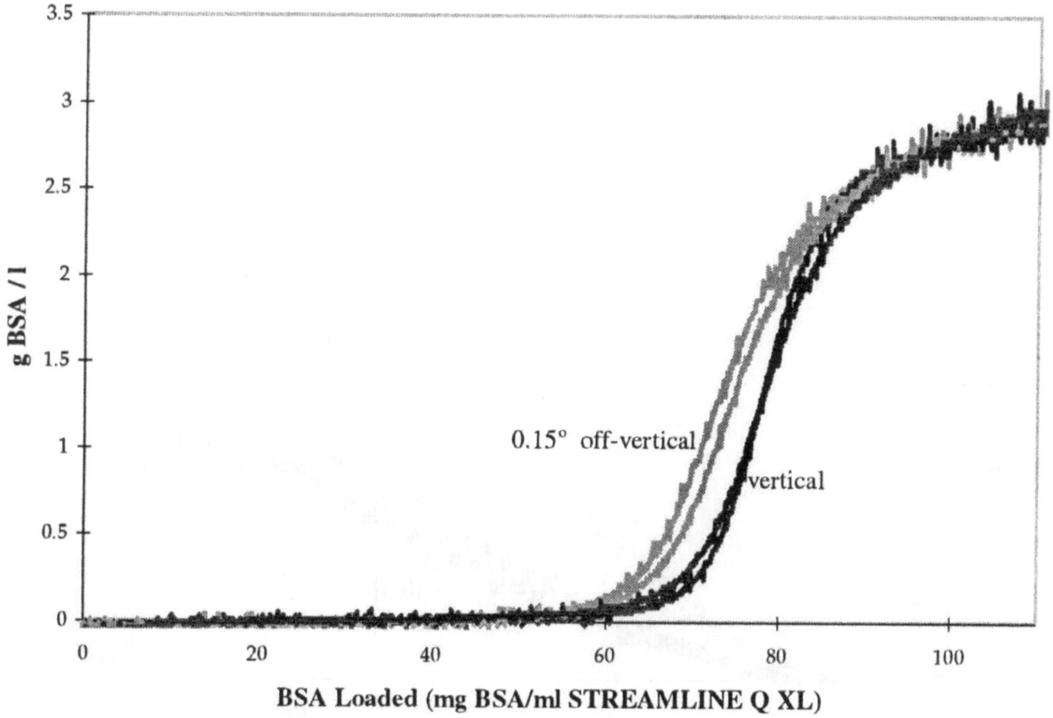

Figure 2. BSA breakthrough on the 1 cm dia. column packed with STREAMLINE Q XL. The settled bed height was ~20 cm, the linear flow rate was 260 cm h^{-1}, and the BSA concentration in the feed was 4 g l^{-1}.

horizontally towards the low wall where they would then move downwards due to the decreased voidage and liquid flow. This settling would likely be limited as the particles would be displaced away from the wall as larger/denser particles lower in the bed themselves move toward the low wall. As the smaller/lighter particles approach the high wall they would be swept upwards by the higher liquid velocity completing their circulation.

Hudson et al. (1996) measured the change in void fraction across the 8 cm dia. of an inclined liquid-solid fluidized bed. They discovered three distinct regions of particle flux: particles moved up near the high column wall, particles moved down near the low column wall, and in the center of the column there was a region where there was no net particle flux. This region of no net particle flux may help explain the relative sensitivity to column inclination of liquid dispersion in the 1 cm dia. column compared to the 5 cm dia. column. In larger diameter columns, this relatively stable region would presumably represent a greater proportion of the column cross sectional area, and the areas of particle movement would remain strongly associated with the regions near the high and low column walls. Liquid dispersion measured for larger diameter columns subjected to the same degree of misalignment as smaller diameter columns would therefore show less change from when the columns were vertical.

BSA breakthrough profiles

Four runs were performed where BSA (>99% pure, pH 7.5) was applied to the 1 cm dia. column. In Figure 2, the resulting breakthrough curves show that earlier breakthrough occurs when the column is not vertical. Variation between the runs at each alignment were small. The main part of the breakthrough curve (from 0.5 to 2.0 g l^{-1} BSA) was steeper for the vertical column, as liquid dispersion was lower in this system. At 10% breakthrough, 7.8% more protein could be applied when the column was vertical compared to the same column misaligned by 0.15°. For both the vertical and the off-vertical column, the final portions of the breakthrough curves were very shallow, as has been observed routinely when BSA is used as an adsorbate in ion exchange systems.

G6PDH breakthrough profiles

Unclarified yeast homogenate was applied to an expanded bed of STREAMLINE DEAE in a 1 cm dia. column that was aligned 0.185° off-vertical and then to a column that was vertical. Figure 3 shows the

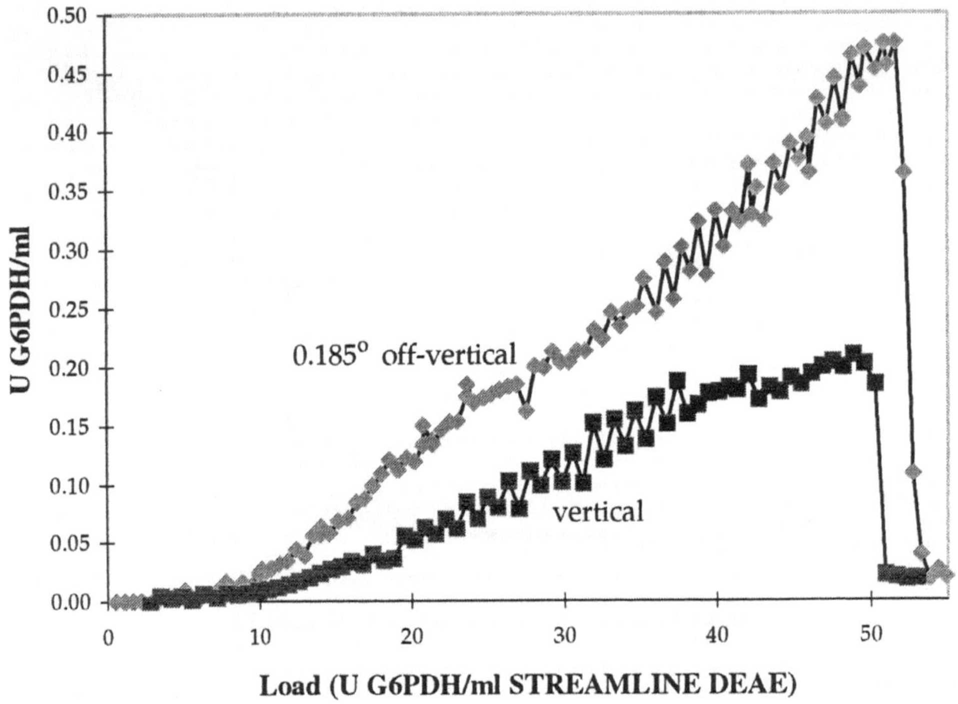

Figure 3. G6PDH breakthrough on the 1 cm dia. column packed with STREAMLINE DEAE. The settled bed height was ~20 cm. Unclarified yeast homogenate (~0.9 U G6PDH ml^{-1}, pH 6) was loaded at 180 cm h^{-1}.

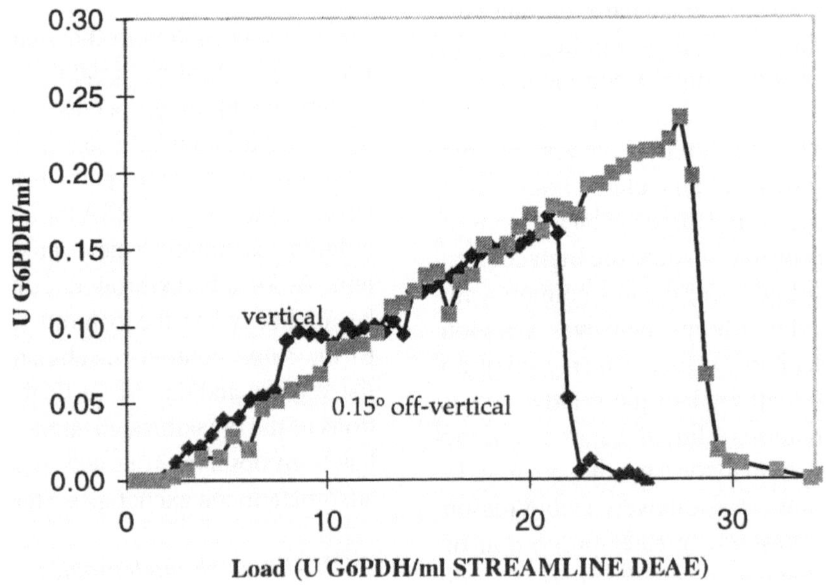

Figure 4. G6PDH breakthrough on the 5 cm dia. column packed with STREAMLINE DEAE. The settled bed height was ~20 cm. Unclarified yeast homogenate (~0.9 U G6PDH ml^{-1}, pH 6) was loaded at 220 cm h^{-1}.

initial part of the gradual breakthrough of G6PDH in the column effluent in each case. As was observed in the single component adsorption experiments, breakthrough occurred earlier for the run performed in the off-vertical column. When the yeast homogenate was applied to the 1 cm dia. vertical column, 10% breakthrough of G6PDH corresponded to a 55% greater load compared to the same column aligned 0.185° off-vertical. This demonstrates the effect of increased dispersion on separation efficiency.

A similar experiment was also performed using the 5 cm dia. column. Unclarified yeast homogenate was applied to an expanded bed that was aligned 0.15° off-vertical and then to a vertical column. Figure 4 shows that the breakthrough profiles of G6PDH in the column effluent for the vertical and 0.15° off-vertical runs were essentially indistinguishable. The relatively small effect of misalignment on dispersion in the 5 cm dia. column results in a negligible effect on separation efficiency. The G6PDH breakthrough curves for the 1 cm and 5 cm dia. columns can not be directly compared as there was a difference in the conductivity of the yeast homogenate prepared for the two scales of separation.

The performance of an EBA separation is influenced by adsorption behaviour, particle side mass transport, liquid side mass transport, liquid dispersion and particle dispersion. Adsorbate breakthrough profiles provide an overall representation of the combined effect of these factors. However, the relative importance of the various transport and kinetic limitations and dispersive effects varies from system to system, thus the *importance* of increased liquid and particle dispersion caused by column misalignment would be expected to be system dependent.

Conclusions

Misalignment of EBA columns results in increased liquid dispersion, as measured by residence time distributions. Liquid dispersion is more strongly affected by alignment in 1 cm dia. columns compared to 5 cm dia. columns.

Column misalignment, even at angles so small as to be undetectable by visual assessment of particle movement, can have a significant effect on separation efficiency. The extent by which misalignment will affect observed separation efficiency is dependent on the relative importance of transport limitations, kinetic limitations and dispersive effects to the system of interest.

Scale down of EBA by using smaller diameter columns has considerable benefits in convenience and reduced costs for optimization and validation work. In the systems examined here (when using 1 cm dia. columns for such scale down studies), it was important to ensure the column was aligned vertically to minimize liquid/particle dispersion. If a separation process is not significantly affected by liquid/particle dispersion, projecting results obtained using a small diameter column to a larger scale would be more straightforward.

References

Bruce LJ and Chase HA (1998) Monitoring of glucose-6-phosphate dehydrogenase breakthrough curves within an expanded bed adsorption column. Bioseparation, this volume.

Hudson C, Briens CL and Prakash A (1996) Effect of inclination on liquid-solid fluidized beds. Powder Tech. 89: 101–113.

Johansson U (1994) Hydrodynamics of a liquid-solid fluidized bed in the low Reynolds number region. Licentiate thesis. Royal Institute of Technology, Stockholm, Sweden.

Karau A, Benken C, Thommes J and Kula M-R (1997) The influence of particle size distribution and operating conditions on the adsorption performance in fluidized bed. Biotech. Bioeng. 55: 54–64.

Levenspiel O (1972) Chemical Reaction Engineering. Wiley, New York.

McCreath GE, Chase HA and Lowe CR (1994) Novel affinity separations based on perfluorocarbon emulsions. J. Chrom. A. 659: 275–287.

Van der Meer AP, Blanchard CMRJP and Wesselingh JA (1984) Mixing of particles in liquid fluidised beds. Chem. Eng. Des. Res. 64: 214–222.

Address for correspondence: H.A. Chase, Department of Chemical Engineering, University of Cambridge, Cambridge, UK (Fax: +44 (0)1223 334796)

Bioseparation **8:** 77–83, 1999.
© 1999 *Kluwer Academic Publishers.*

Evaluation of the effect of in-bed sampling on expanded bed adsorption

L.J. Bruce & H.A. Chase*
Department of Chemical Engineering, University of Cambridge, Cambridge, UK

Received 7 July 1998; accepted in revised form 29 September 1998

Key words: expanded bed adsorption, fluidized bed adsorption, on-line monitoring

Abstract

An expanded bed adsorption (EBA) column (5 cm diameter) has been modified to allow the abstraction of liquid samples from various positions along the height of an expanded bed. As the adsorbent particles were fluidized, in-bed monitoring of key component concentrations during feedstock application, washing and elution was achieved by the withdrawal of liquid samples from the voids within the expanded bed through ports along the wall of the column. Component levels in the withdrawn streams can be assayed using on-line analytical chromatography or samples can be collected and assayed off-line. On-line monitoring can be used to control the duration of the loading stage and as a tool to provide information about the hydrodynamic and adsorption/desorption processes that occur during expanded bed adsorption. Studies of residence time distributions indicated that the modifications to the column do not significantly affect liquid dispersion. Using the adsorption of glucose-6-phosphate dehydrogenase from yeast homogenate on Streamline DEAE as a model system, comparison of breakthrough curves for runs when in-bed monitoring was and was not performed also suggested that separation efficiency is not appreciably affected by in-bed sampling.

Introduction

Expanded bed adsorption (EBA) is a novel processing technique for the purification of proteins which combines clarification, concentration and initial purification in one step (Chase, 1994; Hjorth, 1997; Thommes, 1997). EBA is related to traditional chromatographic purification processes where the feed solution is loaded on a bed of adsorbent packed in a column. In EBA, however, the adsorbent is not tightly packed. Instead, the flow rate up through the bed of adsorbent particles is controlled to expand the bed, and consequently the process can tolerate particulate material in the feed stream, eliminating the need for separate clarification and concentration steps. Bed expansion is controlled to minimize liquid and solid phase dispersion, allowing the adsorption process to retain some chromatographic action.

* Address for correspondence: H.A. Chase, Department of Chemical Engineering, University of Cambridge, Cambridge, UK. Tel +441 223 334781; Fax +441 223-334796; E-mail: hac1000@cam.ac.uk

On-line monitoring has been used to provide timely information about the state of bioprocesses as varied as fermentation (Nilsson et al., 1994), selective flocculation (Habib et al., 1997) and traditional chromatography (Chase, 1986). On-line monitoring of EBA separations is possible as a result of advances in techniques used for the rapid analysis of samples containing particulate material (Richardson et al., 1996; Habib et al., 1997). With adsorption separations, information about the level of key components in the effluent allows for tighter control and results in improved process efficiency. During operation of an EBA column, this information can be used to optimize feedstock loading so that the adsorbent is not overloaded, resulting in loss of the product in the column effluent, and so that the adsorbent is not underloaded, which also results in reduced productivity. This is especially important with biological feedstocks which often exhibit considerable run-to-run variability. After loading, wash buffer consumption and possible associated product loss can be minimized by using on-line monitoring to ensure sufficient removal of contaminant components from within the bed. Monitoring of the effluent for the product during elution enables collection of the product in the minimum volume, and if the effluent is monitored for other eluting components, collection of contaminant components can be avoided and high purity product fractions can be ensured.

Studies of the fluid mechanics of traditional packed bed chromatography have included the use of nuclear magnetic resonance techniques (Lightfoot et al., 1995) and slicing frozen gel beds (Yamamoto et al., 1986). Chase et al. (1998) hypothesized that the higher void fraction and the fluidized nature of the bed should make EBA more amenable to liquid withdrawal from within the bed, compared to a packed column. On-line monitoring of components *within the bed*, or in-bed monitoring, has two additional advantages over on-line monitoring of the column effluent which make it an important development tool: (1) tighter control of loading, washing and elution is possible as information about the progress of the adsorption is attained earlier, and (2) additional insight into the hydrodynamic and adsorption/desorption processes can be gleaned from in-bed data.

Chase et al. (1998) demonstrated the potential of on-line monitoring of components within an EBA column for both a single component system and the adsorption of lysozyme from skimmed cows' milk. They noted that sample withdrawal from within the bed should be designed so as to minimize any increase in liquid or particle dispersion and resulting reduction in separation efficiency.

In this paper, the effect of in-bed sampling on liquid dispersion, as represented by residence time distributions (RTDs), is examined. Column performance is also compared for runs with and without in-bed monitoring using the adsorption of glucose-6-phosphate dehydrogenase (G6PDH, a relatively stable intracellular enzyme) from yeast homogenate as the model system. The effluent breakthrough curves provide evidence that the effect of in-bed monitoring on adsorption separations is minimal. The in-bed breakthrough profiles are presented and discussed as well.

Materials and Methods

Modified EBA column and system

The expanded bed adsorption experiments were carried out using the STREAMLINE 50 system (Pharmacia Biotech, Uppsala, Sweden). For the runs where in-bed monitoring was performed, a 100 cm long, 5 cm diameter glass column, with 0.5 cm diameter holes at 5 cm intervals along the height of the column was substituted for the glass column supplied with the STREAMLINE 50 system, while the rest of the STREAMLINE 50 system was used.

Sintered glass sample ports were created at the 0.5 cm diameter holes by attaching 0.8 cm outer diameter glass tubing, as shown in Figure 1, to the column using silicone sealant. The end of the tubing had a plug of glass sinter (nominal pore size 160–220 μm). When inserted at the port holes, the end of the sinter was flush with the column wall. The glass sinter prevented the withdrawal of adsorbent from the column, but allowed for the removal of the feedstock, including any cell debris. Samples of the liquid phase within the column were withdrawn using a Cole-Parmer peristaltic pump. Figure 2 shows the equipment configuration employed. A solenoid-actuated valve switched withdrawal between the sample ports (located 10, 25 and 40 cm above the distributor) for one minute each in a pre-programmed sequence. The withdrawn liquid was sent to a fraction collector during the breakthrough experiments.

Residence time distributions

RTDs were used to assess the liquid dispersion occurring in the columns. The STREAMLINE DEAE adsorbent was fluidized at 75 $\pm$4 ml min^{-1} with 50 mM NaH_2PO_4 and the expanded bed was allowed to stabilize for at least 30 minutes. Prior to the RTD run, a 20% v/v acetone tracer was loaded in the 21 ml sample loop (2.5% of the column volume) and the top adapter was moved to within a few mm of the top of the expanded bed. The tracer was applied to the column and the UV absorbance of the column effluent was recorded. Tracer elements of a liquid flowing through a column of adsorbent will not have equal residence times as axial mixing of the liquid will occur. This is caused by both molecular diffusion along the column length and non-uniformity of linear velocity in the mobile phase over the cross section of the column.

Moment analysis of the output concentration data is used to characterize the resulting distribution as described by Levenspiel (1972). The first moment, $\bar{t}$, describes the mean residence time, and the second moment, σ^2, represents the variance of the distribution. The normalized variance, σ_θ^2, is:

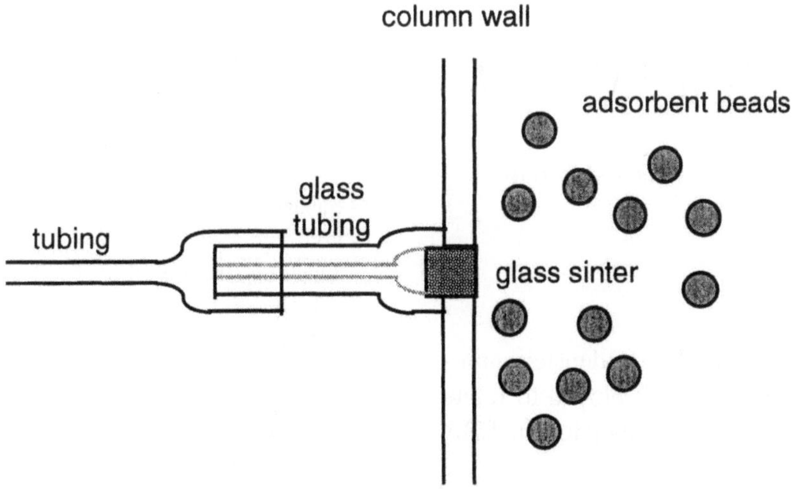

Figure 1. Sintered glass sample port. The plug of glass sinter (0.5 cm diameter, 160–220 μm nominal pore size) prevents withdrawal of adsorbent from the column.

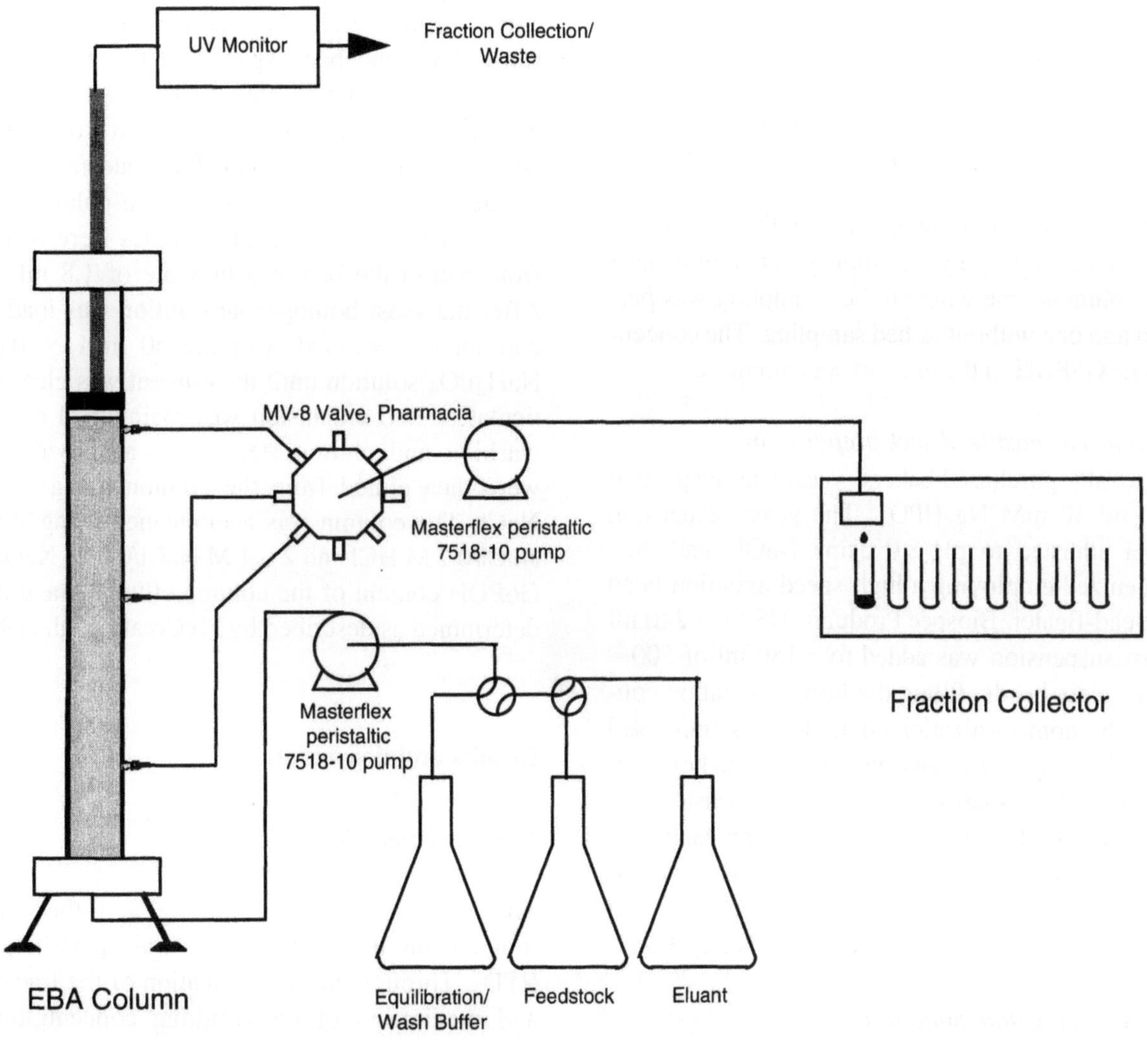

Figure 2. Equipment configuration. Sample ports were located 10, 25 and 40 cm above the bottom of the column.

$$\sigma_\theta^2 = \frac{\sigma^2}{\bar{t}^2} = 2\frac{D_{axl}}{uL} - 2\left(\frac{D_{axl}}{uL}\right)^2\left(1 - e^{\frac{-uL}{D_{axl}}}\right)$$

The axial dispersion, D_{axl}, can be reduced to

$$D_{axl} = \frac{\sigma_\theta^2 uL}{2}$$

when plug flow is approached. The Bodenstein number (Bo) is a dimensionless term employed to relate convective transport of liquid to dispersion in EBA columns:

$$B_o = \frac{uL}{D_{axl}}$$

where u is the linear flowrate, L is the length of the fluidized bed and D_{axl} is the axial dispersion coefficient. Bo decreases as the amount of mixing in the system increases.

Off-line monitoring of glucose-6-phosphate dehydrogenase breakthrough

The effect of in-bed sampling on breakthrough curves was evaluated by applying crude yeast homogenate to two columns, one where in-bed sampling was performed and one without in-bed sampling. The concentration of G6PDH in the effluent was monitored.

Homogenized yeast feedstock preparation
890 g locally purchased bakers' yeast was suspended in 890 ml 50 mM Na_2HPO_4. The yeast suspension was pH adjusted to pH 8.0 using NaOH and then homogenized employing a high speed agitation bead mill (Bead-Beater, Biospec Products, USA). ~240 ml of yeast suspension was added to ~180 ml of 500 – 750 μm glass beads, filling the homogenization container. The homogenization container was immersed in an ice bath. The yeast suspension was then homogenized for eight 20 s periods. The yeast suspension was allowed to cool for at least 2 min between each 20 s homogenization period. The yeast homogenate was diluted to a concentration of ~20 g solids (dry) L^{-1} and a pH of 6.0 using 50 mM Na_2HPO_4 and NaH_2PO_4.

Column loading and monitoring
The settled bed height of the Streamline DEAE was 19.5 cm. The bed was expanded to approximately twice the settled bed height and equilibrated using 3 L of 50 mM Na_2HPO_4 / NaH_2PO_4, pH 6.0. When the expanded bed was stable and had reached a constant height, the feedstock was applied to the column. Table 1 summarizes the load conditions for the two runs. Since the bed expands as the homogenate is loaded, due to the higher viscosity of the feedstock, the top adapter was set high (at 52 cm) so that loading could proceed at a constant flow rate for the length of the run. Throughout the run the column effluent fractions were collected and samples were withdrawn from within the bed at a flowrate of 1.8 ml min^{-1}. After the yeast homogenate solution was loaded, the column was washed with the 50 mM Na_2HPO_4 / NaH_2PO_4 solution until the effluent was clear of particulates. Bed expansion was maintained during the washing and elution steps. The adsorbed proteins were then eluted from the column using 2 L 1 M NaCl. The column was also cleaned using 2 L 1 M NaCl/0.2 M HCl and 2 L 1 M NaCl/0.5 M NaOH. The G6PDH content of the column effluent fractions was determined as described by McCreath et al. (1994).

Table 1. Yeast homogenate load conditions. The 5 cm diameter column was packed with STREAMLINE DEAE to a settled bed height of ~20 cm. Prior to loading, the bed was expanded and equilibrated with 3 L of 50 mM Na_2HPO_4 / NaH_2PO_4, pH 6.0

	No in-bed sampling	In-bed sampling
Flowrate (ml min^{-1})		
- equilibration	75	77
- load	75	77
- sample off-take rate	n.a.	1.8
Feedstock		
- G6PDH (U ml^{-1})	0.62	0.86
- Volume loaded (L)	12.3	10.3
- G6PDH loaded (U)	7620	8850

Results and discussion

Residence time distributions

The effect of in-bed sampling and the sampling flowrate on liquid dispersion was quantified using RTDs. Throughout the application of the tracer pulse and monitoring of the resulting concentration distribution in the effluent, samples were continuously withdrawn from the column from ports 10, 25 and 40 cm up the height of the column for one minute each,

Table 2. The relationship between sampling off-take rate and *Bo*. The 5 cm diameter column was packed with STREAMLINE DEAE to a settled bed height of ~20 cm. The bed was expanded at 230 ± 12 cm h^{-1} with 50 mM NaH_2PO_4. 20% v/v acetone was the tracer used.

Off-take rate (ml/min)	Bo	Standard Deviation
0	75	2.3
0.8	68	0.7
1.3	62	1.5
1.8	30	0.3

in sequence. The relationship between the off-take rate and the resulting *Bo* is summarized in Table 2.

The low standard deviation values show that run-to-run reproducibility was good. Higher off-take rates resulted in lower *Bo* values, indicating increased mixing. As liquid was withdrawn through the sample ports, adsorbent particles were observed to 'pack' in a small area (~1 cm diameter) around the sample port. When sample withdrawal was switched to another port, the 'packed' adsorbent dropped off the wall and quickly dispersed. As the off-take rate increased from 0.8 to 1.8 ml min^{-1}, the 'packed' area increased from a diameter of approximately 0.7 cm to 1.5 cm. The resulting deformation of the smooth column inner surface likely affects liquid dispersion in the column; however, at low off-take rates bed stability appears only slightly affected.

Increased liquid dispersion caused by sample withdrawal would be reduced in systems involving beds of larger diameter or by minimizing the off-take rate required. The latter could be achieved by decreasing the volumes associated with the sampling equipment (tubing, filter volumes, etc.); however, this equipment must still be capable of handling particulate material from the feed solution. As well, sample off-take does not need to be continuous. Sample withdrawal is only necessary to fill the sample loop before each assay.

G6PDH breakthrough curves

Off-line analysis of the samples withdrawn from three different ports throughout the load, wash and elution stages provided breakthrough curves for the three positions up the length of the bed, as shown in Figure 3. The progression of G6PDH along the length of the bed can be seen during loading. The concentration of G6PDH in the liquid phase was greater nearer the bottom of the bed than at higher positions during loading as the adsorbent was exhausted earliest at positions closest to the bed inlet. The G6PDH concentration increased most rapidly at the port nearest the distributor and least rapidly at the port near the top of the bed.

During the washing phase, the in-bed sampling system followed the progressive decreases in the liquid phase concentration of G6PDH along the length of the bed. As expected, the concentration fell first at sampling positions nearer to the bed inlet as the washing buffer swept non-adsorbed material up and out of the expanded bed. The profiles for the elution stage show the G6PDH peak moving through the column.

Although on-line analysis of the in-bed samples was not done in this case, rapid monitoring of the in-bed samples has been performed using rapid chromatographic assays on an FPLC system where samples are filtered prior to analysis. Chase et al. (1998) describe using a one minute, on-line chromatographic assay for lysozyme and Clemmitt et al. (1998) developed an on-line chromatographic assay (with an in-line filter to remove particulates from the samples and a four minute cycle time) for glutathione S-transferase with a polyhistine tail.

The effect of in-bed monitoring on separation efficiency in a realistic situation was evaluated by applying a crude yeast homogenate to expanded beds where in-bed monitoring was and was not employed, then comparing the product breakthrough curves for the two runs. Fractions of the effluent collected during these runs were analyzed for G6PDH. Figure 4 shows that the resulting breakthrough curves for the runs where in-bed sampling was and was not employed were essentially indistinguishable.

Conclusions

Studies of residence time distributions and protein breakthrough curves indicated that in-bed sampling does not significantly affect column performance.

The in-bed monitoring procedure is shown to have advantages over simple monitoring of column effluent; the ability to study the development of breakthrough curves can provide additional insight into the underlying hydrodynamic and adsorption/desorption behaviour, and this information is provided in a more timely manner, allowing for tighter process control. On-line, in-bed monitoring is especially useful for expanded bed procedures as the feedstocks are typically highly

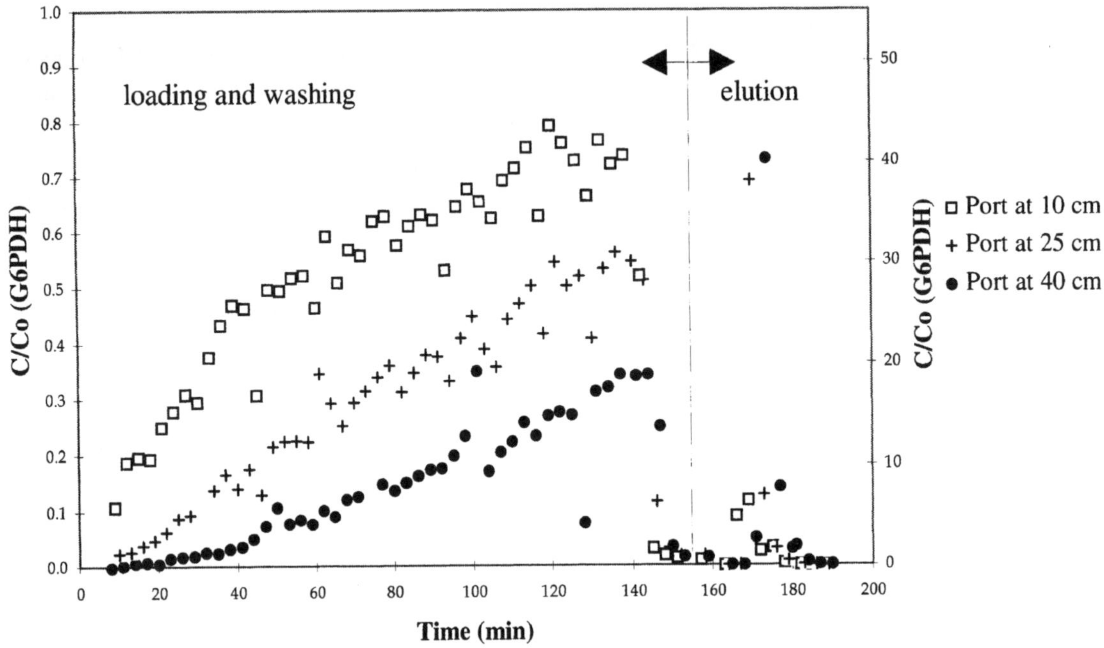

Figure 3. G6PDH breakthrough profiles during loading, washing and elution. The 5 cm diameter column was packed with STREAMLINE DEAE to a settled bed height of ~20 cm. Unclarified yeast homogenate was loaded at ~235 cm h^{-1}. The off-take rate for the in-bed sampling was 1.8 ml min^{-1}, 1 min withdrawal at each port in sequence.

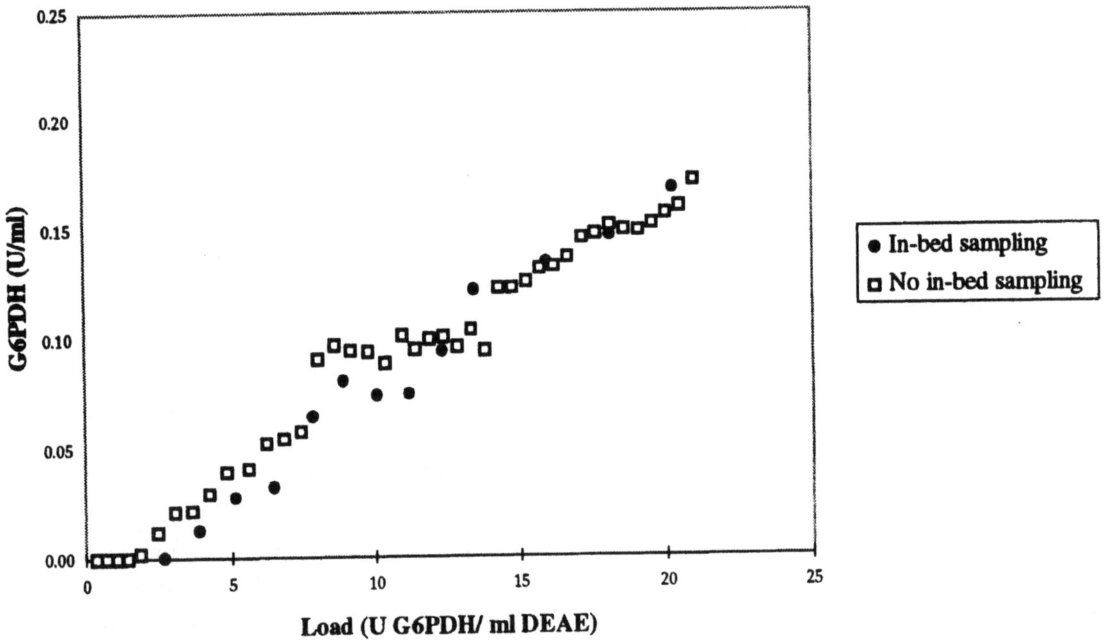

Figure 4. G6PDH breakthrough in column effluent. The 5 cm diameter column was packed with STREAMLINE DEAE to a settled bed height of ~20 cm. Unclarified yeast homogenate was loaded at ~230 cm h^{-1}. The off-take rate for the in-bed sampling was 1.8 ml min^{-1}.

complex and subject to considerable batch-to-batch variation.

References

Chase HA, Nash DC and Bruce LJ (1998) On-line monitoring of breakthrough curves within an expanded bed adsorber. Bioprocess Eng. In press.

Chase HA (1994) Purification of proteins by adsorption chromatography in expanded beds. Trends Biotechnol. 12: 296–303.

Chase HA (1986) Rapid chromatographic monitoring of bioprocesses. Biosensors. 2: 269–286.

Clemmitt RH, Bruce LJ and Chase HA (1998) On-line monitoring of the purification of GST-$(His)_6$ from an unclarified *Escherichia coli* homogenate within an immobilised metal affinity expanded bed. Bioseparation. submitted for publication.

Habib GB, Holwill I and Hoare M (1997) Rapid piloting of a selective flocculation process for product purification. J. Biotechnol. 59: 91–101.

Hjorth R (1997) Expanded bed adsorption in industrial processing: Recent developments. Trends Biotechnol. 15: 230–235.

Levenspiel O (1972) Chemical Reaction Engineering. Wiley, New York.

Lightfoot EN, Athalye AM, Coffman JL, Roper DK and Root TW (1995) Nuclear magnetic resonance and the design of chromatographic separations. J. Chrom. A. 707: 45–55.

McCreath GE, Chase HA and Lowe CR (1994) Novel affinity separations based on perfluorocarbon emulsions. J. Chrom. A. 659: 275–287

Nilsson M, Vijayakumar AR, Holst O, Schornack C, Hakanson H and Mattiasson B (1994) Online monitoring of product concentration by flow-ELISA in an integrated fermentation and purification process. J. Ferment. Bioeng. 78: 356–360.

Richardson P, Molloy J, Ravenhall R, Holwill I, Hoare M and Dunnill P (1996) High speed centrifugal separator for rapid online sample clarification in biotechnology. J. Biotechnol. 49: 111–118.

Thommes J (1997) Fluidized bed adsorption as a primary recovery step in protein purification. Adv. Biochem. Eng. Biotech. 58: 185–230.

Yamamoto S, Nomura M and Sano Y (1986) Scaling up of medium-performance gel filtration chromatography of proteins. J. Chem. Eng. Japan. 19: 227–231.

Bioseparation **8:** 85–91, 1999.
© 1999 *Kluwer Academic Publishers.*

EBA columns with a distribution system based on local stirring

Elias Zafirakos & Allan Lihme
UpFront Chromatography A/S, Lersoe Parkalle 42, DK-2100 Copenhagen OE, Denmark

Received 8 September 1998; accepted in revised form 15 June 1999

Key words: expanded bed elution, mixed zone influence, stirred distribution, volumetric scale up

Abstract

A new type of liquid distribution system for expanded bed columns has been developed. The construction differs from traditional distribution designs by not having any small pores (like filters or distribution plates) in the flow path of the crude feedstock. A stirrer at the bottom of the column distributes the incoming feedstock. Due to the stirring, jet streams are prevented and a stable expanded bed is formed above a mixed zone. This article describes the new column design, and investigates the performance of the stirred distribution concept experimentally by measuring theoretical plate number and breakthrough profile. Furthermore, the possibilities of scaling up the concept will be discussed based on theoretical plate measurements.

Introduction

An important feature in successful chromatographic purification is to have plug flow through the chromatographic bed. To achieve plug flow efficient distribution of the incoming liquid is necessary. Furthermore, it is important that the bed is homogenous and stable during the process.

In traditional chromatography columns, a plate with lots of small pores distributes the liquid stream into the column. Due to the small sized pores the incoming liquid is distributed. The pore size gives a limitation to the nature of the incoming feed liquid. If the feed contains particulate impurities or if small air bobbles are present in the feed, then the distribution may be disturbed due to clogging of the pores.

The idea behind the stirred distribution is to avoid limitations caused by having small pores in the flow path. The feed stream enters the column directly through holes with a minimum 2-mm diameter at the bottom of the column. In the column, a stirring device placed at the bottom by the inlet distributes the incoming liquid. The rpm of the stirrer can be regulated to meet the needs from different viscosities or flow velocities.

Materials

Expanded Bed Matrices

General physical properties

Beads based on a non-porous glass sphere coated with agarose, density range: 1.3–1.5, size range 100–300 μm.

UFC DEAE resin (breakthrough study).

UFC Mimo (case study).

The columns:

Expanded bed columns of diameter: 2, 5, 20 and 40 cm, UpFront Chromatography A/S.

The basic principle is similar over the range of column diameters (Figure 1).

At the bottom there is an outlet for packed bed operation, above this there is a resin support net on to which the resin can pack. Above the resin support net is an expanded bed inlet and above this inlet is a stirring device. The liquid enters the column through the EBA inlet, while the packed bed outlet is closed. Once inside the column a stirrer distributes the liquid, and creates a mixed zone.

The column top is fitted with a pipe having 3/4 length of the column. The liquid leaves the column through a simple hole (the outlet pipe). The diameter of this hole is equal to or bigger than the inlet diameter.

The pipe functions as the top inlet/outlet and the height is adjustable. It is lowered or raised to midnimise the liquid headspace above the chromatographic bed.

Other equipment
UV-monitor, Pharmacia Biotech system UV1, adsorption at 280 nm. UV recorder, Pharmacia Biotech Recorder 101. Peristaltic Pump, Verder Pericor SF70 and Verder SR 240.

Chemicals
Acetone solution 0.25% v/v. Bromine phenol blue solution. BSA solution from Biofac. Red split lentils.

Methods

Operating the column

Start up
Resin volume is measured by dispersing an approximate resin volume in water. Transfer it to the column and sediment for 5 min. The packed bed height is then measured and the resin volume calculated.

The column is adjusted to vertical position to avoid gravity influence on the beads in non-vertical direction during expansion. Finally the flow is turned on and the stirrer started. The rpm of the stirrer is adjusted in order to give an even distribution and to minimise the height of the mixed zone. When a stable expansion level is reached, the bed is inspected visually. If channels are observed, stirrer speed is increased until the channels disappear and the expanded bed stabilises.

The position of the outlet is adjusted to minimise the liquid head above the bed. Adjustment of the pipe should be done carefully, as submerging the pipe into the expanded bed will result in loss of chromatographic beads. The column principle is illustrated in Figure 1.

Residence Time Distribution (RTD)
The RTD studies were done using Pharmacia guidelines on expanded bed chromatography (ref. 1). The bed was expanded to approximately 2 h_0 (h_0 being the packed bed height) and the flow rate was measured. The liquid headspace above the bed was limited to 5 cm by adjusting the top outlet.

When the baseline on the UV recorder was stable, the acetone solution was applied (0.25% v/v). When the positive step input signal was stable with the acetone solution water was reapplied to the column. Plain

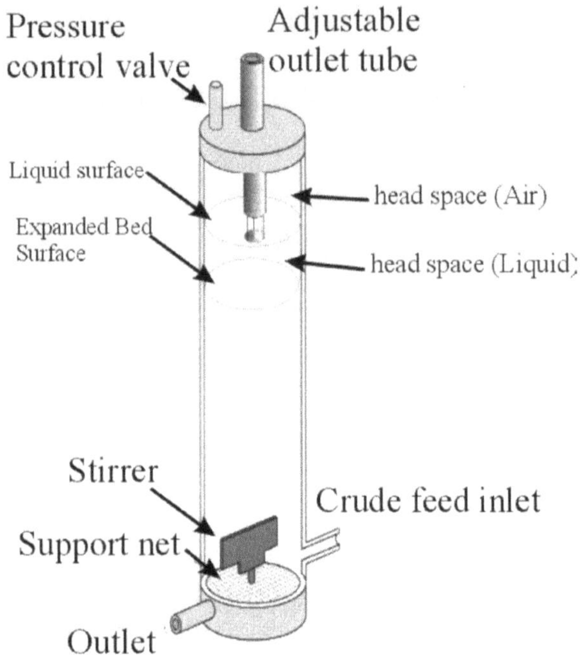

Figure 1. A schematic illustration of the column concept. The support net and the packed bed outlet are only used for packed bed operation and do not interfere with the expanded bed operation.

water was applied until the negative step signal returned to the baseline level. Finally, the number of theoretical plates was calculated from the negative step input signal:

$$N = t^2/\sigma^2$$

t = the mean residence time, σ = the standard deviation.

The recorded UV signal obtained by using the method described above is hereafter referred to as 'the RTD curve'. The RTD was measured for the full range of columns to test if the concept was scalable.

Mixed zone influence
The mixed zone influence was examined in 2 and 5-cm columns, by measuring the RTD in various experiments.

For convenience, a new factor, h_{mix} is introduced. h_{mix} is defined as the percentage of the expanded bed that appears mixed by visual inspection of the stabilised expanded bed, i.e. h_{mix}= (height of mixed zone/total expanded bed height)*100. The height of the mixed zone is easy to identify visually. Theoretical plate number was determined per metre packed bed height. Where nothing else is indicated the experiments were made with approximately 20 cm packed bed height, h_0, and the degree of expansion, h/h_0 ~

2. The mixed zone was increased/decreased by, respectively, increasing or decreasing the rpm of the stirrer.

Breakthrough profile

A model system with BSA and DEAE ion exchanger was used to obtain a breakthrough profile on the 2 cm column. Experimental set up was as follows:

2.5 mg BSA/ml in the feed, linear velocity 5 cm/min and 10 cm/min, stirrer rpm $\sim$ 300. Fractions of $\sim$ 30 ml was collected and measured at UV 280 nm.

Case study

This 'case study' is included in the article to illustrate the capabilities of the expanded bed concept on very crude feedstock.

A 20-cm column with stirred distribution was used to purify *Lens culinaris agglutinin* from an extract of crushed red split lentils. The feed stock contained a high amount of particulate impurities; therefore, to avoid precipitation the feed was under continuous stirring during application. The entire process was done in expanded bed mode including the elution. The stirrer rpm was approx. 50.

The crude feed extract appeared milky/orange. To distinguish the white resin from the feed 1 litre of red-dyed, inactive resin was mixed with the active (white) Mimo resin. Therefore, the resin appears red on the pictures.

Scale up possibilities

To test the scale up potential RTD was measured in the 20-cm and 40-cm diameter columns. The packed bed height was approximately 40-cm corresponding to 12.5 and 50 litres of resin, respectively. The 20-cm column was tested with stirrer speed of approximately 50 rpm and the 40-cm with 30 rpm and the linear velocity was set to 300 cm/h in both experiments, corresponding to 94 and 377 litre/h, respectively.

To visualise the stable expanded bed colour experiments were performed by injecting Bromine Phenol Blue into the columns.

Results and discussion

The mixed zone influence

As illustrated on the graph, increasing the height of the mixed zone has a significant impact on the number of theoretical plates.

Figure 2 shows that increasing h_{mix} with a constant resin volume results in a decrease in the number of theoretical plates. If a theoretical plate number above 170 N per metre is needed then the mixed zone should not exceed 10–15% of the expanded bed in the tested columns.

For the 2-cm column a reduction of h_{mix} from 25% to 12.5% increases the theoretical plate number 4 times (diamonds on Figure 2). A reduction of h_{mix} from 30% to 15% increases the theoretical plate number 2.5 times, in the 5-cm column. This difference may be due to the difference in diameter.

Another explanation could be that the concentration of chromatographic beads in general is higher in the lower part of the expanded bed. If, for instance, the lowest 25% of the expanded bed contains 50% of the beads, then even a small decrease of h_{mix} (e.g. from 20 to 15%) will add a lot of beads to the stable part of the expanded bed.

All the above experiments resulted in RTD curves indicating stable expansion; also tracer injected colour indicated stable expansion and plug flow above the mixed zone (Figure 4). Therefore, the number of theoretical plates per metre packed bed, N/m, is not the only factor to consider when establishing whether the expanded bed is stable in the UpFront column. A certain h_{mix} will correspond to a certain N/m in a given stable expanded bed. If h_{mix} is high (e.g. 50) then a relatively low number of N/m is expected even if the bed above the mixed zone is stable. Yet if h_{mix} is low then a low value of N/m indicates an unstable bed.

In a traditional expanded bed column, stable expansion corresponds to a theoretical plate number of 20–30 N per 15 cm packed bed height (ref. 1). This number is not directly transferable to the UpFront column. Stable expansion has been obtained in the UFC column with plate numbers down to 100 per metre.

For an efficient process it is important that there is a minimum number of theoretical plates and that all the resin is in contact with the feed stream. Given that all the adsorbent in the mixed zone is in contact with the feed stream, the number of theoretical plates above the mixed zone assures an efficient adsorption. The necessary number of theoretical plates per metre depends on the specific process. Perhaps 25–30 theoretical plates in the stable expanded bed part above the mixed zone is sufficient for almost all applications.

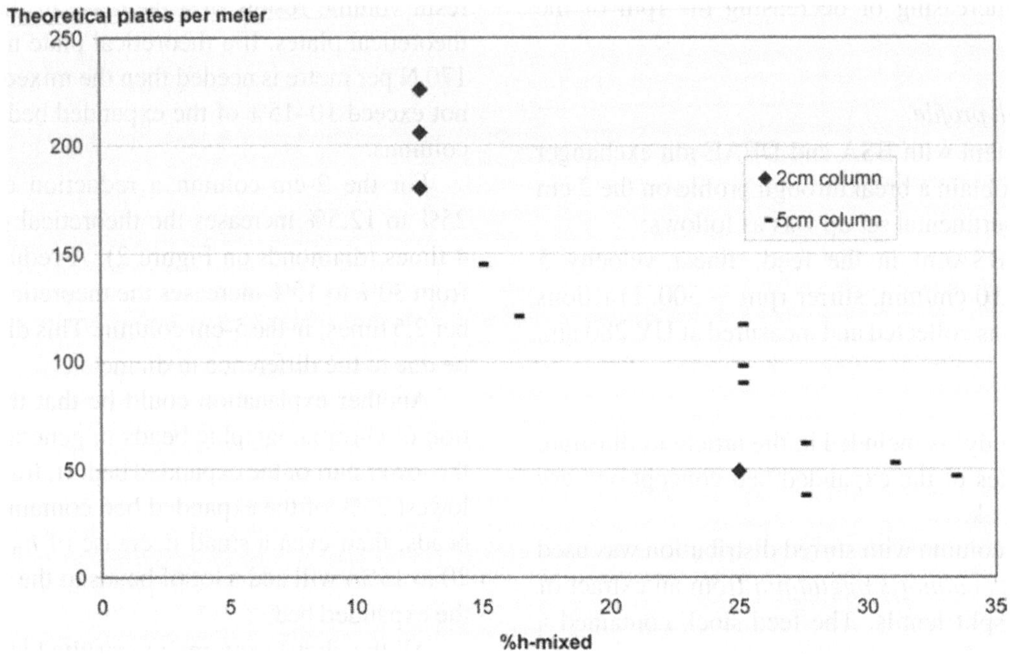

Figure 2. Theoretical plate number per metre packed bed height, N/m, as a function of h_{mix}.

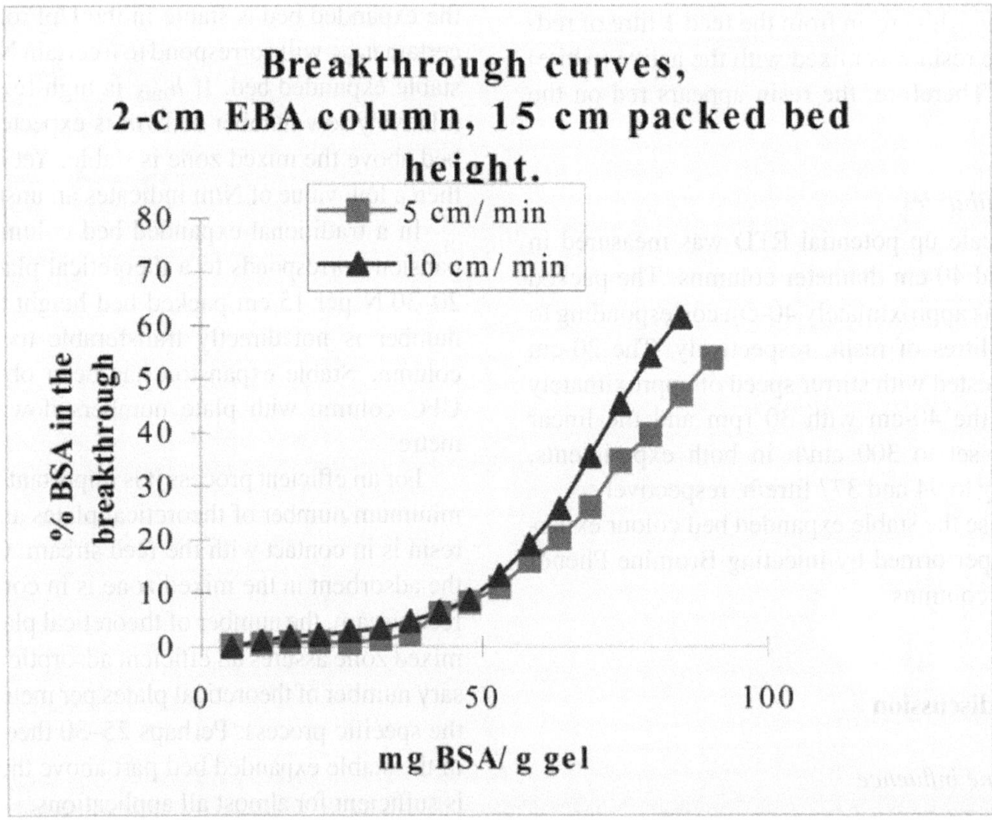

Figure 3. Breakthrough curves for the 2-cm column using DEAE ion exchanger and BSA solution.

Figure 4. A stable expanded bed in the 40-cm column, where a trace of Bromine Phenol Blue has been fed to the column. Notice the sharp front, no jet streams are observed.

Case study

The study indicates that plug flow is obtainable with viscous crude feed stock containing particulate impurities as well as with pure buffer. As the process pictures show, the incoming feed stream creates a sharp front at all levels of the purification process, indicating plug flow. No clogging during the entire process further supports the concept of the column used on crude feed stock.

Elution in expanded bed mode yielded a concentrated eluate corresponding the volumes obtained with packed bed elution (Zafirakos et al., unpublished data). Using expanded bed mode during the entire process

(A) (B) (C) (D) (E)

Figure 5. Expanded bed process pictures: (A) Sample application; (B, C) expanded bed wash; (D, E) expanded bed elution.

Table 1. Scale up possibilities. The data given in the table below is average values for experiments in the different column diameters. The expansion was approx. 2

Column diameter, cm	Packed bed height, cm	Mixed zone, h_{mix}	Theoretical plates	Theoretical plates per metre
2	22	12.5	45	206
5	27	15	36	135
20	48	9	96	200
40	40	20	43	107

therefore seeems to be a very interesting alternative to the combined expanded/packed bed operation.

Scale up

For the 2 and 20 cm columns, a theoretical plate number above 200 N/m, and above 100 N/m for the 5 and 40 cm column was obtained (Table 1). The variation in h_{mix} range from 8 to 20% reflects this range.

The experiments demonstrate that it is possible to obtain a stable expanded bed, using stirred distribution in columns ranging from 2 to 40 cm in diameter.

The number of theoretical plates per metre increases as the percentage of mixed zone is decreased. The task is therefore to minimise h_{mix} to a level which is acceptable for the process.

The higher the expanded bed the higher theoretical plate number can be obtained (ref. 3). Using the 'all expanded bed' technique, where all the steps are performed in expanded bed (as in the LCA case study), the height of the bed can in principle be infinite. When one has estimated the number of theoretical plates needed for a given process volumetric scale up can be done either in width or in height to meet this need.

Conclusion

The concept of stirred distribution seems to work well for column diameters up to 40 cm. Stable expansion above the mixed zone was obtained in all the experiments. Theoretical plate numbers from 100 to 200 N/m indicates that efficient adsorption is obtainable in the tested columns.

It is clear that h_{mix} has a high influence on the theoretical plate number, the higher h_{mix} the lower number of N/m. This is also what could be expected; if h_{mix} was 100%, and given an ideal mixing, the theoretical plate number should be one. When using stirred distribution one has to estimate the 'necessary number' of N/m for the process. It could be argued that a stable expanded bed corresponding to, for instance, 15 cm packed bed on top of a mixed zone is sufficient to give a good adsorption/desorption process

The optimal h_{mix} is an issue that needs further investigation. It will probably vary from different processes; a large h_{mix} (and a low theoretical plate number) is acceptable in processes with fast equilibrium and vice versa. The h_{mix} value is controlled by the design/speed of the stirrer and by the column design. Further, it is controlled by the packed bed height, increasing the packed bed height decreases h_{mix}. In traditional chromatography, the height of the bed is limited due to the increase in backpressure. When performing entire expanded bed processes the height is no longer a limiting factor, and the 'all expanded bed' process opens up for new opportunities with respect to scale up.

The possibility to scale up in height using expanded bed elution means that modular column designs may substitute very large column diameters in industrial purification plants.

Visual inspection of the bed revealed a stable expansion above the mixed zone in all the experiments. Furthermore, the RTD curves indicated stable expansion. Finally the system is very robust towards disturbance from air bubbles and particulate impurities. Therefore the stirred distribution system seems to be a serious alternative to traditional systems when dealing with Fluid/expanded bed reactors.

Acknowledgements

The author would like to thank the laboratory technicians Brian Olsen and Anja Dideriksen and Stud. Scient Rene Oehlenschlager for valuable assistance.

References

Pharmacia Biotech, Edition AA (1997) Expanded Bed Adsorption, Principles and Methods, 5–16.

Lihme A, Zafirakos E, Hansen MB and Olander M (1999) Simplified and more robust EBA processes by elution in expanded bed mode. Bioseparation, this volume.

Thömmes J, Weiher M, Karau A and Kula MR (1995) Hydrodynamics and performance in fluidized bed adsorption. Biotechnol. Bioeng. 48: 367–374.

Address for correspondence: Elias Zafirakos, UpFront Chromatography A/S, Lersoe Parkalle 42, DK-2100 Copenhagen OE, Denmark. Tel.: +45-39-27-37-63; Fax: +45-39-27-31-11;. E-mail: ez@upfront-dk.com

Bioseparation **8:** 93–97, 1999.
© 1999 *Kluwer Academic Publishers.*

Simplified and more robust EBA processes by elution in expanded bed mode

Allan Lihme[1,2], Elias Zafirakos[1], Marie Hansen[1] & Morten Olander[1]
UpFront Chromatography A/S, Lersøpark Allé 42, DK-2100 Copenhagen Ø, Denmark.
Homepage: www.upfront-dk.com ([2]E-mail: al@upfront-dk.com)

Received 8 September 1998; accepted in revised form 26 December 1998

Key words: EBA, packed bed elution, expanded bed elution, elution profiles, elution volume, EBA columns, matrix density, degree of expansion

Abstract

This paper illustrates the feasibility of eluting EBA columns in the expanded bed mode as an alternative to the generally used method of packed bed elution. It is shown that at linear flow rates of 1 – 3 cm/min the difference in total elution volume between expanded bed elution and packed bed elution is less than 20%. It is suggested that expanded bed elution offers a range of significant advantages, while the drawbacks will be insignificant in most applications. The key to the success of this method seems to be the use of EBA matrices with a relatively low degree of expansion (i.e. a high density) at the linear flow rates employed for elution of bound product.

Introduction

The standard method for elution of bound products from EBA matrices has so far been packed bed elution. The choice of this method has been based on a wish to maximise product concentration and minimise consumption of buffer. However, performing the process using packed bed elution adds significantly to the complexity and cost of the process as well as the complexity and cost of the EBA column. It is suggested that these drawbacks may be avoided by performing all steps in expanded bed mode. Our expectation is that a low degree of expansion during the elution phase will minimise the required elution volume. In this study we therefore employ EBA matrices of relatively high density (i.e. d = 1.3–1.5 g/ml) which has a low degree of expansion at flow rates between 1 and 5 cm/min..

When applying expanded bed processes there are a number of drawbacks associated with elution in packed bed mode, which can be eliminated by expanded bed elution:

Packed bed elution – drawbacks:

- Packed bed elution is a complicated and slow procedure due to the need for flow reversal, packing and re-expansion of the bed during each cycle.
- Aggregation of adsorbent particles when packed during elution may impair re-expansion of the bed and significantly harm the efficiency of CIP procedures.
- The operational packed bed height is limited by the pressure stability of the matrix and the back pressure obtained during packed bed operation. This restrains the design of the column to short and broad columns and thereby limits the number of theoretical plates.
- Packed bed elution complicates and increases the cost of the EBA column due to the need for a support screen and tightly sealed plungers. The presence of a support screen, seals and nets further makes it more difficult to clean the column.

Expanded bed elution – advantages:

- The EBA process is simplified and faster as all operations are performed in the same flow direction.

- Aggregation of the adsorbent particles is minimised because the matrix is kept suspended during operation. Packing during storage does not give rise to aggregation because the matrix is clean at this point.
- The operational bed height can be very high resulting in more theoretical plates and thereby higher dynamic capacity.
- A shift to expanded bed elution allows for extremely simple and low cost column design.
- Automation of the EBA process becomes significantly less complicated when all steps are performed in the same flow direction.

In this presentation we investigate whether it is generally possible to perform expanded bed elution without experiencing the hitherto expected drawback of significantly increased elution volume.

Experimental

EBA adsorbent:

Elution experiments were performed with an EBA ion exchanger of the DEAE-type, cat. no.: 500-0400, UpFront Chromatography A/S, Denmark.

The UpFront EBA matrices (Lihme et al., 1991a) are based on spherical cross-linked agarose beads comprising a solid glass core, d = 1.3 – 1.5 g/ml and a size range of Ø = 100 – 300 μm.

EBA column:

All experiments were performed with an UpFront EBA column (Ø = 2 cm) (2,3), cat. no.: 700–0100, UpFront Chromatography A/S, Denmark, See also Figure 1.

BSA model system:

Purified bovine serum albumin (BSA) was bound to 1 litre of the DEAE-ion exchanger by batch incubation at pH 7 (10 litre 2 mg/ml BSA solution to one litre DEAE-ion exchanger). After washing the DEAE-ion exchanger was divided into samples of 65 ml which were eluted successively in either packed bed or expanded bed mode. Binding and washing buffer: 10 mM potassium phosphate pH 7.0 Elution buffer: 10 mM potassium phosphate + 1 M NaCl sodium chloride pH 7.0 The elution experiments were performed at different linear flow rates ranging from 1.0 – 4.5 cm/min. The course of the elutions were monitored

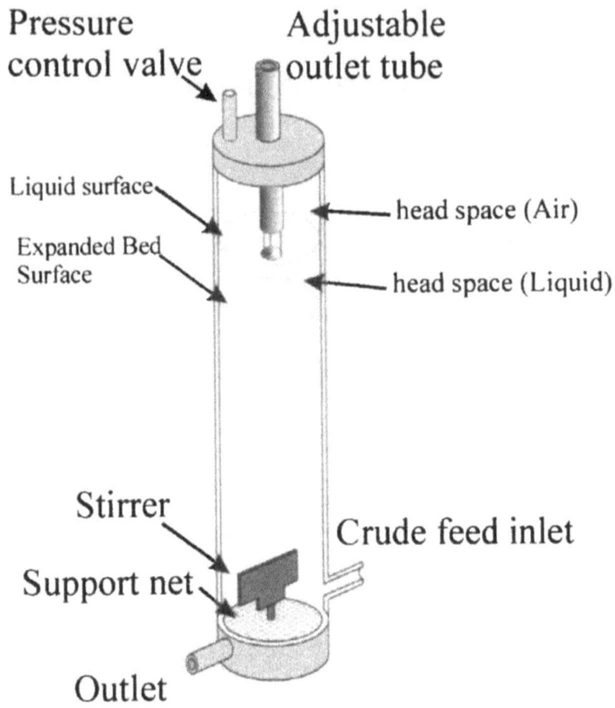

Figure 1. The UpFront EBA column (Zafirakos & Lihme, 1998) has no distribution plates that clog up with crude raw materials. A mechanical stirrer ensures even distribution and plug flow. The support net and the bottom outlet is used for packed bed elution only. Stirring speed is adjusted according to flow rate and viscosity of the raw material. The liquid head space above the expanded bed is controlled by the position of the adjustable outlet tube and the air head space is controlled via the pressure control valve.

using a UV-1 monitor from Amersham Pharmacia Biotech, Sweden. The elution volume was determined by the UV-tracks obtained and defined as the volume eluted between the beginning of the elution peak and the point where 98% of the BSA was eluted.

Results and discussion

Typical elution profiles

Figure 2 shows examples of typical UV-profiles obtained during the elution experiments. The linear flow rate used in this example is 1 cm/min for both packed bed and expanded bed elution. As can be seen the peak from packed bed elution emerges at about 60 ml, while the expanded bed elution peak emerges at about 70 ml. The expanded bed elution peak is also slightly broader than the packed bed elution peak. Generally, however, the peak tailing seems to be more pronounced and irregular for the packed bed elution experiments.

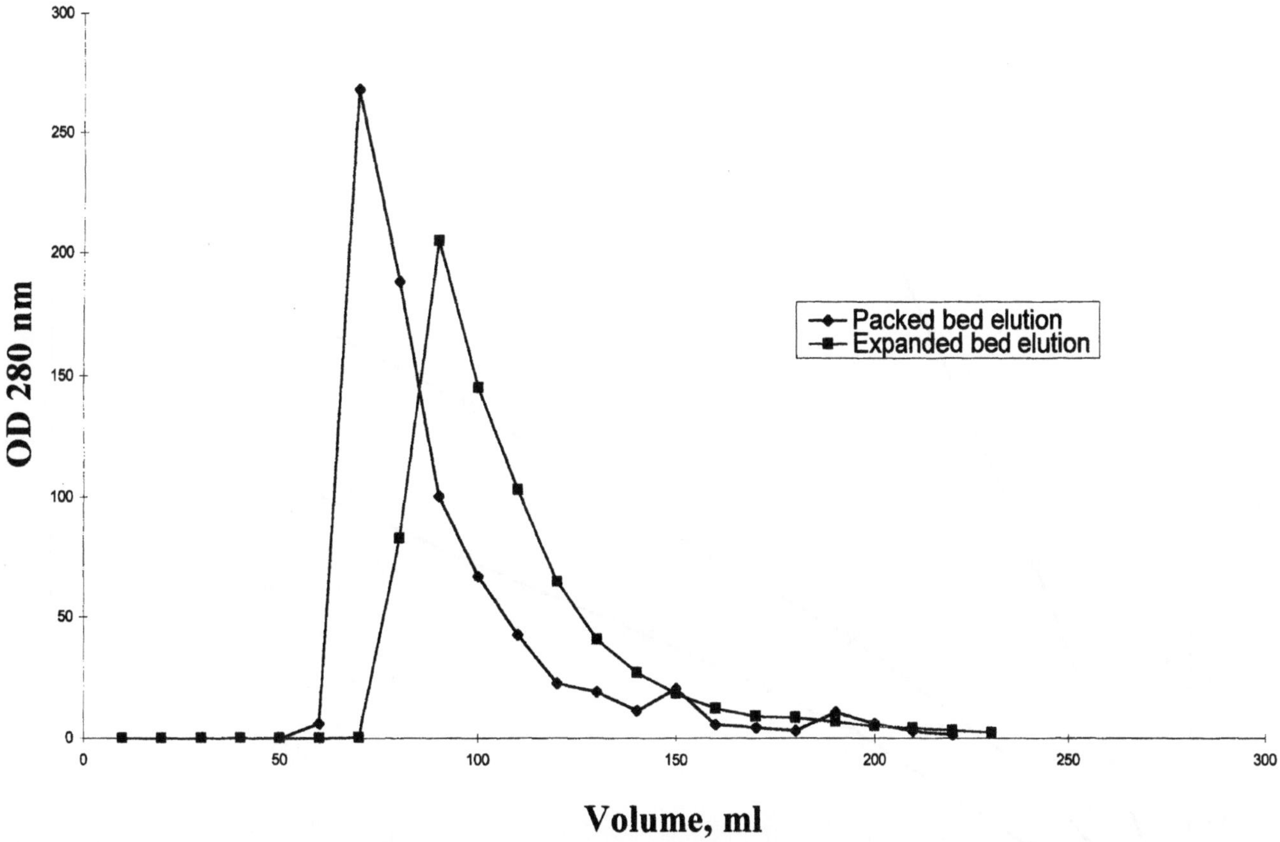

Figure 2. Typical elution profiles obtained by elution of BSA from a DEAE-ion exchanger (65 ml) in either packed bed or expanded bed. Linear Flow rate: 1 cm/min Column: UpFront EBA column Ø = 2 cm.

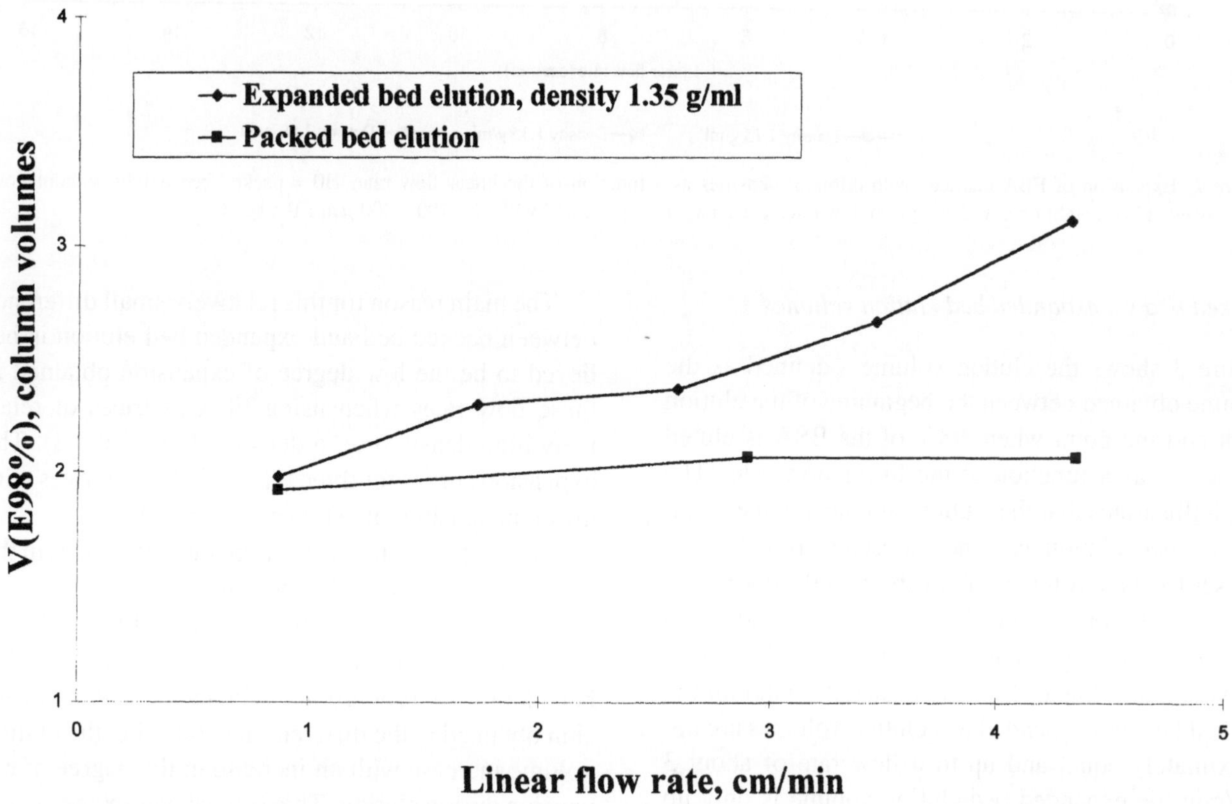

Figure 3. Packed bed vs. Expanded bed elution volumes ($V_{E98\%}$). Elution volumes are expressed as 'column volumes' i.e. relative to the volume of matrix (65 ml) employed in the experiments

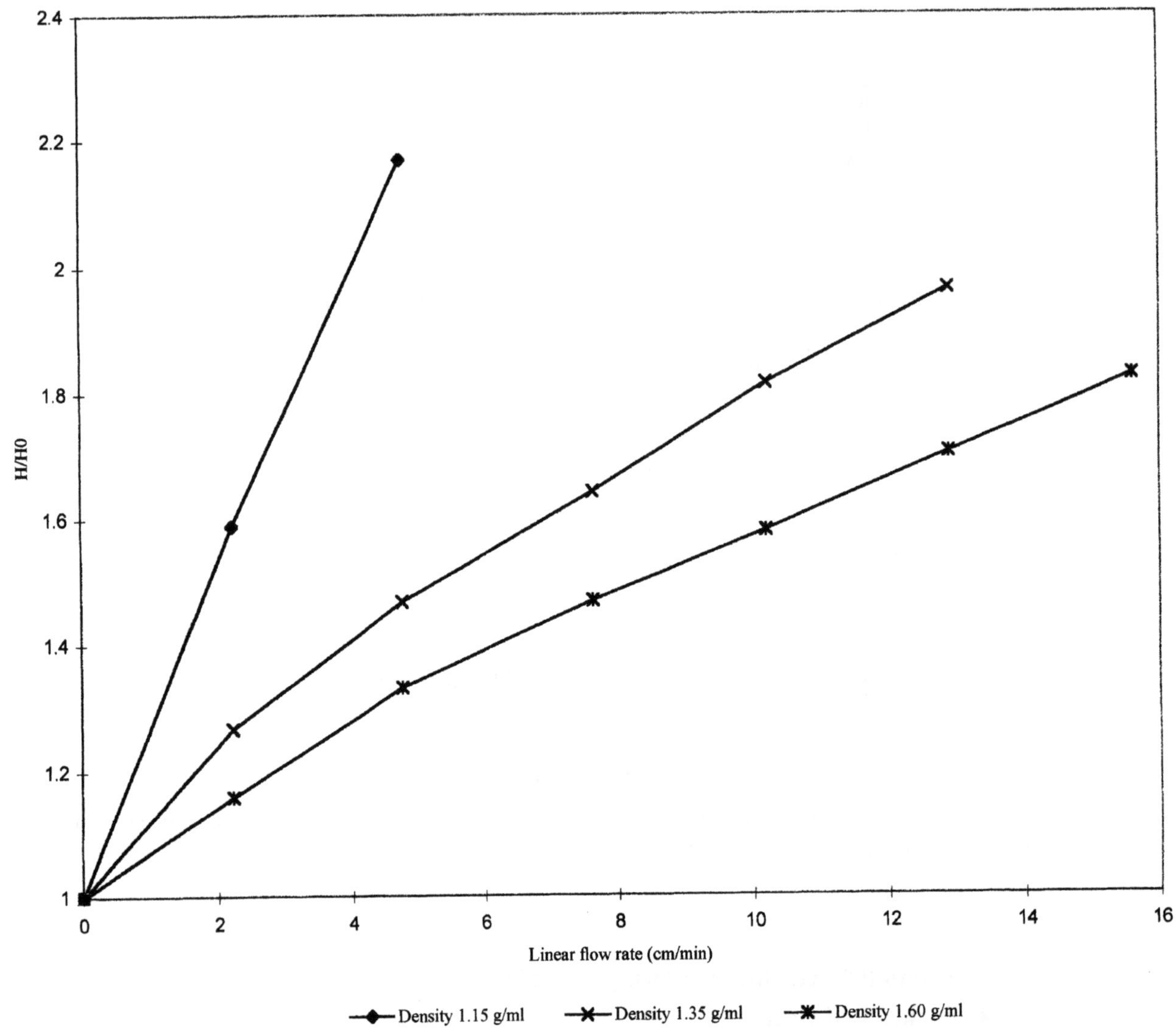

Figure 4. Expansion of EBA matrices with different densities as a function of the linear flow rate. H0 = packed bed height without flow, H = expanded bed height observed at a given flow rate. The data are all produced with Ø = 100 – 300 μm EBA beads.

Packed bed vs. expanded bed elution volumes

Figure 3 shows the elution volume - defined as the volume obtained between the beginning of the elution peak and the point where 98% of the BSA is eluted ($V_{E98\%}$) - as a function of the linear flow rate. The graph illustrates that the elution volume obtained with packed bed elution is rather constant around twice the settled bed volume within the tested flow range, while the elution volume obtained using expanded bed elution increases gradually with increasing flow rates. However, at a linear flow rate of about 1 cm/min the packed bed and expanded bed elution volumes are approximately equal and up to a flow rate of about 3 cm/min the expanded bed elution volume is only up to about 20% higher than the volume obtained using packed bed elution.

The main reason for this relatively small difference between packed bed and expanded bed elution is believed to be the low degree of expansion obtained at these flow rates when using EBA matrices of relatively high density (i.e. a density of at least 1.3). The expansion curves for three types of EBA matrices with different densities are shown in Figure 4.

The graph illustrates the dramatic differences in the degree of expansion obtained by varying the density of the adsorbent. By comparing with Figure 3 it is clear that the elution volumes obtained by expanded bed elution are proportional with the degree of expansion obtained at the different flow rates i.e. the elution volume increase with an increase in the degree of expansion during elution. This is what was expected and it seems that the optimal EBA matrix to employ for expanded bed elution is a matrix which has a low degree of expansion (e.g. less than 1.2–1.3) at the linear flow rate used for elution.

At linear elution flow rates of 1–3 cm/min optimal results are therefore expected when an EBA adsorbent with a density of 1.3–1.5 g/ml is employed (given that the particle size range is about 100 – 300 μm in diameter. If the particles are smaller the density must be higher). According to our experience, a further advantage of using matrices of high density is that a more robust EBA process is achieved. It is less vulnerable towards variations e.g. density and viscosity differences between buffers and raw materials or the passing of large air bubbles through the bed.

A density of 1.3–1.5 g/ml seems to be the best compromise between the degree of expansion, process robustness and total binding capacity per litre adsorbent. For certain applications e.g. involving highly viscous raw materials or ultra high flow rates it may however still be relevant to use adsorbents with even higher densities.

In conclusion we find that the differences found with respect to elution volume between packed bed and expanded bed elution (i.e. less than 20% at flow rates of 1–3 cm/min) are too small to justify the significant mentioned drawbacks associated with packed bed elution. The potential advantages obtained by expanded bed elution in terms of process simplification and cost reductions - as described in the introduction - seem to be much more significant, and should be a significant inspiration for the development of the next generation of EBA equipment.

Acknowledgements

The authors wish to thank technicians Anja Dideriksen, Tine Møllegaard, Brian Olsen, Anne Skovgaard, and Parvin Tehrani for valuable technical assistance during this project.

References

Lihme A, Nielsen, CS, Bøg-Hansen TC (1990a) Substance carrying conglomerate. Patents: EP 0538.350, EP 0607998.

Lihme, A, Nielsen, CS, Bøg-Hansen TC (1990b) Method of distributing aliquid in a fluid bed reactor. Patent application: EP 0722771 A1

Zafirakos E and Lihme A (1998) "UpFront Expanded Bed Columns – Performance and Scale Up", submitted for publication in Bioseparation.

Address for correspondence: UpFront Chromatography A/S, Lersøpark Allé 42, DK-2100 Copenhagen Ø, Denmark. Homepage: www.upfront-dk.com

Bioseparation **8:** 99–109, 1999.
© 1999 *Kluwer Academic Publishers.*

Cell/adsorbent interactions in expanded bed adsorption of proteins

Jan Feuser[2], Joachim Walter[2], Maria-Regina Kula[1] & Jörg Thömmes[1]
[1]*Institut für Enzymtechnologie, Heinrich-Heine Universität Düsseldorf, Stetternicher Forst, 52426 Jülich, Germany;* [2]*Boehringer Ingelheim Pharma KG, Birkendorfer Straße 65, 88397 Biberach an der Riss, Germany*

Received 17 August 1998; accepted in revised form 30 October 1998

Key words: expanded bed adsorption, fluidisation, initial recovery, protein purification

Abstract

Expanded bed adsorption (EBA) is an integrated technology for the primary recovery of proteins from unclarified feedstock. A method is presented which allows a qualitative and quantitative understanding of the main mechanisms governing the interaction of biomass with fluidised resins. A pulse response technique was used to determine the adsorption of various cell types (yeast, Gram positive and Gram negative bacteria, mammalian cells and yeast homogenate) to a range of commercially available matrices for EBA. Cells and cell debris were found to interact with the ligands of agarose based resins mainly by electrostatic forces. From the adsorbents investigated the anion exchange matrix showed the most severe interactions, while cation exchange and affinity adsorbents appeared to be less affected. Within the range of biologic systems under study *E. coli* cells had the lowest tendency of binding to all matrices while hybridoma cells attached to all the adsorbents except the protein A affinity matrix. The method presented may be employed for screening of suitable biomass/adsorbent combinations, which yield a robust and reliable initial capture step by expanded bed adsorption from unclarified feedstock.

Introduction

Adsorption in liquid fluidised beds has been introduced as an integrated technology for the initial recovery of proteins from unclarified feedstock (Chase, 1994; Hjorth, 1997; Thömmes, 1997). By fluidising suitable adsorbents a stable bed is formed whose increased interstitial volume allows the application of particle containing samples without the risk of blocking the bed. Stabilisation of the fluidised bed by choosing adsorbents with an appropriate distribution in density and/or particle size paired with a well designed fluid distribution unit at the column inlet yields a plug flow pattern, which is supposed to produce a performance comparable to protein capture in packed beds. Adsorption in fluidised beds, which are stabilised by stratified or classified fluidisation, is termed expanded bed adsorption (EBA). During EBA in theory all particles contained in the feed shall pass freely through the expanded bed while the target compound is bound. Washing procedures will remove the residual particles and unbound proteins from the void volume and the product of interest can be eluted subsequently. Thus a combination of solid/liquid separation and adsorptive purification is obtained in a single new unit operation. Numerous examples have been described in literature where this principle has been applied successfully for the purification of proteins from yeast (Noda, 1996), mammalian (Batt et al., 1995; Thömmes et al., 1996) and bacterial cells (Hansson et al., 1994) as well as from homogenates (Barnfield-Frej et al., 1994; Chang et al., 1995).

The system parameters governing the performance of EBA have also been investigated based on experiments with single component adsorption from aqueous buffers (Hjorth et al., 1995; Chang and Chase, 1996; Karau et al., 1997). From the behaviour in ideal systems the following conclusions were drawn:

- Stratified (classified) fluidised beds show reduced axial mixing; provided a minimum sedimented bed height is chosen the performance of EBA will not be limited by dispersion
- Adsorption of proteins to fluidised porous matrices is mainly limited by internal transport processes

– Contrary to packed bed protein adsorption fluid side transport should be considered as a factor potentially contributing to the overall performance.

Based on model systems these investigations have concentrated on the attempt to show that an expanded bed has similar performance as a packed bed. Application of EBA, however, implies that biomass is introduced to the expanded bed. Unfortunately there is only limited information available from literature, how cells or cell debris interfere with fluidisation and adsorption during the purification of proteins from whole broth by EBA. In 1994 Erickson et al. described sticking of mammalian cells to a protein A coated controlled pore glass adsorbent during EBA of antibodies from whole CHO cell culture broth (Erickson et al., 1994). These interactions led to the formation of aggregates from cells and adsorbent and reduced the efficiency of the overall process. Barnfield-Frej et al. (1994) investigated the influence of *E. coli* homogenate concentration in the feed during adsorption of recombinant annexin V to an expanded anion exchange resin. Biomass concentration in the sample had to be restricted for two reasons. Firstly increased sample viscosity due to biomass concentration higher than 7–8% dry weight caused hydraulic transport of adsorbent beads out of the 1 m column (15 cm sedimented bed height were used). The second effect observed was the formation of flow channels in the bed, an event taking place at biomass loads of more than 8% dry weight. Although this was not further investigated, interactions between homogenate particles and adsorbent can be assumed to be the major cause for deteriorated fluidisation. Thömmes et al. described cell breakthrough during EBA of IgG2a on a rProteinA modified matrix (Thömmes et al., 1996). No retardation of cells in the bed was found and the overall cell concentration in the eluate was reduced more than 100 fold, indicating the absence of cell/adsorbent interactions. Similar results were found by Feuser et al. (1998) during EBA of IgG1 with the same adsorbent. Using a cation exchange resin, however, hybridoma cells were severely retarded during sample application. This was attributed to adsorption of cells to the stationary phase under the conditions of sample application (pH 5.5). Besides reducing the stability of the expanded bed, the presence of cells may also have an influence on the adsorbents' static binding capacity for the target protein. Chase and Draeger (1992a, b) performed BSA adsorption to matrices of varying ligand chemistry in the presence of different types of cells. Charge effects dominated the influence on static capacity while small organisms reduced the capacity to a higher extent than larger ones. This was attributed to penetration of cells into the porous adsorbent structure. Based on results reported in literature as well as on own experience the following consequences of biomass/adsorbent interaction during EBA can be anticipated:

– The bed stability may be reduced due to formation of channels and stagnant zones during sample application
– The sorption performance of beds with channels and stagnant zones can deteriorate
– The static capacity may be regarded in case of penetration of small cells or debris into porous adsorbents
– During elution in packed bed mode resin particles, to which biomass has adsorbed, may show hindered sedimentation. The resulting poor packing quality may lead to distorted elution peaks
– The product solution can be contaminated with biomass, which is co-eluted under desorption conditions
– Adsorbent can be elutriated from the column due to reduction in the specific weight of the particles after adsorption of biomass
– Harsh regeneration conditions will be needed in order to release tightly bound cellular material
– The life expectancy of the stationary phase may be reduced due to irreversible fouling effects and the harsh regeneration conditions.

During development of EBA processes the adsorption conditions in general are optimised using clarified supernatants and after extensive fine tuning of the conditions the protocol is transferred to a whole broth procedure. In the case of severe biomass/adsorbent interactions and subsequent process failure the whole optimisation has to be repeated for different adsorption conditions. In some cases due to the disappointment of the operators the whole EBA system may regarded as not suited for protein purification purposes. Thus a fast and simple method is needed, which can be used in parallel to the development of sample application conditions in order to check certain combinations of biomass and adsorbent for potential interactions. Only conditions which favour target protein binding AND suppress cell adsorption should be chosen for the whole broth operation.

In this paper a method fulfilling these criteria will be presented and its usefulness for quantification of biomass/adsorbent interactions shall be demonstrated. The method is based on a pulse/response experiment where the adsorption of biomass to the stationary phase during passage through a stable expanded bed is quantified. By setting a threshold value of maximum cell adsorption during this experiment a criterion can be obtained, which allows to decide which process conditions will be favourable for bed stability in terms of biomass/adsorbent interaction and which conditions should be avoided in the EBA protocol.

Materials and methods

Cells

Freeze-dried yeast (*Saccharomyces cerevisiae*) was purchased from Goldhand Vertriebsgesellschaft mbH (Düsseldorf, Germany). Prior to the experiments the freeze dried yeast was suspended in 10 mM potassium phosphate buffer at pH 7.0 and allowed to re-hydrate for 30 min with gentle stirring of the suspension. Yeast homogenate was obtained by disruption of a 40% [w/w] yeast suspension in 10 mM potassium phosphate buffer at pH 7.0 in a continuously operated bead mill with 0.5 mm glass beads.

E. coli (strain DSM 498 was obtained from DSM, Braunschweig, Germany) was cultivated in 5 l shake flasks at 35 ° C in an incubator at a rotation speed of 100 rpm. The medium consisted of 5.0 g L^{-1} peptone and 3.0 g L^{-1} meat extract and the cells were harvested after 4 h at the end of the exponential growth phase.

Hybridoma cells (mouse-mouse hybridoma cell line) were grown in continuous culture under standard conditions as described elsewhere (Born et al., 1996).

Staphylococcus carnosus TM 300 (Schleifer and Fischer, 1982) was cultivated in a synthetic medium containing glucose as carbon source.

Adsorbents

All matrices used are listed in Table 1 and belonged to the STREAMLINE series of adsorbents (Amersham Pharmacia Biotech, Uppsala, Sweden). In particular the following media were used: STREAMLINE AC (base matrix, no ligand), STREAMLINE SP, STREAMLINE DEAE, STREAMLINE Phenyl, STREAMLINE rProteinA.

Equipment and conditions for expanded bed experiments

The experiments were conducted using an in-house designed expanded bed column with 0.02 m internal diameter, a fluid distribution unit at the bottom, and a movable unit at the column top, which allowed the adjustment of the top adapter directly above the expanded bed. The experimental set-up is shown in Figure 1: the stationary phase to be investigated was filled into the column to a sedimented bed height of 0.12 m and the bed was expanded with the respective mobile phase at a fluid velocity of $7.5\text{–}8.3{\cdot}10^{-4}$ m s^{-1}. The condition for the formation of a stable expanded bed were a constant bed height at approximately 2.5 fold expansion as well as the absence of flow channels and stagnant zones in the bed after visual inspection. After bed stabilisation a pulse of biomass was applied into the mobile phase using a three way injection port. Pulse duration was three minutes, resulting in a total pulse volume of 42–47 mL. Cell concentration in the pulse before and after passage through the expanded bed was detected by optical density (OD) in a flow spectrometer (Knauer, Berlin, Germany) at 600 nm. Data were transferred to a PC equipped with a software suited for peak integration. The biomass peaks were integrated and the fraction of cells transmitted was obtained by dividing the area of the response signal at 600 nm by the area of the input signal at 600 nm. A linear correlation of the signal at 600 nm with the concentration of biomass in the samples was obtained in the range of concentration under study as discussed below.

Sample preparation

The concentration of the cells in the pulses applied had to be adjusted in order to obtain a linear correlation between OD 600 nm and the concentration of cells. The concentrations given here correspond to the upper end of the linear range of calibration of the flow through detection units used in this study. If not otherwise stated the dilution buffer used for all experiments was 10 mM potassium phosphate at pH 7.0. Yeast cells and yeast homogenate were suspended in dilution buffer to a final concentration of 0.03% [w/w] dry weight. *E. coli* cultivation broth was diluted 1:3 resulting in a final cell concentration of approximately 0.37% [w/w] wet weight. *Staphylococcus carnosus* broth was diluted 1:20 yielding a final biomass concentration of 0.1% [w/w] wet weight. Hybridoma cells

Table 1. Qualitative summary of cell adsorbent interactions in the systems under investigation (+ more than 80% cell transmission in pulse response experiment; – less than 80% cell transmission in pulse response experiment; -- bed collapsed during pulse application; NA system not investigated)

Biomass	STREAMLINE AC	STREAMLINE Phenyl	STREAMLINE DEAE	STREAMLINE SP, pH 7.0	STREAMLINE rProteinA
S. cerevisiae	+	+	–	+	NA
S. cerevisiae homogenate	+	+	–	+	NA
E. coli	+	+	+	+	NA
S. carnosus	+	+	–	+	NA
Hybridoma	+	--	--	+	+

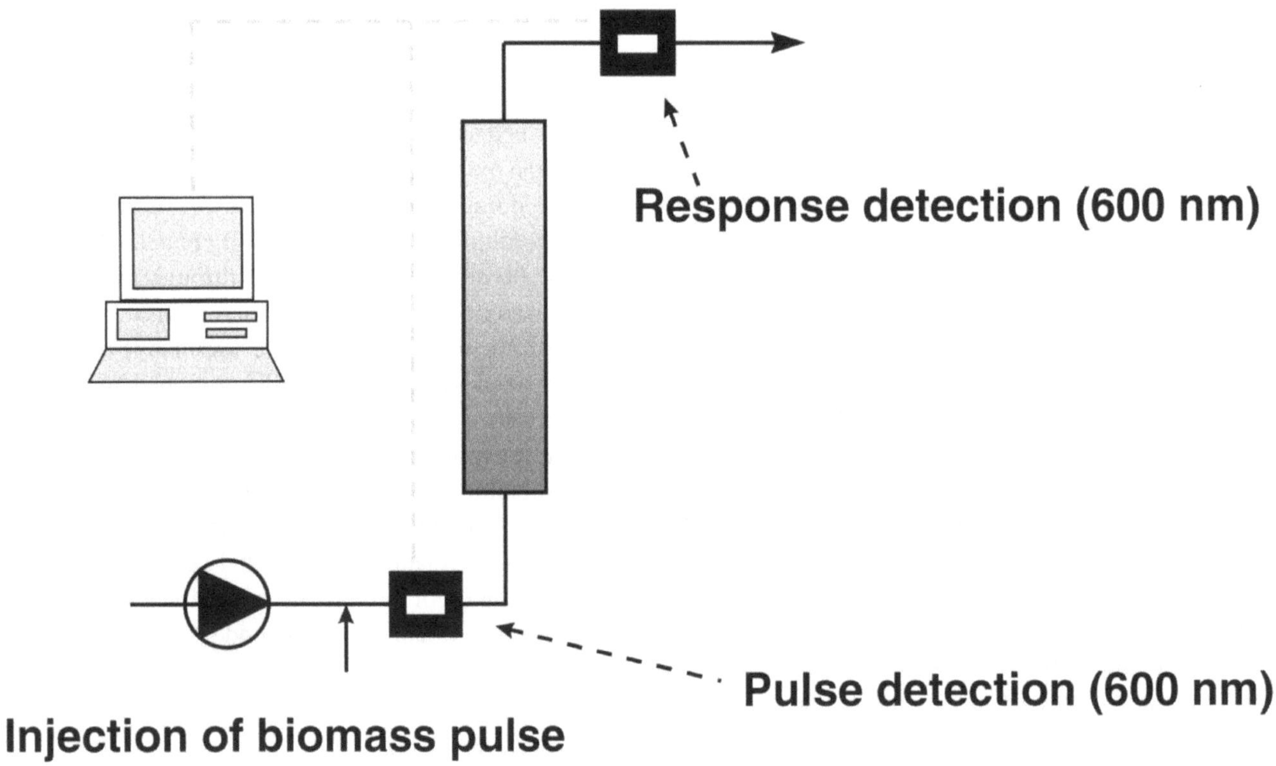

Figure 1. Experimental set-up.

were used undiluted with a mean cell concentration of $2{\cdot}10^6$ cells/ml.

Error of measurement

All experiments were performed at least in triplicate, the confidence interval of the data obtained depended on the biomass used. For the whole cell systems adsorption of biomass could be estimated with an accuracy of $\pm10\%$. For the yeast homogenate the reproducibility depended strongly on the reproducibility of the cell disruption. The extent of adsorption of particles to fluidised matrices certainly is a function of the size of the biomass particles in the sample, which in turn is determined by the method of cell disruption. Furthermore the adsorption of biomass will be mediated by the presence of nucleic acids, in particular by the length of the nucleic acid strands. Variations in the cell disruption will result in variations in both these parameters. This has to be considered when experiments with homogenates are analysed. In our case the confidence interval could be estimated as $\pm$ 20%.

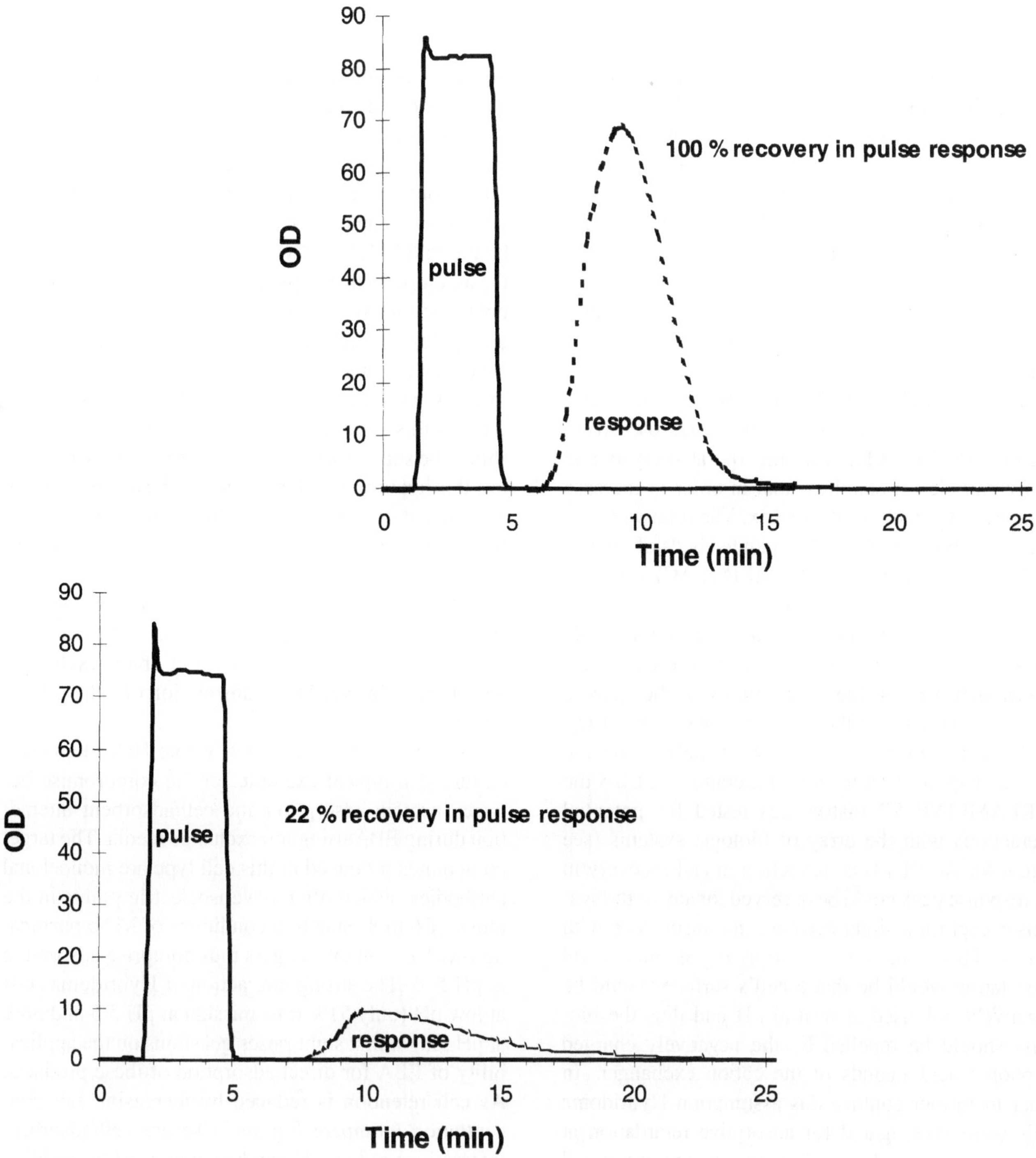

Figure 2. Pulse/response signals. (a) No interaction (Hybridoma pulse to STREAMLINE SP at pH 7.0); (b) strong interaction (Hybridoma pulse to STREAMLINE SP at pH 5.0).

Results and Discussion

Retention of cells in a fluidised bed can be caused by mechanical retardation in the beds' void fraction or by adsorption of cells to the fluidised resin. If a pulse of cells is applied in the experimental set-up described above mechanical retardation will result in a distorted response peak with a significantly increased mean residence time (compared to a not retarded tracer molecule). Additionally extended peak tailing will be found. The total response peak area, however, will not differ from the input pulse, as no adsorption takes place and each cell applied to the column will pass through sooner or later. Cell adsorption will be detected by a reduction in the overall peak area, thus it was decided to define the percentage of cell transmission (response peak area/input pulse peak area) through the bed as a quantitative signal for biomass/adsorbent interaction under the conditions studied. This quantity is expressed in Figures 2–8 as recovery in pulse response (%). Figure 2 shows two representative outcomes of pulse response experiments. In Figure 2a an experiment without measurable cell adsorption is shown, while Figure 2b depicts significant adsorption of cells as evidenced by the smaller area of the response peak.

Before investigating the influence of ligand chemistry on the adsorption behaviour to fluidised adsorbents the affinity of various organisms to the base matrix of the STREAMLINE series was studied. These adsorbents consist of a cross-linked agarose structure of $1 \cdot 10^{-4}$ m radius with incorporated crystalline quartz particles or stainless steel in order to increase the apparent density of the matrix. The relative transmission of biomass through expanded beds of the unmodified base matrix (STREAMLINE AC) is shown in Figure 3. As can be clearly seen, none of the organisms investigated adsorbed to this material at pH 7.0, so it can be concluded that any adsorptive effects shown later are not due to an affinity to the agarose backbone but rather to the respective ligand chemistry.

As cation exchange processes appear to be the dominant application in current literature on EBA the STREAMLINE SP matrix was tested for potential interactions with the array of biologic systems (see Figure 4). At pH 7.0 no reduction in cell recovery in the response peak could be observed for any of the systems except for a slight decrease in transmission with *E. coli*. This result was not surprising as the general expectation would be that a cell's surface should be negatively polarised at neutral pH and thus the biomass should be repelled by the negatively charged sulphonic acid ligands of the cation exchanger. In order to further confirm this assumption Hybridoma cells were investigated for adsorptive retardation at various pH values. Figure 5 shows an increase in cell adsorption to the cation exchanger with decreasing pH. This is in accordance with the expectation of an increasing number of positive charges on a cell surface at reduced pH, which in turn will favour binding to negatively charged ligands. Obviously, when cation exchangers are employed in EBA, reducing the pH in order to improve target protein binding may lead to an alteration of the cell surface charge and thus cell binding to a cation exchanger may be promoted.

Looking at the behaviour of the organisms studied with the anion exchange matrix STREAMLINE DEAE at pH 7.0 (Figure 6) we find a severe interaction of all biologic systems but *E. coli*. Hybridoma cells did not pass through a fluidised bed of this matrix and even caused the bed to collapse, which most likely is due to the formation of an aggregated network from cells and adsorbent particles. Whole *S. cerevisiae* was transmitted only to 28% while approximately half of the yeast homogenate particles and *S. carnosus* cells were bound to STREAMLINE DEAE. This again supports the assumption that at pH 7.0 most cells are negatively polarised and are thus prone to adsorption to positively charged anion exchangers. From the data shown it appears that the interaction between cells and adsorbents in EBA were mainly dominated by electrostatic effects. If this were true, cell retention in fluidised beds should be suppressed at high ionic strength. In order to test this hypothesis cell retention in the pulse/response experiment at varying conductivity was investigated for the system *S. cerevisiae*/STREAMLINE DEAE, which had shown very strong adsorption of cells (only 28% transmission in the pulse/response). As demonstrated in Figure 7, addition of salt in fact reduces the affinity of *S. cerevisiae* to the anion exchanger and at 200 mM NaCl the transmission of cells is fully restored.

Isolation of products from whole Hybridoma cell culture is a typical example for the compromise between protein adsorption and cell/adsorbent interaction during EBA using ion exchange media. The target compounds produced in this cell type are monoclonal antibodies, which often have isoelectric points in the range of 6 to 8. Standard conditions of MAb purification with cation exchangers thus comprise adsorption at pH 5–6. The strong interaction of Hybridoma cells at low pH (only 51% transmission at pH 5.5 and 38% at pH 5.0) at first sight raises questions on the applicability of EBA for direct adsorption of these products. As cell retention is reduced by increasing salt concentration (compare Figure 7) severe cell/adsorbent interactions at low pH can be suppressed by addition of a charged mediator. As increasing ionic strength also reduces the static protein binding capacity of ion exchangers, the user is faced with an optimisation problem: a compromise has to be found between a reasonable capacity for the target compound at low conductivity and low pH and a stable fluidisation at high pH and high ionic strength. The successful examples of MAb purification on cation exchangers from whole broth may serve as a good example of such a compromise: EBA of monoclonal antibodies at pH 5–6 often is operated with undiluted cell culture broth

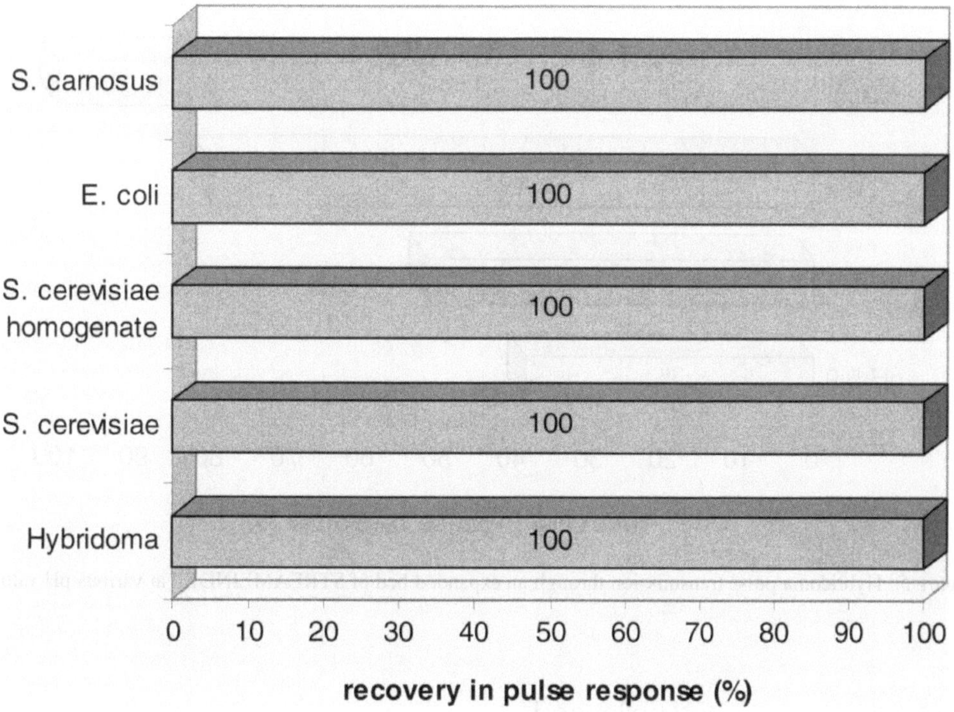

Figure 3. Cell pulse transmission through an expanded bed of the base matrix (STREAMLINE AC).

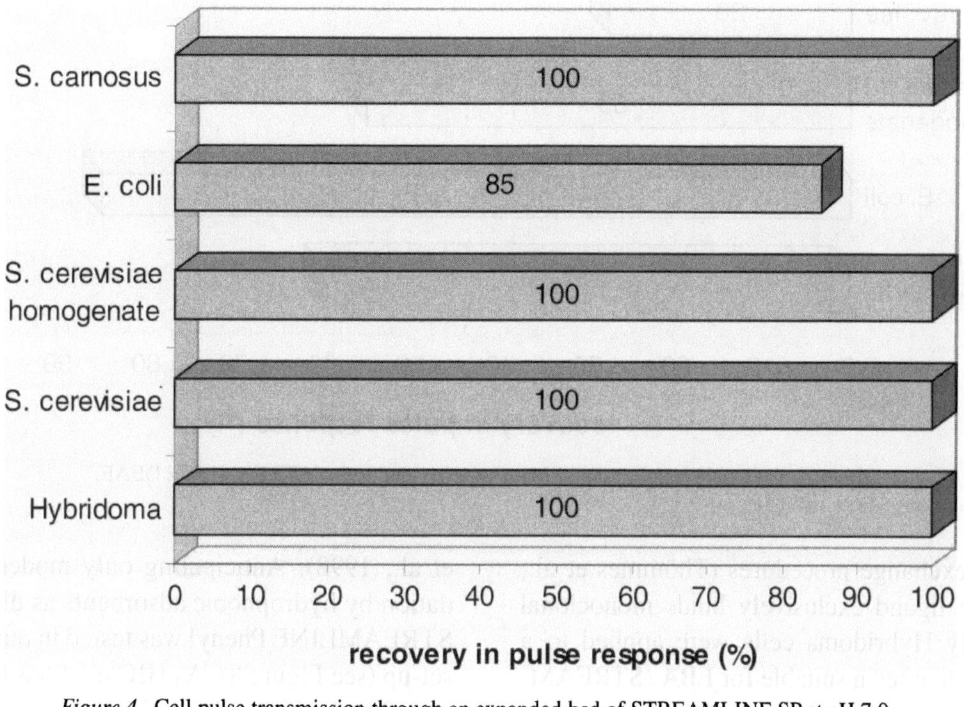

Figure 4. Cell pulse transmission through an expanded bed of STREAMLINE SP at pH 7.0.

(Erickson et al., 1994; Batt et al., 1995), where the conductivity is equal to a 150 – 180 mM NaCl solution (13–15 mS cm^{-1}). The increased ionic strength in the medium is able to suppress cell adsorption and makes operation at pH 5–6 possible without hampering the sorption efficiency by unstable fluidisation.

Under the assumption of charge dominated biomass adsorption to expanded stationary phases affinity matrices as well as media for hydrophobic interaction should have no or at least only little tendency to bind either of the organisms tested in this work. EBA of monoclonal antibodies to adsorbents modified with protein A has proven to be a valuable alter-

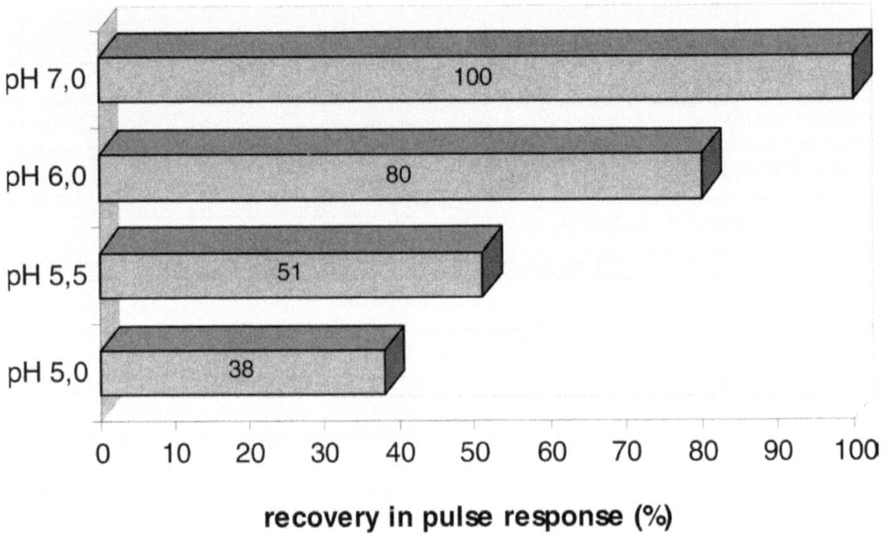

Figure 5. Hybridoma pulse transmission through an expanded bed of STREAMLINE SP at various pH values.

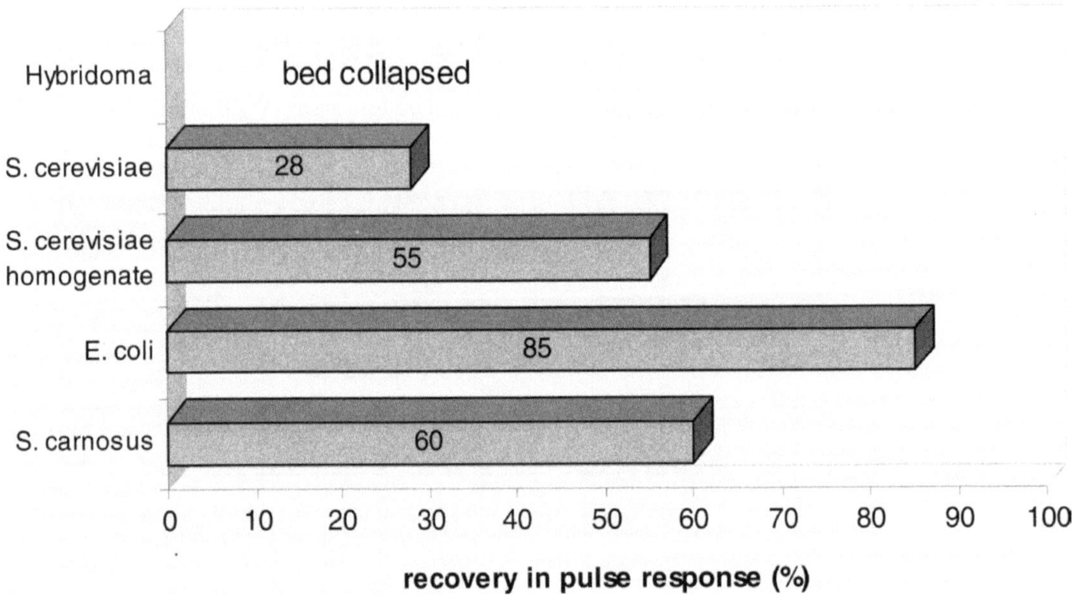

Figure 6. Cell pulse transmission through an expanded bed of STREAMLINE DEAE.

native to ion exchange procedures (Thömmes et al., 1996). As the ligand exclusively binds monoclonal antibodies only Hybridoma cells were applied to a bed of a respective resin suitable for EBA (STREAMLINE rProtein A, porous agarose matrix with particle diameter of $1.3 \cdot 10^{-4}$ m, apparent density increased by incorporation of stainless steel particles). In the respective pulse/response experiments 100% transmission of Hybridoma cells during passage through the expanded bed was found indicating the total absence of biomass/adsorbent interactions. Adsorption based on hydrophobic interaction in fluidised beds has only been described recently (Engberg et al., 1998; Noppe et al., 1998). Anticipating only moderate cell retardation by hydrophobic adsorbents as discussed above STREAMLINE Phenyl was tested in our experimental set-up (see Figure 8). As HIC methods usually are operated under high salt conditions ammonium sulphate was added to the mobile phase to a final concentration of 1 M (90 mS cm^{-1}). Under these conditions Hybridoma cells were completely retarded in the fluidised bed and even caused a collapse of the bed, similar to the behaviour in combination with the anion exchange resin. It has to be considered, however, that the very high ionic strength irreversibly damaged the Hybridoma cells with concomitant release of in-

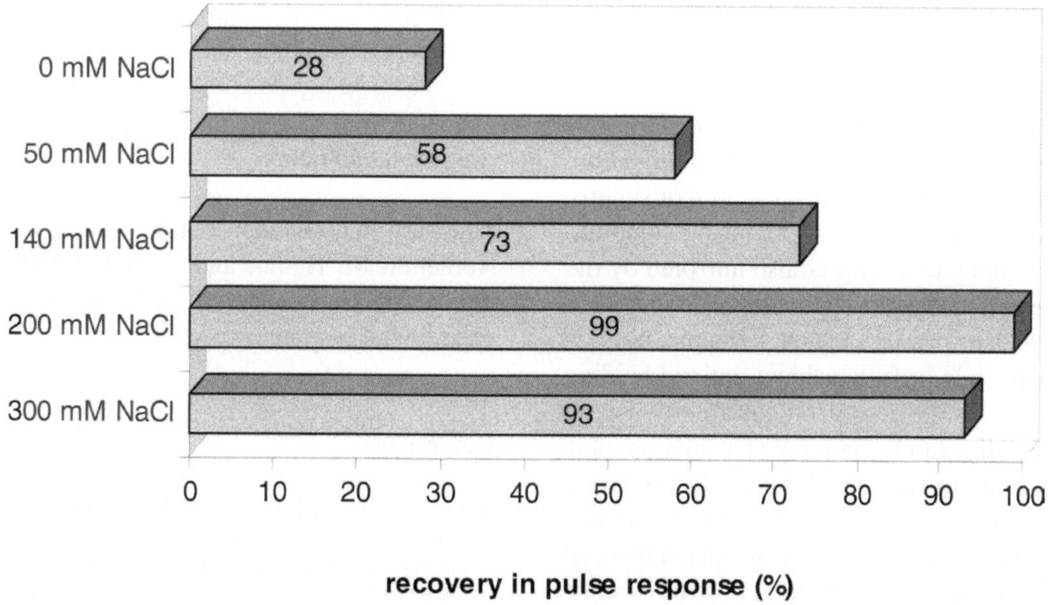

Figure 7. S. cerevisiae pulse transmission through an expanded bed of STREAMLINE DEAE at varying NaCl concentration in mobile phase.

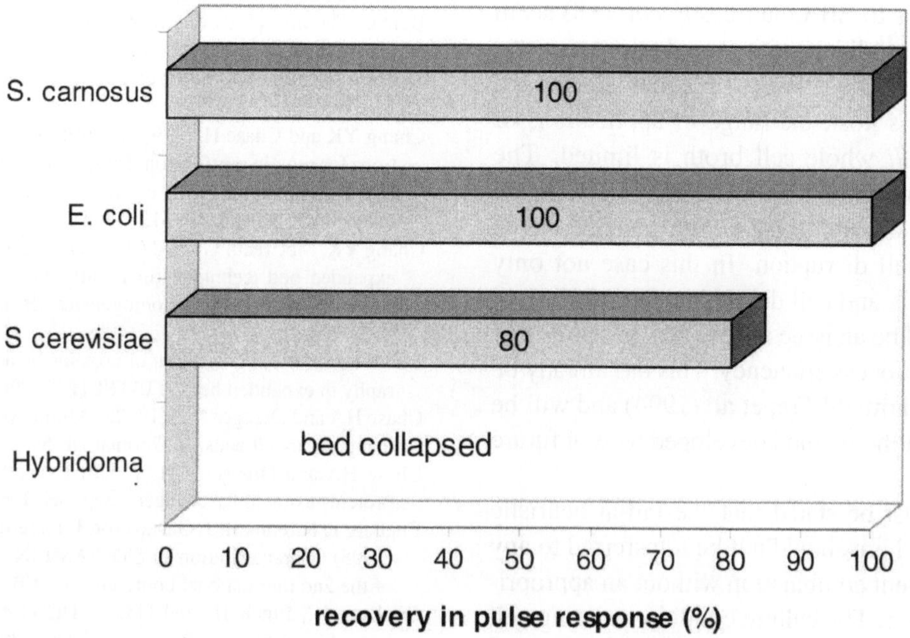

Figure 8. Cell pulse transmission through an expanded bed of STREAMLINE Phenyl under high salt conditions (1 M ammonium sulphate; 90 mS cm^{-1}).

tracellular material, especially nucleic acids. Thus the interaction of Hybridoma cells with STREAMLINE Phenyl under high salt conditions may have been promoted by the presence of long chain nucleic acids. The other organisms can be characterised as not or only moderately interacting systems even under high salt conditions, which makes hydrophobic interaction in EBA systems a good candidate for reliable and robust operation.

Conclusions

Based on the results shown here an initial guideline can be established, which may be used to differentiate between suitable and inappropriate combinations of biomass and adsorbent in EBA from unclarified broth. Table 1 summarises this information according to the following rules: non or only moderately interacting systems are characterised by the '+' sign, the borderline criterion being a transmission of cells in the pulse/response experiment of at least 80%. Strong interaction is depicted by the '–' sign whereas collapse of the bed is indicated by '– –'.

Employing ion exchange processes, which are very popular as a first capture step in protein purification, in an EBA mode cation exchange at neutral or slightly acidic pH seems to be the most promising combination for stable operation. This is also mirrored by the large number of successful applications described in literature, which are based on such protocols. If pH is reduced too much in order to achieve optimal binding for certain proteins, however, cell adsorption may be significant. From a stability point of view hydrophobic interaction seems to be a good candidate for initial recovery in EBA with the exception of hybridoma cells. The complete absence of interactions of this cell type with the rProtein A adsorbent, however, offers a good alternative for direct capture of MAb from whole broth. Comparing the overall affinity of the different biological systems to EBA matrices *E. coli* cells seem to be the easiest candidate with regard to process stability. As there is only a limited number of secreted products from this host, the range of applications of EBA from *E. coli* whole cell broth is limited. The majority of recombinant products from *E. coli* is accumulated in the cytosol or the periplasma and has to be released by cell disruption. In this case not only adsorption of cells and cell debris but also binding of nucleic acids will be an issue with regard to stability of fluidisation and process efficiency. This has already be pointed out by Barnfield-Frej et al. (1994) and will be investigated with the method developed here in future work.

Finally it must be stated that the initial heuristic rules established here should not be transferred to any other cell/adsorbent combination without an appropriate critical attitude. The culture conditions during cell propagation will have a considerable influence on the physiological status of any biological system and will thus determine the affinity of a cell surface to EBA matrices. The purpose of this contribution is NOT to define fixed rules but rather to point out a very simple technique which can be applied by potential users of EBA in order to characterise their specific system for cell/adsorbent interactions. Experimental conditions for EBA processes (cell line, adsorbent, pH, conductivity) may be screened for indications of biomass adsorption based on the methodology outlined here. Optimising process conditions with regard to both protein adsorption AND cell/adsorbent interaction will take into account that EBA is a truly integrated process which combines the separation of cells from the initial broth with a bona fide adsorptive operation.

Acknowledgements

The very helpful contributions of C. Janknecht, F. Nothelfer, M. Halfar, and A. Karau both during the practical realisation as well as during the design of the experiments is gratefully acknowledged.

References

Barnfield-Frej A-K, Hjorth R and Hammarstroem A (1994) Pilot scale recovery of recombinant annexin V from unclarified *E. coli* homogenate using expanded bed adsorption. Biotechnol. Bioeng. 44: 922–929.

Batt BC, Yabannavar VM and Singh V (1995) Expanded bed adsorption process for protein recovery from whole mammalian cell culture broth. Bioseparation 5: 41–52.

Born C, Thömmes J, Biselli M, Wandrey C and Kula M-R (1996) An approach to integrated antibody production: coupling of fluidized bed cultivation and fluidized bed adsorption. Bioproc. Eng. 15: 21–29.

Chang YK and Chase HA (1996) Development of operating conditions for protein purification using expanded bed techniques: The effect of the degree of bed expansion on adsorption performance. Biotechnol. Bioeng. 49: 512–526.

Chang YK, McCreath GE and Chase HA (1995) Development of an expanded bed technique for an affinity purification of G6PDH from unclarified yeast homogenates. Biotechnol. Bioeng. 48: 355–366.

Chase HA (1994) Purification of proteins by adsorption chromatography in expanded beds. TIBTECH 12: 296–303.

Chase HA and Draeger NM (1992a) Affinity purification of proteins using expanded beds. J. Chromatogr. 597: 129–145.

Chase HA and Draeger NM (1992b) Expanded bed adsorption of proteins using ion exchangers. Sep. Sci. Technol. 27: 2021–2039.

Engberg P, Farenmark J, Gustavsson J, Lagerlund I and Sandberg L (1998) Characterisation of STREAMLINE Phenyl. Proceedings of the 2nd international conference on EBA. Napa, USA.

Erickson JC, Finch JD and Greene DC (1994) Direct capture of recombinant proteins from animal cell culture media using a fluidized bed adsorber. In: B Griffiths, RE Spier and W Berthold (eds) Animal Cell Technology: Products for today, prospects for tomorrow (pp. 557–560). Butterworth & Heinemann, Oxford.

Feuser J, Halfar M, Lütkemeyer D, Ameskamp N, Kula M-R and Thömmes J (1998) Interaction of mammalian cell culture broth with adsorbents in expanded bed adsorption of monoclonal antibodies. Proc. Biochem. *(In press)*.

Hansson M, Stahl S, Hjorth R, Uhlen M and Moks T (1994) Single-step recovery of a secreted recombinant protein by expanded bed adsorption. BIO/TECHNOLOGY 12: 285–288.

Hjorth R (1997) Expanded bed adsorption in industrial bioprocessing: recent developments. TIBTECH 15: 230–235.

Hjorth R, Kämpe S and Carlsson M (1995) Analysis of some operating parameters of novel adsorbents for recovery of proteins in expanded beds. Bioseparation 5: 217–223.

Karau A, Benken C, Thömmes J and Kula M-R (1997) The influence of particle size distribution and operating conditions on the adsorption performance in fluidized beds. Biotechnol. Bioeng. 55: 54–64.

Noda M (1996) Process for purifying recombinant human serum albumin. European Patent Application EP 0 699 687 A2.

Noppe W, Haezebrouck P, Hanssens I and De Cuyper M (1998) A new approach to alpha-Lactalbumin purification using STREAMLINE Phenyl. Proceedings of the 2nd international conference on EBA. Napa, USA.

Schleifer KH and Fischer U (1982) Description of a new species of the genus Staphylococcus: Staphylococcus carnosus. Int. J. Syst. Bacteriol. 32: 153–156.

Thömmes J (1997) Fluidized bed adsorption as a primary recovery step in protein purification. Adv. Biochem. Eng. 58: 185–230.

Thömmes J, Bader A, Halfar M, Karau A and Kula M-R (1996) Isolation of monoclonal antibodies from cell containing hybridoma broth using a Protein A coated adsorbent in expanded beds. J. Chromatogr. A 752: 111–122.

Address for correspondence: Jörg Thömmes, Institut für Enzymtechnologie, Heinrich-Heine Universität Düsseldorf, Stetternicher Forst, 52426 Jülich, Germany. (Fax: 2461 612490; e-mail: j.thoemmes@fz-juelich.de).

Bioseparation 8: 111–114, 1999.
© 1999 Kluwer Academic Publishers.

Validation issues related to expanded bed technology

Gail Sofer
BioReliance, 9900 Blackwell Road, Rockville, MD 20850, USA

Received 27 July 1998; accepted in revised form 9 December 1998

Abstract

Expanded bed adsorption technology is being implemented in manufacturing processes for biotherapeutics. In order to market a product, validation must be performed to document that the process performs its intended function. The key considerations for validation of expanded bed technology are presented.

Introduction

Validation is an issue that everyone in industry must address, but often firms try to ignore it until it is too late to design it into a process in a cost-effective manner. Lately, validation of chromatographic processes has become more of a target for regulatory inspectors. Expanded bed adsorption technology is a chromatographic technique that allows purification of very crude feedstreams, including those with particulates. Special attention should be paid to cleaning and to establishing control parameters that ensure consistent performance. With expanded bed steps now being incorporated into new processes, it is important to consider what the validation requirements are and how to phase them into a reasonable time plan.

Validation of expanded bed adsorption is not very different from validating other types of chromatographic or isolation techniques. The major issues related to expanded bed adsorption include qualification of the column and system, raw materials qualification, process validation, and cleaning validation. Santization routines and media lifetime evaluations are also part of the validation considerations.

Column and system qualification

Column qualification includes an installation qualification (IQ) and an operational qualification (OQ).[1] The IQ ensures that each component of the column is specified and meets design criteria. The OQ tests that the column functions properly. Whereas for standard chromatography it is common to measure HETP and/or A_s, expanded bed adsorption column qualification is performed by using a test molecule such as acetone or benzyl alcohol and testing for residence time distribution and measuring bed expansion. The formal qualification of the column is not performed until clinical trial material is being prepared, usually prior to Phase III trials, but it is important to determine optimal bed height during process design and correlate the acceptable bed height ranges with desired product quality. This information should be included in a development report that justifies how the process was designed. All changes made during process development should be documented. In addition to determining that the bed heights in both expanded and packed modes are within acceptable limits that provide the requisite product quality, it is important to define acceptable ranges of conditions, such as pH and ionic strength, that ensure that the expanded bed adsorption media are properly equilibrated.

During the formal system qualification, IQ and OQ are performed for equipment, such as valves and pumps.[2] If the system is automated, then IQ and OQ

[1] PDA Biotechnology Task Force on Purification and Scale-up. Industry Perspective on the Validation of Column-Based Separation Processes for the Purification of Proteins. Technical Report No. 14. PDA, Bethesda, MD, 1992.

[2] Sofer, G. and Hagel, L. Handbook of Process Chromatography: A Guide to Optimization, Scale-up, and Validation., Academic Press, 1997.

are also performed for both hardware and software.[3], [4] This is a rather arduous task for those not familiar with such work. For example, one commercial software package for process chromatography, called UNICORN, requires about 80 pages of tests. Today, software validation is one of the targets of regulatory inspectors. It is particularly important that security issues are addressed in the system qualification. As with the column, this formal qualification is not performed until clinical trial material is made.

It is important to point out that the approach for column and system qualification described here is only one way of approaching the task. Some activities can be combined to save time. The end result should be that there is documented evidence that the expanded bed column and system meet design criteria and consistently function properly.

Raw materials

Raw materials used in expanded bed adsorption processes include buffers, additives, expanded bed adsorption media, and the feedstream. These must all be qualified. Identity tests and certificates of analysis may be sufficient for purchased raw materials, provided that the user company has certified the vendor.[5] Buffer quality and the quality of additives must be defined. In some cases, detergents may contain unknown impurities that lead to inconsistent performance. Whenever an additive is used, it should be one for which there is a validatable assay.

For expanded bed adsorption media, certificates of analysis and regulatory support files (RSF) which describe extractables and both chromatographic and chemical stability should be available from the vendor. Although it is impossible to audit all vendors prior to obtaining a license to market a product, it is important to have a written plan for vendor certification. It is also critical to have some kind of acceptance criteria for raw materials even in early process development, since inconsistent results due to variability in raw materials can be very difficult to interpret.

The column feedstream should be defined within a range of acceptable parameters. These parameters may include total protein, DNA, endotoxin, and viscosity. For expanded bed adsorption, variability in feed viscosity can influence the degree of expansion and lead to inconsistent performance.

Process performance qualification

The process performance qualification is where the major validation effort occurs. This effort is directed towards ensuring consistency. Some regulatory documents describe the need to demonstrate validation of the purification process.[6], [7] For expanded bed adsorption, it is important to understand the goals of the step. For example, what impurities are being removed? If viral clearance is to be claimed for this step, then viral clearance evaluation studies must be performed. If DNA is to be removed, then the amount of removal must be validated during preparation of consistency batches. The product purity and yield must be reproducible within a defined range, and the product as well as impurities must elute consistently under a range of specified conditions, i.e., pH, conductivity, and volume.

Leachables should be known, and this information is available in vendor-produced Regulatory Support Files or Drug Master Files. If a protein ligand is employed, however, the user should validate the removal of any protein leachables in the remainder of the processing steps. This can usually be performed as a validation study during preparation of consistency batches during Phase III clinical trials. Leakage levels are usually highest during cleaning, sanitization and storage. It is essential, therefore, to removal all storage solutions and any concomitant leakage products.

Regeneration conditions should be defined, e.g., pH and conductivity for ion exchangers. And storage conditions must also be defined and demonstrated to inhibit microbial growth. All of the parameters that are validated during the process performance qualification should be established in broad ranges during process development, and then the ranges tightened as further understanding of the process is gained with experience.

During the validation runs, which are generally 3–5 consecutive batches at pilot or full scale, it is

[3] Grigonis, G.J., Subak, E.J., and Wyrick, M.L., Validation Key Practices for Computer Systems Used in Regulated Operations, Pharm. Technol. 21 (6) (1997) 74-97.

[4] PDA Committee on Validation of Computer-Related Systems. Technical Report No. 18. PDA J. Pharm. Sci. Technol. 49, S1 (1995).

[5] Supplier Certification Task Force. Supplier Certification—a model program. PDA J. 43 (43 (1989) 151-157.

[6] U.S. FDA, Guidance for Industry for the Submission of Chemistry, Manufacturing, and Controls Information for a Therapeutic Recombinant DNA-Derived Product or a Monoclonal Antibody Product for In Vivo Use. August 1996.

[7] U.S. FDA, Points to Consider in the Manufacture and Testing of Monoclonal Antibody Product for Human Use. February 1997

important to have established in-process control parameters. These may include sample load specifications for the total product mass and certain impurities, as well as viscosity. The acceptable operating temperatures for both feedstream and buffers should be defined. These specifications as well as those for flow rates and bed height must be adhered to during routine manufacturing of licensed product.

Cleaning, sanitization, and maintenance

Cleaning validation is one of the most important issues for expanded bed adsorption. Media are always dedicated to one product, but equipment may be used for multiple products. In the latter case, equipment cleaning validation is necessary to demonstrate absence of carryover.[8] This may require product-specific assays as well as general assays such as those for total organic carbon (TOC) and protein. The cleaning agents must be compatible with the media and hardware, as well as any in-line components such as filters. Ideally, the cleaning protocol is designed during development.

Bioburden and endotoxin assays should be performed routinely after cleaning and sanitization, and also after storage just prior to use. In fact, several firms have recently had questions from the U.S. FDA related to lack of endotoxin and bioburden testing after storage of chromatography columns. Increased focus on good manufacturing practices (GMPs) during post-approval inspections may also lead to questions from regulatory investigators who may perceive discolored columns to be unclean. The best approach is to be prepared with data that demonstrate the column is clean and the discoloration is not causing changes in product purity.

In addition to cleaning, a complete maintenance program requires consideration of utilities, which can have a significant effect on column lifetime (see below). Humidity and temperature control, as well as water and air quality, should be specified. Humidity control, especially in cold rooms, is needed to inhibit mold growth. While increased temperature decreases viscosity and may enable you to achieve a more stable bed, it is important to keep in mind that temperature fluctuations may affect protein precipitation and cause a more difficult cleaning problem. The quality of the water should be such that it is consistent and does not introduce impurities. Water for injection (WFI) is

[8] PDA Biotechnology Cleaning Validation Committee, Cleaning and Cleaning Validation: A Biotechnology Perspective. PDA, Bethesda, Md. 1996.

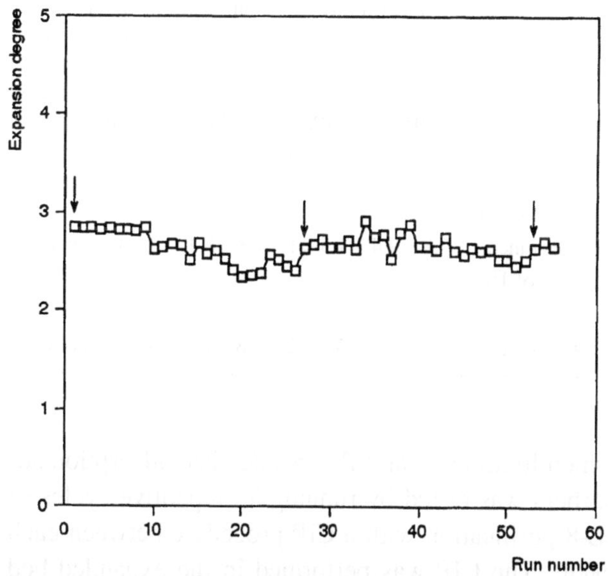

Figure 1. Lifetime study for an expanded bed of SP-SL used to purify Interleukin 8 expressed in *E. Coli* inclusion bodies. The lifetime of the expanded bed adsorbent was tested by running 52 repetitive cycles of interleukin 8. A cleaning procedure was performed between each cycle. The cleaning procedure consisted of a 4-hour contact time with 1% (w/v) DARACLEAN 8471 (Grace Dearborn Ltd.) using upward flow of 30 cm/h, followed by 30 mM $NaPO_4$, pH 6.5 for 4 sedimented volumes with upward flow of 100 cm/h. The dynamic capacity for lysozyme at 300 cm/h was measured before the first cycle, after 25 cycles, and after 50 cycles. Before the first cycle, the capacity was 69 mg/ml adsorbent; after 25 cycles, it was 71 mg/ml adsorbent; and after 50 cycles, it was 74 mg/ml adsorbent.

usually used in chromatography, especially after the initial steps in the process. The use of purified water, rather than WFI, in early processing steps such as expanded bed adsorption, can certainly be justified for feedstreams such as those from *E. coli*. Filtered air is important in controlling microbial contamination. The class of air will depend greatly on the degree of exposure of the product to the environment. For purification areas, Class 10,000 (Class C) is fairly standard. Since expanded bed adsorption is performed in the recovery, not the purification, suite, a lower air classification should be acceptable, especially with fermentation feedstreams.

Media lifetime

Expanded bed adsorption media lifetime studies should be performed at both small scale and concurrently with pilot and full scale production. Where possible, the feedstream should be from manufacturing scale. Data should be available prior to submitting a market application. Cleaning and lifetime are often addressed together in a process validation study. Figure 1 illustrates a cleaning-in-place (CIP) study in

Table 1. A comparison of viral clearance in logs$_{10}$ by new resin and end-of-use resin. Data courtesy of Amersham Pharmacia Biotech.

	BHV		BVD		EMC		HIV	
	a	b	a	b	a	b	a	b
New Resin								
Clearance	7.8	7.4	1.5	<1	>8	>8.1	>6.4	>6.4
End-of-Use								
Resin								
Clearance	7.3	>7.9	3	2.4	8	7.9	>6.5	>6.6

which lifetime on an SP-expanded bed adsorption adsorbent was tested by running 52 repetitive cycles of IL-8 purification with a CIP procedure between each cycle. The CIP was performed in the expanded bed mode with the adaptor at twice the sedimented bed height. As noted in this figure, the contact time with the cleaning agent was important. Once contact times for cleaning are established in development studies, they should be written into standard operating procedures. This figure illustrates the use of a rather stringent cleaning agent, but it is noted that the components of the cleaning agent are all defined so that an assay can be performed to demonstrate the absence of cleaning agent residues. In this particular study, a standard protein, lysozyme, was used to evaluate the consistency in dynamic capacity after more than 50 cycles. This is a practical approach for a vendor, but as noted above, the actual product feedstream should be used to obtain data to submit for a license application.

When viral clearance is claimed for an expanded bed adsorption column, there should be some evidence that there is consistent virus removal over the lifetime of the expanded bed adsorption media. This is an issue that is important to products derived from cell culture, animal transgenics, or natural products such as plasma. In Table 1, data from a virus evaluation study performed for a plasma fractionation process on Fast Flow ion exchangers illustrates the ability of chromatography media to consistently remove 4 different viruses for more than 400 cycles. Similar studies on expanded bed adsorption columns are not yet in the public domain, but Table 1 illustrates the type of data which should be obtained when viral clearance claims are to be made using expanded bed adsorption.

Regulatory questions related to chromatography, including expanded bed adsorption

Recently worldwide regulatory authorities have been focusing more on chromatography than they had in the past. The reason for this is not totally clear, but is most likely because with more biotechnology and biological processes approved and in clinical trials, there is more exposure to the use of chromatography to obtain the requisite product quality. Regulatory inspectors' questions to firms related to the use of chromatography have included requests for more information on processing times, holding points, validation of column lifetime and performance prior to use, loading capacity, recoveries, and removal of impurities.

Questions related directly to expanded bed adsorption include requests for dimensions of the bed in expanded and packed mode, evidence that cells pass through the column in expanded mode, cleaning procedures, and lifetime of the expanded bed adsorption media.

The first FDA-approved product using expanded bed adsorption has now been approved and placed on the market. The validation issues were satisfactorily addressed. These issues are not very different from those for traditional chromatography steps.

Summary

Validation of expanded bed adsorption should include qualification of the column and system, as well as raw materials such as the expanded bed adsorption media and processing chemicals. The process performance qualification should address viral clearance, where applicable, and the demonstration of the consistent removal of impurities. Cleaning validation should be performed and with expanded bed adsorption this is a major issue that should be addressed during process design. Sanitization routines should be shown to be consistently effective and the media lifetime claims validated. Validation should ensure consistent performance and it should be designed into a purification process to ensure final product quality and prevent failed batches.

Bioseparation **8:** 115–120, 1999.
© 1999 *Kluwer Academic Publishers.*

Highly dense beaded sorbents suitable for fluidized bed applications

N. Voute & E. Boschetti*
BioSepra s.a., 35, Avenue Jean Jaurès, 92395 Villeneuve la Garenne, France

Received 28 August 1998; accepted in revised form 5 February 1999

Key words: chromatography, fluidized bed, mineral oxides, proteins, sorbents

Abstract

New highly dense beaded sorbents suitable for fluidized bed applications of protein separations are presented. They are prepared using porous mineral oxides supporting functional hydrogels responsible for protein interaction. Beads of small diameter (70μm) are selected to reduce mass transfer resistance. Zirconium oxide was the preferred mineral material due to its high density (5.9g/ml) allowing high fluidizing liquid velocities (600 cm/h) into columns with a moderate bed expansion (lower than 3). Composite mineral - hydrogel sorbents are evaluated for their ability to rapidly adsorb proteins in fluidized bed and to separate with an appropriate resolution macromolecule mixtures in packed bed. Lysozyme dynamic capacities of 68 and 53 mg per ml of sedimented bed were obtained at fluidizing velocities of 450 and 900 cm/h.

Introduction

Capture and separation of biologicals from crude feed stocks by solid-liquid fluidized bed or expanded bed have been described in the last few years (Chang et al., 1995; Draeger and Chase, 1991; Raymond et al., 1998; Spence et al., 1994). Contrary to classical packed beds, fluidization of solid adsorbents provides a practical option to process very crude materials containing particles in suspension such as protein aggregates, cells or cell debris. In this separation mode, microbeads are lifted inside a column by an upward liquid stream generated by buffers and sample solutions. As a result of the increased interparticle porosity, fluidized beds allow the treatment of feed stocks with particles in suspension while packed beds only accept clear feed stocks. To the contrary of packed beds, fluidized beds do not generate any significant back pressure (Lochmuller et al., 1988) and they do not require any specific packing performance (Wells et al., 1987).

This mode of separation is attractive for biochemical processes because it reduces the number of preliminary steps of the downstream processing. It obviates the requirement for solid-liquid separation necessary for packed bed adsorption process, resulting in an increase of the overall productivity and recovery as well as in a decrease of the capital investment.

The ability of fluidized bed technology to quickly process large volume of dilute feed stock is interesting for the development of new biologicals. Monoclonal antibody production (Batt et al., 1995; Thömmes et al., 1996), plasmid production (Lyddiatt and O'Sullivan, 1998), and retroviral vector production (Braas et al., 1996) are typical examples of biotechnological processes, with low expression level and concomitant large volume of feed, for which fluidized bed technology is a valid alternative. Performing fluidized bed adsorption at high linear velocity is important to improve the column productivity. The linear velocity at which a fluidized bed can be operated is essentially dependent on the physical properties of the solid phase. However, upper limits are set on one hand by the terminal velocity of the particle and on the other hand, by the kinetics of sorption. When the column linear velocity exceeds the particle terminal velocity, the bed expansion increases sharply and the beads leave the column or form a packed bed at the upper distribution plate of the column, where clogging can occur. In addition, high linear velocity results in very short residence times that are detrimental for the kinetics

* Corresponding author.

of sorption, generally dominated by intraparticle mass transfer resistance. Several particle physical properties can be adjusted in order to maximize the performances of a fluidized bed process. As suggested by the Stokes' law, ways to prevent a too large expansion factor under high upward flows are the increase of the bead diameter or the increase of the solid density. However, the increase in the bead diameter is incompatible with the decrease in residence time resulting from a high upward flow. Increasing the bead diameter enlarges the characteristic diffusion length, which in turn requires a higher residence time for the sorption to occur efficiently.

Therefore substantial efforts have been devoted to the development of matrixes with increased specific weight compared to packed bed. Examples of materials with increased density include agarose quartz composite (Hjorth et al., 1995), perfluoropolymer derivative (McCreath et al., 1995), and fluoride modified zirconium oxide particles (Griffith et al., 1997). The latter material possesses the required density to afford high fluidizing velocity combined with a low particle diameter but is somehow hampered by low protein capacity.

In this study, a novel approach in the design of solid phase particles usable in fluidized bed mode is described. It produces high levels of dynamic binding capacity at elevated velocities thanks to the combination of small diameter, high density bead with a high capacity hydrogel filling the bead porosity.

Materials and methods

Chemicals

Zirconium oxide powders, salts and buffers were from Aldrich (Brussels, Belgium), monomers and crosslinkers were from Polysciences (Warrington, PA, USA) and proteins were from Sigma Chemicals (Saint Louis, MO, USA). Streamline sorbents and Streamline columns were from Amersham Pharmacia Biotech (Orsay, France).

Preparation of porous mineral oxide beads

Porous mineral beads were prepared by agglomeration of finely dispersed non-porous mineral oxide particles (Unger 1979). Briefly, an aqueous slurry containing 20 to 30% by weight of mineral oxide particles (0.1-2 μm particle size) was sprayed into a drying column. Water was very rapidly eliminated by evaporation and porous beads obtained. Primary beaded mineral oxides were then sintered at 600 to 800 °C for two hours. The distribution of bead diameters was between 35 and 120 μm.

Functionalization of mineral oxide beads

Mineral oxide beads were chemically modified to obtain ion exchangers according to the method of monomer intrusion followed by a polymerization (Boschetti and Coffman 1997). In practice a given weight of dry porous mineral beads was contacted with an acrylic monomer mixture composed of a functionalized monomer (methacryloylaminopropyltrimethylammonium chloride for the preparation of hydrophilic anion exchangers or acrylamidomethylpropane sulfonic acid for the preparation of hydrophilic cation exchangers), and a bifunctional acrylic monomer as crosslinker (N,N' methylene bis methacrylamide). Total concentration of monomers was 20% (w/v) and crosslinking ratio was 1%. Polymerization was started using a radical initiator (azobisisobutyronitrile). After polymerization, the resulting beads containing the functionalized hydrogel within their pore volume were extensively washed with solutions of 0.1 M sodium hydroxide, 0.1 M hydrochloric acid and distilled water.

Determination of bead physical properties

Apparent density of dry zirconia particles was measured by weighting a graduated glass cylinder before and after loading a sample. The cylinder was tapped for at least 1 min to allow uniform packing of the particles before measuring the volume of the mineral beads.

Particle density (effective particle density during fluidization) was determined using the Archimedes principle. This density refers to the density of the combined mineral shell and material filling the bead porosity.

Pore volume of dry porous zirconia beads was measured by extensive impregnation in paraffin oil.

Particle porosity was calculated using pore volume and material bulk density, determined by pycnometry.

Dynamic binding capacity

Dynamic binding capacity was measured with lysozyme for cation exchangers and with bovine serum albumin for anion exchangers. Protein concentration used was 5 mg/ml; buffer used were 50 mM acetate buffer pH 4.5 for cation exchangers and 50 mM Tris-HCl buffer pH 8.5 for anion exchangers.

Dynamic binding capacities of packed and fluidized beds were measured by frontal analysis at different flow rates in columns of 1.1 cm I.D. and 5 to 6 cm in settled bed length. Calculations were made at 10% breakthrough.

Fluidization properties

Expansions of cation exchangers were performed in a 2.5 cm I.D. Streamline column, using a 50 mM acetate buffer pH 4.5 as fluidizing solution. 25% v/v glycerol was added to the acetate buffer to increase the solution viscosity up to 2.3 mPa.s. Different flow rates were used and bed expansion measured after the bed stabilization was achieved.

Packed bed protein separation properties

Properties related to the separation of proteins were determined in packed bed columns. A 1.1 cm I.D. column was filled with 8 ml of cation-exchangers. The protein mixture used was composed of 5 mg/ml ovalbumin, 5 mg/ml β-lactoglobulin, 2.5 mg/ml cytochrome *c* and 2.5 mg/ml lysozyme. The amount of sample injected was 1 ml. Column equilibration and washing were performed in 50 mM acetate buffer at pH 4.5. Elution was carried out with a sodium chloride gradient up 1 M in 60 min at a flow rate of 125 ml/h.

Results and discussion

Zirconium oxide beads were analyzed for their porosity and apparent density. Porosity is an important parameter as it determines the amount of hydrogel that can be incorporated within the media and therefore, it controls the binding capacity of final sorbent. By varying the concentration of the slurry as well as the sintering temperature, porosities ranging from 0.16 to 0.57 were obtained. The apparent density is related to the bead porosity with the following equation:

$$\rho = (1 - \varepsilon_0)(1 - \varepsilon_p)\rho_s \qquad (1)$$

with ρ; apparent density, ε_o: interparticle porosity, ε_p: bead porosity, ρ_s: solid density.

The linear relationship between bead porosity and apparent density is shown on Figure 1. The slope of the line allows to estimate a value of 5.9 g/ml for the bulk density of zirconia, This data falls within ranges previously reported for the monoclinic and tetragonal forms of zirconia (Nawrocki et al., 1993).

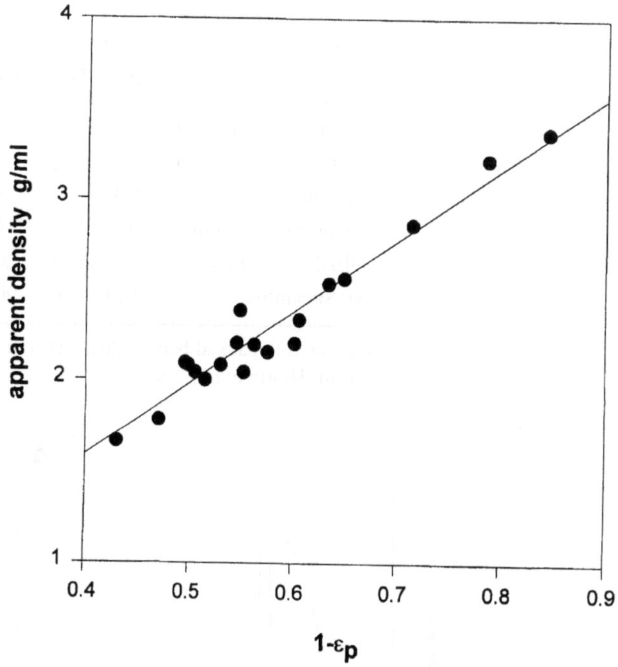

Figure 1. Dependence of zirconia apparent dry density on bead intraparticle porosity (ε_p)

Maximizing the bead porosity is beneficial for the capacity, but it is also detrimental to the media density, which in turn contributes to the particle terminal velocity and consequently, the fluidization properties.

In the attempt to maintain as high as possible the protein binding capacity, the targeted porosity value was set to 0.55. When using zirconium oxide porous beads exhibiting this level of porosity to support ionizable hydrogels, excellent ion exchangers were obtained. Main properties of these new ion exchange sorbents are shown in Table 1. Bead porosity influences the dynamic sorption capacity of both anionic and cationic zirconia resins. A reduction of 33% of porosity leads to a similar reduction in protein binding capacity.

High porosity zirconia based ion-exchange sorbents show marginally higher binding capacities, compared to commercially available fluidized bed media (SP and DEAE Streamline). However, more efficient mass transfer at very low column residence time is expected for zirconia sorbents due to their lower particle diameter: 70 μm compared to 200 μm for Streamline.

Separation properties of investigated ion exchangers were determined by packed bed chromato-

Table 1. Comparative properties of fluidized-bed media

Type of sorbent	Bead porosity	Bead diameter (μm)	Particle density g/ml	Dynamic binding capacity (mg/ml)
Cationic (Q) Zirconia	0.38	70	3.9	43*
Cationic (Q) Zirconia	0.55	70	3.2	68*
Anionic (S) Zirconia	0.48	70	3.6	87*
Anionic (S) Zirconia	0.55	70	3.2	105*
DEAE Streamline	Unknown	200	1.2	55**
SP Streamline	Unknown	200	1.2	73**

*measured in packed bed mode at 100 cm/h
**from (Hjorth et al., 1995)

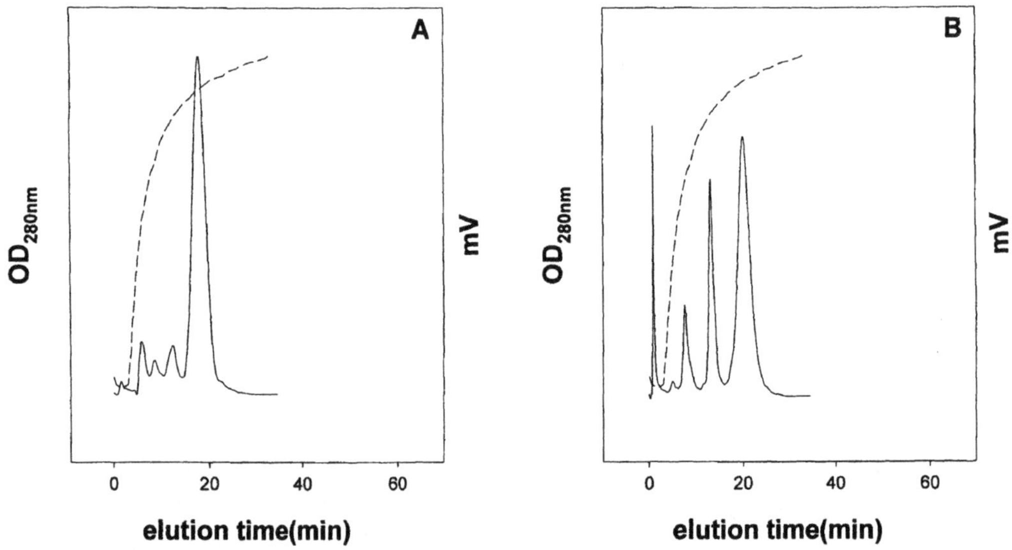

Figure 2. Packed bed separation of standard proteins (1.1 cm I.D.column, 8 cm packed bed length - 1 ml load: 5 mg/ml ovalbumin, 5 mg/ml β-lactoglobulin, 2.5 mg/ml cytochrome c and 2.5 mg/ml lysozyme - Equilibration and wash buffer: 50 mM acetate buffer pH 4.5 - Linear gradient elution up to 1 M NaCl in 60 min at 125 ml/h). Full line: UV trace - Dot line: output in mV from conductivity meter. A: SP Streamline (particle diam. 200 μm) - B: S Zirconia (porosity 0.55 -particle diam. = 70 μm).

graphic experiments. Cation exchange zirconia materials demonstrated that protein mixtures can easily be resolved. Proteins were separated according to their net charge as per classical ion exchange separation as shown on Figure 2. SP Streamline ion exchangers were also evaluated under similar separation conditions. Separation efficiency of the Streamline material was significantly lower due to a larger particle size. The impact of particle diameter on separation efficiency is explained by the large contribution of intraparticle diffusion to the overall mass transfer rate for protein chromatography (Li et al., 1995). The use of small beads significantly improves the separation power of a chromatography column by increasing its efficiency and its resolution.

The underlying justification of the large particle diameter for classical fluidized bed media is related to their relatively low density. In order to obtain an operational fluidizing velocity compatible with an industrial application, the low density of classical media is balanced by a large particle diameter, allowing to increase the terminal particle velocity, as illustrated by the Stokes' law:

$$U_t = \frac{(\rho_p - \rho_l) g d_p^2}{18\mu} \tag{2}$$

where: U_t: terminal particle velocity, ρ_p: particle density, ρ_l: liquid density, μ: liquid viscosity, g: gravity, d_p: particle diameter.

A shortcoming of this approach is the concomitant increase of intraparticle diffusional resistance leading to lower efficiency.

Table 1 also shows the significant difference in the particle density between zirconia derived sorbents and

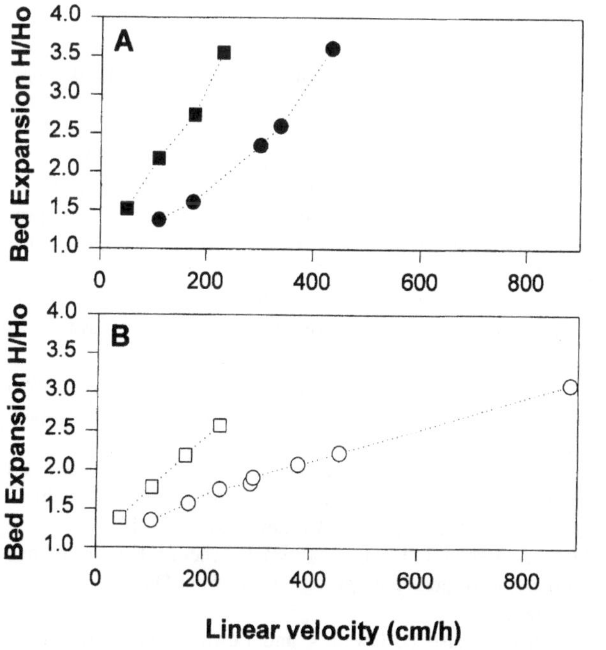

Figure 3. Bed expansion of cation exchange media as a function of fluidizing velocity performed in a 2.5 cm I.D. Streamline column, using a 50 mM acetate buffer pH 4.5 (circle) and 25% v/v glycerol in acetate buffer (square). A: SP Streamline (particle diam. 200 μm) - B: S Zirconia (porosity 0.55 -particle diam. = 70 μm).

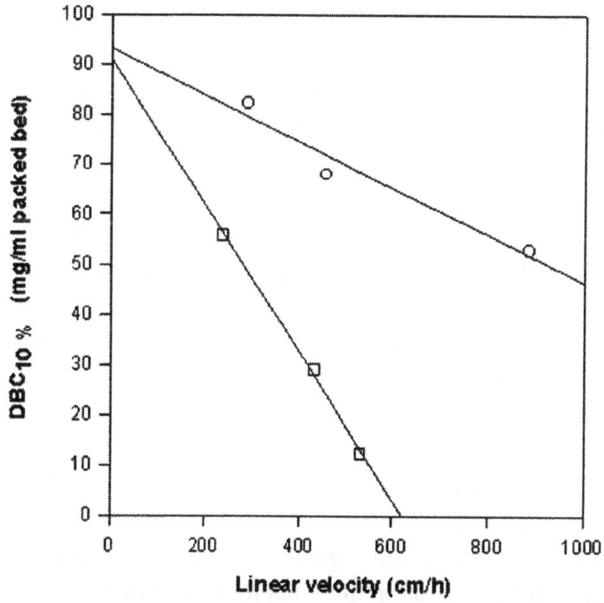

Figure 4. Dependence of dynamic binding capacity on fluidizing velocity. Lysozyme binding capacity measured at 10% breakthrough by frontal analysis on a 1.1 cm I.D. column - Settled bed height 6 cm. Square: SP Streamline (particle diam. 200 μm) - Circle: S Zirconia (porosity 0.55 -particle diam. = 70 μm).

Streamline media. The larger density of the new sorbent was targeted in order to reduce particle diameter without jeopardizing the terminal particle velocity.

Fluidization property studies were performed in a comparative manner and main results are shown on Figure 3. Bed expansion behaviors of a commercially available resin and new cation exchange zirconia material are reported at increasing linear velocities and for two fluidizing solution viscosities. A bed expansion of 2.5 was obtained at 590 cm/h and 220 cm/h with solution viscosity of respectively 1 and 2.3 mPas for S zirconia ion exchanger. In similar conditions, fluidizing velocities of 320 cm/h and 150 cm/h were obtained for SP Streamline. As a result, a substantial reduction in the time needed for the treatment of a given volume of feed is expected, leading to a productivity increase for fluidized bed separation based on zirconia beads. Thanks to the large density of zirconia based media, a low bed expansion is obtained at large linear velocity (600 cm/h), despite a relatively small particle diameter. This situation is extremely favorable for reduction of buffer consumption and dilution of purified fraction when elution is performed in fluidized bed mode.

A comparative dynamic adsorption study was performed in fluidized bed mode. Figure 4 shows the dynamic uptake of lysozyme for fluidized beds of cationic ion exchangers: S zirconia and SP streamline. For both media a decrease of dynamic binding was observed when increasing the fluidizing velocity. However, the extent of the reduction is significantly lower for the 70μm diameter zirconia beads, compared to the 200 μm SP Streamline. This result is explained by the increase in the external surface area and the reduction in diffusion path length in small particles that improve the transport efficiency (Li et al., 1995).

In fluidized bed mode, an efficient capture of the targeted molecule is a prerequisite to reduce loss in the flowthrough and minimize the volume of the column. Efficient capture is illustrated by steep breakthrough curves leading to comparable values of 10% breakthrough capacity and equilibrium static capacity. For hydrogel filled pore material, such as S zirconia, efficient binding is observed due to the small particle diameter but also due to specific property of the ionizable gel that is known to exhibit rapid solid diffusion mass transfer mechanism (Boschetti and Coffman, 1997; Weaver and Carta, 1996).

An intensive investigation is under way to demonstrate the applicability of this new material at preparative scale.

Conclusion

The hypothesis of using solid phase particles of high density and small diameter for high speed fluidized bed applications is documented and feasibility demonstrated by a number of experimental data. Such beads can be transformed in high capacity ion exchange sorbents by filling their macroporosity with functionalized hydrogel responsible for the reversible capture of biologicals. In particular, zirconia porous beads functionalized with an anionic hydrogel can be fluidized at high linear velocity. Moreover, frontal analysis experiments indicate that minor reduction in the binding efficiency is observed at high linear velocity due to a short characteristic diffusion length. In addition, these beads show good selectivity to separate components of protein mixtures in packed bed.

Acknowledgments

This work was partially supported by the French Ministry of Research and Eureka organization.

Streamline is a trade name of Amersham Pharmacia Biotech.

References

Batt B, Yabannavar V and Singh V (1995) Expanded bed adsorption process for protein recovery from whole mammalian cell culture broth. Bioseparation, 5: 41–52.

Boschetti E and Coffman J (1997) Enhanced diffusion chromatography and related sorbents for biopurification. In Subramanian G (ed) Bioseparation and Bioprocessing, (pp. 157–198) Wiley VCH, Weinheim.

Braas G, Searle P, Slater N and Lyddiatt A (1996) Strategies for the isolation and purification of retroviral vectors for gene therapy. Bioseparation, 6: 211–228.

Chang Y, McCreath G and Chase H (1995) Development of an expanded bed technique for an affinity purification of G6PDH from unclarified yeast cell homogenates. Biotechnol. Bioeng., 48: 355–366.

Draeger N and Chase H (1991) Liquid fluidized bed adsorption of protein in the presence of cells. Bioseparation, 2: 67–80.

Griffith C, Morris J, Robichaud M, Annen M, McCormick A and Flickinger M (1997) Fluidization characteristics of and protein adsorption on fluoride-modified porous zirconium oxide particles. J. Chromatogr. A, 776: 179–195.

Hjorth R, Kampe S and Carlsson M (1995) Analysis of some operating parameters of novel adsorbents for recovery of proteins in expanded beds. Bioseparation, 5: 217–223.

Li Q, Grandmaison E, Hsu C, Taylor D and Goosen M (1995) Interparticle and intraparticle mass-transfer in chromatographic separation. Bioseparation, 5: 189–202.

Lochmuller C, Ronsick C and Wigman L (1988) Fluidized-bed separators reviewed: a low pressure drop approach to column chromatography. Preparative Chromatogr., 1: 93–108.

Lyddiatt A and O'Sullivan D (1998) Biochemical recovery and purification of gene therapy vectors. Current Opin. Biotech., 9: 177–185.

McCreath G, Chase H, Owen R and Lowe C (1995) Expanded bed affinity chromatography of dehydrogenases from bakers yeast using dye ligand perfluoropolymer supports. *Biotechnol. Bioeng.*, 48: 341–354.

Nawrocki J, Rigney M, McCormick A and Carr P (1993) Chemistry of zirconia and its use in chromatography. J. Chromatogr. A, 657: 229–282.

Raymond F, Rolland D, Gauthier M and Jolivet M (1998) Purification of a recombinant protein expressed in yeast: optimization of analytical and preparative chromatography. J. Chromatogr. B, 706: 113–121.

Spence C, Schaffer C, Kessler S and Bailon P (1994) Fluidized-bed receptor-affinity chromatography. Biomedical chromat., 8: 236–241.

Thömmes J, Bader A, Halfar M, Karau A and Kula M (1996) Isolation of monoclonal antibodies from cell containing hybridoma broth using a protein A coated adsorbent in expanded beds. J. Chromatogr. A 752: 111–122.

Unger K (1979) Porous silica, Elsevier Scientific Publ. Co. Amsterdam–Oxford–New York.

Weaver L, and Carta G (1996) Protein adsorption on cation-exchangers: comparison of macroporous and gel-composite media. Biotechnol. Progress, 12(3): 342–355.

Wells C, Lyddiatt A and Patel K (1987) Liquid fluidized bed adsorption in biochemical recovery from biological suspensions. In Verall M (ed), Separation for Biotechnology, Ellis Harwood, Chicester, 217-224.

Bioseparation **8**: 121–129, 1999.
© 1999 *Kluwer Academic Publishers.*

Characterization of very dense mineral oxide–gel composites for fluidized-bed adsorption of biomolecules

Nicolas Voute, Damien Bataille, Pierre Girot & Egisto Boschetti
BioSepra S.A., 35 Avenue Jean Jaurès, 92390 Villeneuve la Garenne, France

Received 18 August 1998; accepted in revised form 13 December 1998

Key words: expanded bed adsorption, fluidized bed, inorganic oxide packing, ion exchange media, protein recovery

Abstract

Efficient design of fluidized-bed biomolecule adsorption from crude feed stock requires particles with elevated density, large adsorption capacity and broad chemical stability. Moreover, combinations of small particle diameters with high densities allow for high fluidization velocities while preserving a rapid mass transfer.

This approach has been implemented by combining stable porous mineral oxide of high density (2.2, 4.7, 5.7, 9.4 g/ml) with functionalized hydrogels. The cross-linked hydrogel derivative fills the internal porosity of the beads and provides a high equilibrium binding capacity.

Various porous mineral oxides (silica, titania, zirconia and hafnia) have been characterized in term of fluidization behavior, surface reactivity and chemical resistance to harsh CIP procedures. Porous zirconia particles were also modified into ion-exchangers by suitable surface modification and intraparticle polymerization of functionalized stable derivatives of acrylic monomers. Back-mixings in fluidized bed columns were analyzed by residence time distribution analysis of inert tracers. 328 and 218 mixing plates per meter were found for respectively, bed expansions of 1.7 and 2.9. The dynamic protein adsorption behaviors of zirconia-based polymeric anion-exchange sorbents were obtained in fluidized-bed, using BSA as model protein. A dynamic binding capacity of 62 mg/ml was observed at a fluidizing velocity of 320 cm/h. These investigations substantiate the favorable physical and chemical characteristics anticipated for dense composite beads for use as fluidized bed adsorbents.

Introduction

Recently fluidized-bed technology has been examined as an alternative for the direct capture and purification of biomolecules from crude feed stocks, such as fermentation broth (Raymond et al. 1998), cell homogenates (Chang et al., 1995), cell culture suspension (Spence et al., 1994; Thommes et al., 1995) and biological fluids such as milk of transgenic animals (Degener et al., 1998) and human plasma fraction (Finette et al., 1998). Fluidized-bed systems are tolerant to suspended materials, and consequently avoid the need for removal of cells, cell debris and aggregates prior to the adsorption step. The elimination of sophisticated clarification techniques reduces the processing time of labile molecules, increases the overall yield and reduces the capital investment and direct costs associated with these systems (Batt et al., 1995).

Commercially available adsorbents designed for use in fluidized beds are based on agarose/crystalline quartz composites. Although, successful applications of these adsorbents in the purification of proteins have been recently described, these types of materials are limited by a low specific weight (1.2 g/ml). In consequence large particle diameters (100–300 μm) are required in order to allow a fluidization velocity high enough to be compatible with industrial applications. A major consequence of this approach is the large characteristic diffusion length within the bead that inherently limits the productivity of these media. In this work, the potential of several mineral oxides, as adsorbent beads, is examined with regard to the process constraints of fluidized-bed adsorption. Mineral oxides offer the possibility of higher density than classical polymeric media, resulting in a more rapid fluidizing velocity with a smaller adsorbent particle diameter. The reduction in the particle size leads to a more rapid biomolecule adsorption due to a shorter characteristic diffusion length within the bead. Additionally, mineral oxides such as titania, zirconia and hafnia are remarkably stable to strong cleaning agent (i.e., sodium hydroxide) and permit repeated CIP cycles. Recently, fluoride surface-modified zirconia particles have been evaluated as protein fluidized-bed adsorbent (Griffith et al., 1997). This type of chemical modification produces mixed-mode interactions and is not permanent. In this study, the advantages of dense mineral beads coupled with a permanently fixed-in place, pore anchored hydrogel that imparts a defined chemical functionality (e.g, ion-exchange) are investigated. Moreover, conditions of this functionalization can be optimized as to shield the surface reactivity of the mineral oxide.

Materials and methods

Chemicals and support materials

Porous ceramic particles based on dioxide of silicon, titanium, zirconium, and hafnium were used in these investigations. These materials were prepared as part of an European Eureka program (EU1647) at BioSepra.

BSA and cytochrome *c*, dextran 70, PEI (polyethyleneimine) were obtained from Sigma. CM dextran and PEG 20000 (polyethylene ethylene glycol) were obtained from, respectively, Pfeifer & Langen and Hoechst. All other chemicals used were of analytical grade.

Preparation of mineral oxide–hydrogel composites

Regular HyperD™ sorbents (BioSepra) are composites of stable derivatives of polyacrylamide hydrogel and porous ceramic moiety (Boschetti, 1994). The soft hydrogel completely fills the pores of the ceramic, and allows sorption to occur throughout the entire pore volume. The media combines the high capacity of a classical polyacrylamide hydrogel with the rigidity and density of the ceramic backbone. Various types of porous mineral oxides are suitable for transformation into gel filled type particles, similar to regular HyperD, including dioxides of silicon (silica), titanium (titania), zirconium (zirconia), and hafnium (hafnia). The fluidized bed media are obtained as followed: gel monomers, dissolved in solvent, are introduced into the beads in liquid form, until the porosity of the beads is filled with the monomer solution, then the polymerization is induced. Cationic hydrogel (Q derivative) was obtained by free radical polymerization of 20% MAPTAC (methacryloylaminopropyltrimethylammonium chloride) monomer solution. The crosslinking agent, N-N methylene bis acrylamide, and the initiator AIBN (azobisisobutyronitrile) were used at a concentration of, respectively, 1% wt and 0.3% wt. Vials containing the appropriate amounts of porous mineral oxide, monomer, crosslinking agent and initiator were purged with nitrogen and then placed in a temperature bath for 24 hours at 85 °C. Upon completion of the polymerization the beads were extensively washed with distilled water, 0.1 N NaOH and 0.1 N HCl to remove unreacted monomers, crosslinking agent and initiator.

The result is a rigid and dense bead whose pores are filled with a crosslinked hydrogel. The gel fills the pores of the mineral oxide and uses the whole particle volume for higher capacity. However, prior to the formation of the hydrogel within the pore volume of the bead it was found beneficial to coat the bare surface of the mineral oxides with appropriate organic polymers. This coating provides a shielding of the mineral surface in order to suppress non-specific interactions with biomolecules (Dunlap and Carr, 1996).

Cationic, anionic and neutral polymer coated porous mineral oxide particles were prepared by adsorption of receptively PEI, CM dextran and PEG 20000/ dextran 70 mixture onto the surface. The adsorbed polymers were subsequently crosslinked with 1,4-butanediol diglycidyl ether.

Physical properties of the beads are summarized in Table 1. Mean particle diameter was measured by laser diffraction spectrometry (Mastersizer S, Malvern). Particle density (effective particle density during fluidization) was determined using Archimed principle. Pore volume was measured by extensive impregnation of paraffin oil. Pore diameter was determined by mercury porosimetry (Porosimeter 2000, Carlo Erba). Particle porosity was calculated using pore volume and material bulk density, determined by pycnometry. Particle surface area was measured by nitrogen BET adsorption (Sorpty 1750, Carlo Erba).

Measurement of bare mineral oxide surface reactivity

The surface reactivity of bare mineral oxides has been investigated using fixed bed column packed with underivatized porous particles. A 8 ml column, previously equilibrated in either pH 4.5, 7.5 or 8.6 buffer

Table 1. Characteristics of porous mineral oxide beads

	Silicon oxide	Titanium oxide	Zirconium oxide	Hafnium oxide
Solid density (g/ml)	2.2	4.2	5.7	9.4
Particle diameter (μm)	150, 205, 304	52	49, 63, 77	30
Particle density (g/ml)	1.3	2.8	3.2	4.2
Pore volume (ml/g)	0.95	0.17	0.16	0.2
Particle porosity (%)	68	42	48	47
Pore diametre (Å)	3000	1000	1500	n.d.
Surface aera (m^2/g)	10	n.d.	1	n.d.

n.d.: not determined.

solution (respectively 50 mM acetate buffer, 50 mM Tris-HCl buffer and 50 mM Tris-HCl) was loaded with 1 ml of 20 mg/ml acid or basic protein tracer (BSA, pI=4.9 or cytochrome C pI=9.4). After washing with the equilibration buffer, the column was eluted with a 1.5 M NaCl solution, followed by a 1 N NaOH cleaning step. The column effluent volume was collected in three fractions. The amount of protein tracer recovered in the flowthrough (non-retained fraction), the amount of protein tracer retained by ionic interactions (NaCl peak), and the amount of protein tracer retained by non-specific interactions (NaOH peak) were determined by spectrophotometry.

Bed expansion

The bed expansion characteristics of mineral oxide beads have been measured in a 25 mm i.d. column (Streamline 25 column, Amersham Pharmacia Biotech) with a 3 g/l NaCl solution at different linear velocities and compared to a commercially available material (Streamline DEAE, Amersham Pharmacia Biotech). The sedimented bed height at the start of the experiments was 12–15 cm. The fluidized bed column was connected to a peristaltic pump (Watson and Marlow). The liquid flow rate was altered in random order. 10 min was allowed after each change in the flow rate for the bed to stabilize. Expanded bed height was measured as a function of the liquid superficial velocity.

Flow hydrodynamic characteristics

Mixing of the liquid phase in expanded bed was determined by measuring the exit stream age distribution following an upstream step input (Levenspiel, 1972). The experiments were performed in a low dead volume 11 mm i.d. column (BioSepra) equipped with a 10 μm stainless steel screen at the bottom distributor. After stable expansion was reached, the upper column plunger was positioned at approximately 0.25 cm above the bed surface. Settled bed height was 11 cm. The fluidizing liquid and the tracer were, respectively, a 3 g/l NaCl solution and a 1% v/v acetone solution. The acetone concentration was followed at the column outlet by UV absorbance at 280 nm. The dimensionless output tracer concentrations were plotted against the dimensionless volume. These normalized results were fitted to the tank-in-series model using the number of mixing plate as model parameter (Walas, 1997):

$$F = 1 - \exp(n\theta) \sum_{j=0}^{n-1} \frac{(n\theta)^n}{j!} \tag{1}$$

where F is the dimensionless outlet concentration, θ the dimensionless column effluent volume, n the number of mixing plate.

Dynamic adsorption characteristics

The adsorption performances of expanded bed of Q zirconia were compared at different degrees of bed expansion. The uptake of 1 mg/ml and 5 mg/ml BSA solution in 50 mM Tris-HCl buffer pH 8.6 was measured by frontal analysis on a low dead volume 11 mm i.d. column (BioSepra), equipped with a 10 μm stainless steel screen at the bottom distributor. Prior to the protein loading the column was equilibrated in the 50 mM Tris-HCl buffer pH 8.6. After stable expansion was reached, the upper column plunger was positioned at approximately 0.25 cm above the bed surface. Settled bed height was 11 cm. The protein concentration at the column exit was recorded by UV absorbance at 280 nm.

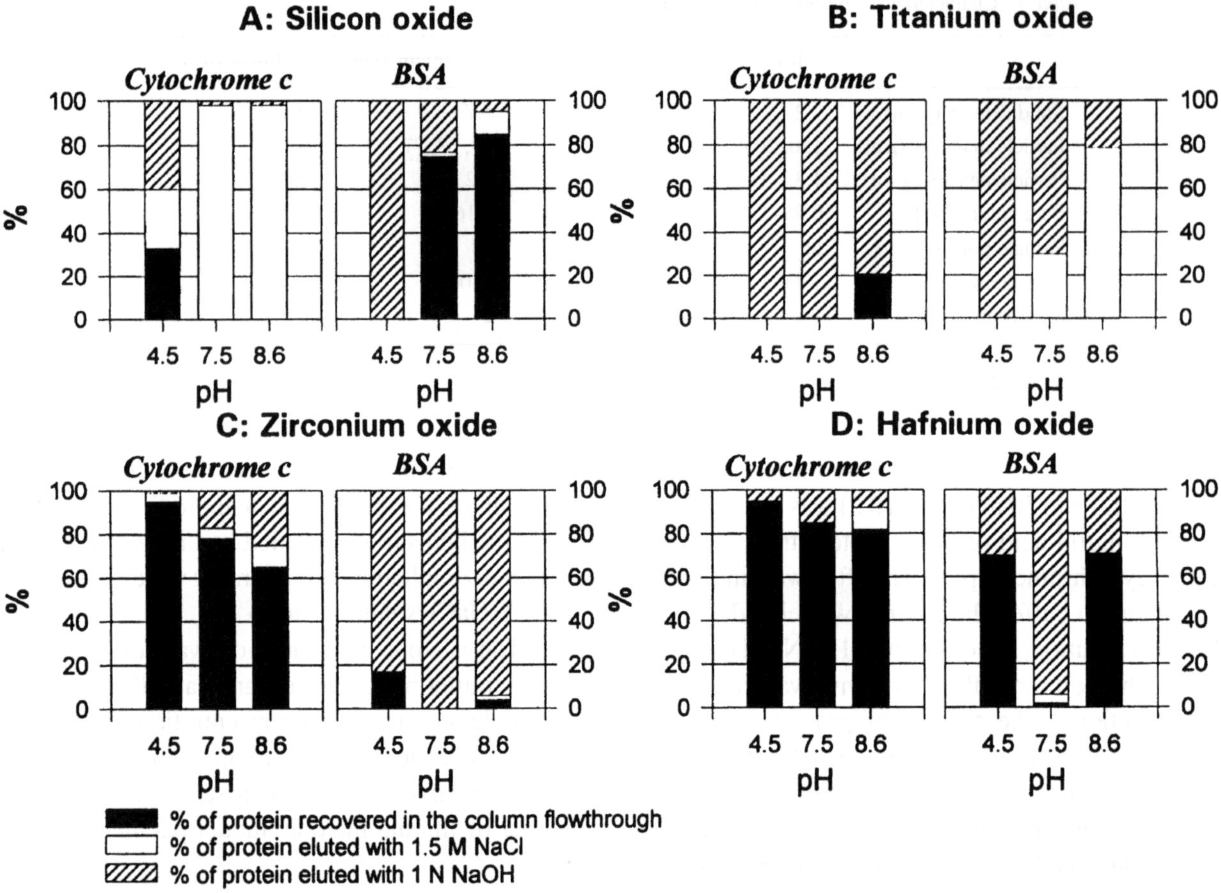

Figure 1. Surface interactions on porous mineral oxide beads (A: silicon oxide, B: titanium oxide, C: zirconium oxide and D: hafnium oxide). Interactions determined by protein adsorption-elution experiments on a 8 ml packed bed column at 3 different pH (load 1 ml of 20 mg/ml BSA or cytochrome *c*, adsorption buffers: 50 mM acetate pH 4.5, 50 mM TrisHCl pH 7.5, 50 mM TrisHCl pH 8.6 - elution buffer: 1.5 M NaCl - cleaning solution: 1 N NaOH).

Results and discussion

Surface reactivity

The adsorption and elution of BSA and cytochrome *c* on surfaces of silicon oxide, titanium oxide, zirconium oxide and hafnium oxide were studied as a function of pH. Figure 1 shows that, as previously reported, silicon dioxide is negatively charged at pH 7.5 and 8.6: BSA, negatively charged at these pH, is not retained, whereas cytochrome *c*, positively charged, binds by coulombic interaction. At pH 4.5, there is a change of the surface charge as indicated by the unretained cytochrome c fraction. However, other interactions exist on the surface of the silicon dioxide as illustrated by the non specific retention of BSA at pH 4.5.

Zirconium and hafnium oxides show different behavior: the non-retention of cytochrome *c* at any pH indicates a cationic surface. However, BSA is non specifically adsorbed pointing out the presence of nonionic interaction sites. This effect is more pronounced on the zirconia surface. Moreover, the incubation of zirconia with a strong Lewis base, 500 mM phosphate buffer, 17 hours, prior to the binding experiment, dramatically changes the behavior of the zirconia surface that acquires anionic properties (data not shown). This result has already been reported (Dunlap and Carr, 1996): it is thought that acid Lewis sites on zirconia surface bind Lewis bases of the solution (e.g., phosphate), which in turn create a negatively charged surface. Titania exhibits a similar behavior with a more pronounced Lewis acidity.

As shown on Figure 2, counteracting these interactions on the zirconium oxide surface is possible by coating the bare surface with organic polymers such as polyethyleneimine, polyethyleneglycol, dextran and CM-dextran. The polyethyeneimine coated zirconia shows cationic properties with a charge density of 45 μeq/ml of packed bed resin. The percentage of protein that is not recovered in the flow-through of the column or not eluted by a salt step is lower than 1%, com-

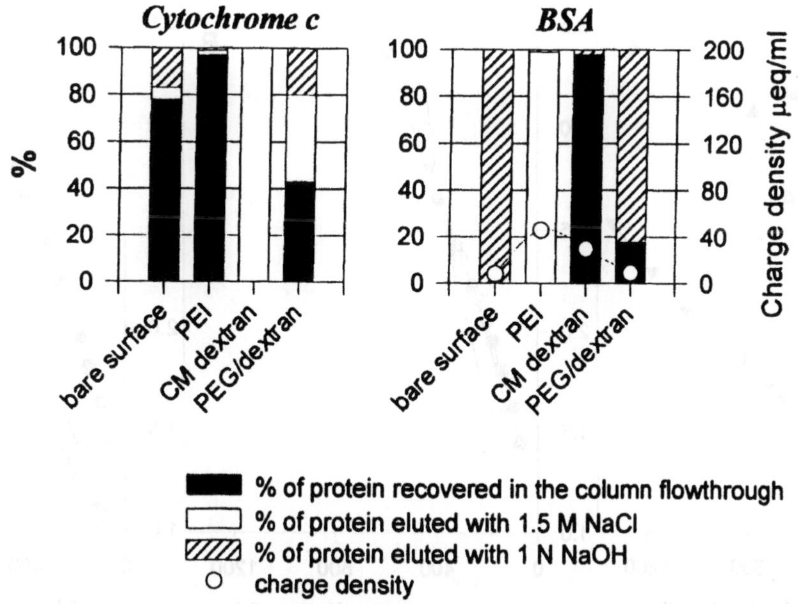

Figure 2. Surface shielding of porous zirconium oxide beads. Surface interactions on bare surface and polymer coated surface (PEI:polyethyleneimine coated surface- CM dextran: CM dextran coated surface and PEG/dextran: polyethylene glycol / dextran copolymer coated surface). Interactions determined by protein adsorption-elution experiments on a 8 ml packed bed column at pH 7.5. Adsorption buffer: 50 mM TrisHCl pH 7.5 - elution buffer: 1.5 m NaCl - cleaning solution: 1 N NaOH Load 1 ml of 20 mg/ml cytochrome *c*, Load 1 ml of 20 mg/ml BSA.

pared to 100% for BSA with the bare surface. Similar shielding efficiency was found for the anionic polymer coating (CM dextran). However, the neutral coating (polyethylene glycol/ dextran copolymer) shows low shielding efficiency, as 82% of loaded amount of BSA required harsh elution condition (1 N NaOH).

Porous zirconia is an attractive option for the production of dense adsorbent for fluidized bed capture. It shows less Lewis acid/base interaction compared to titania, its large scale availability and cost are less problematic than hafnia, and it is more stable in alkaline condition than silicon dioxide. It was also shown that charged polymer coating provides an efficient approach to shield the non-ionic interaction of zirconia surface. Furthermore the low specific surface area of these particles (approximately 1m^2/g) facilitates the shielding of the interacting sites.

Bed expansion

Stable fluidizing conditions have been observed for all the media tested, independently of the particle diameter and density. The combined effect of the particle diameter and particle density on the dimensionless bed expansion is illustrated on Figure 3. The silicon dioxide based material shows that the increase in the superficial velocity at a given bed expansion varies with the square of the particle diameter, as predicted by theoretical model (Richardson and Zaki, 1954; Foscolo et al. 1983). Figure 3 also shows that the relatively low particle diameter of the zirconia material is well compensated by its higher density. Its velocity at any bed expansion is much higher than those of larger but less dense material. It is not true for titania and hafnia material, their densities do not compensate for their smaller diameter, resulting in a similar bed expansion behavior as the polymeric media. However, the smaller particle diameter of the mineral oxide beads will lead to an improved mass transfer compared to the light and large polymeric beads.

The data can also be plotted as a contour plot (velocity vs. particle diameter and density) for a given level of bed expansion (*H/Ho* = 2), and compared (Figure 4) with the prediction of the model of Foscolo (Foscolo et al., 1983) which relates the liquid superficial velocity, u and the interstitial porosity ϵ, as

$$u = u_t \frac{\epsilon^4}{3.33(1-\epsilon)+\epsilon^3} \quad (2)$$

for laminar flow conditions ($Re_p < 0.2$), where u_t is the terminal velocity of the particle.

The graph indicates that a substantial improvement in the linear velocity at 100% bed expansion can be achieved by reducing the particle diameter and increasing the density, whereas the increase in the

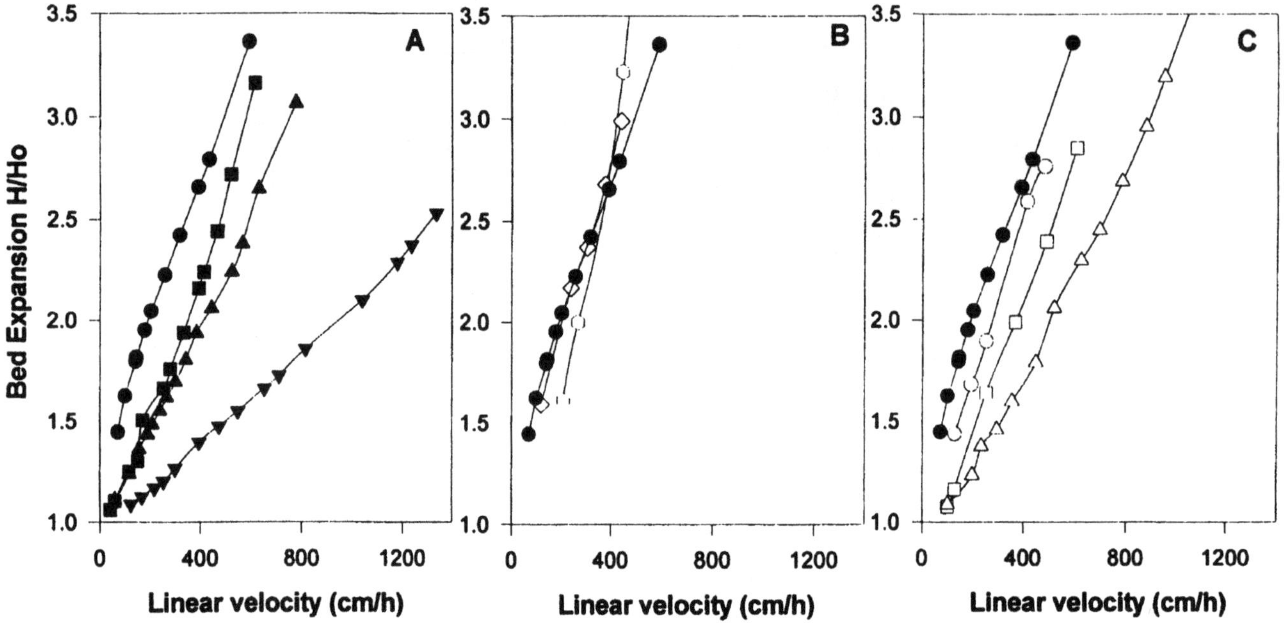

Figure 3. Bed expansion as a function of linear velocity for porous mineral oxide particles and agarose/crystalline quartz composite particules. Settled bed height: 12–15 cm, column i.d. 25 mm. Fluidizing solution: 3 g/l NaCl. A: (●) Deae Streamline, (■) silicon oxide 150 μm, (▲) silicon oxide 200 μm, (▼) silicon oxide 300 μm. B: (●) Deae Streamline, (◇) titanium oxide 52 μm, (○) hafnium oxide 30 μm. C: (●) Deae Streamline, (◯) zirconium oxide 49 μm, (□) zirconium oxide 63 μm, (△) zirconium oxide 77 μm.

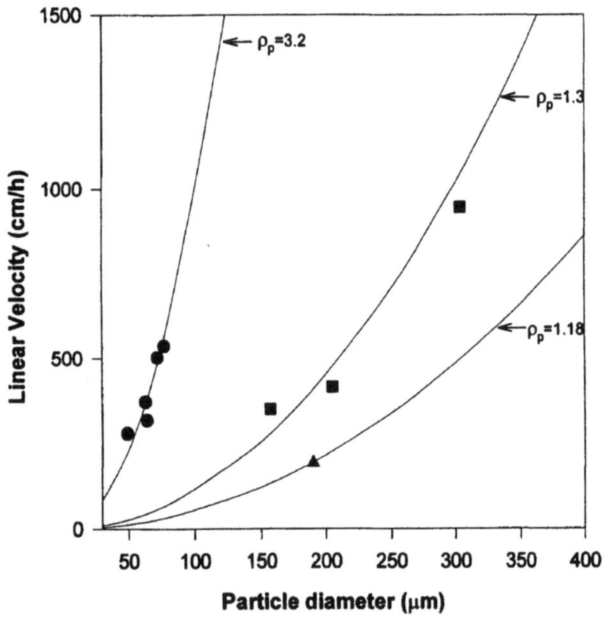

Figure 4. Fluidizing velocity as a function of particle diameter and particle density for a bed expansion of 2. Comparison of experimental data with the prediction of Foscolo's model, (●) porous spherical zirconium oxide particles, (■) porous spherical silicon oxide particles, (▲) porous agarose/quartz composite particles. (—) model prediction (equation 2).

external surface area and the reduction in the characteristic diffusion length will significantly improve the mass transfer.

Flow hydrodynamic characteristics

Results shown in Figure 5 indicate that axial dispersion in the liquid phase of expanded bed of Q zirconia is moderate and should not compromise the capture efficiency of the media. Large number of mixing plates per meter of settled bed height were found in these experiments, 328 and 218 plates/m for, respectively, bed expansions of 1.7 and 2.9. These values are similar to the ones previously reported for less dense polymeric based particles (Batt et al., 1995), and indicate that the particle density does not significantly affect axial mixing in the fluidized bed column. The liquid can be assumed to flow through the bed in a plug flow and mixing in the liquid phase should not significantly contribute to the spreading of the breakthrough profile.

Dynamic adsorption characteristics

The performances of fluidized-bed adsorptions of Q zirconia were examined by measuring breakthrough profiles in frontal chromatography. The normalized results are reported on figure 6 for several fluidizing velocities. Dynamic capacities at 10% breakthrough point of 60 mg/ml and 62 mg/ml are observed for Q zirconia for, respectively, 5 mg/ml and 1 mg/ml BSA solutions in fluidized bed mode at a flow velocity of 320 cm/h.

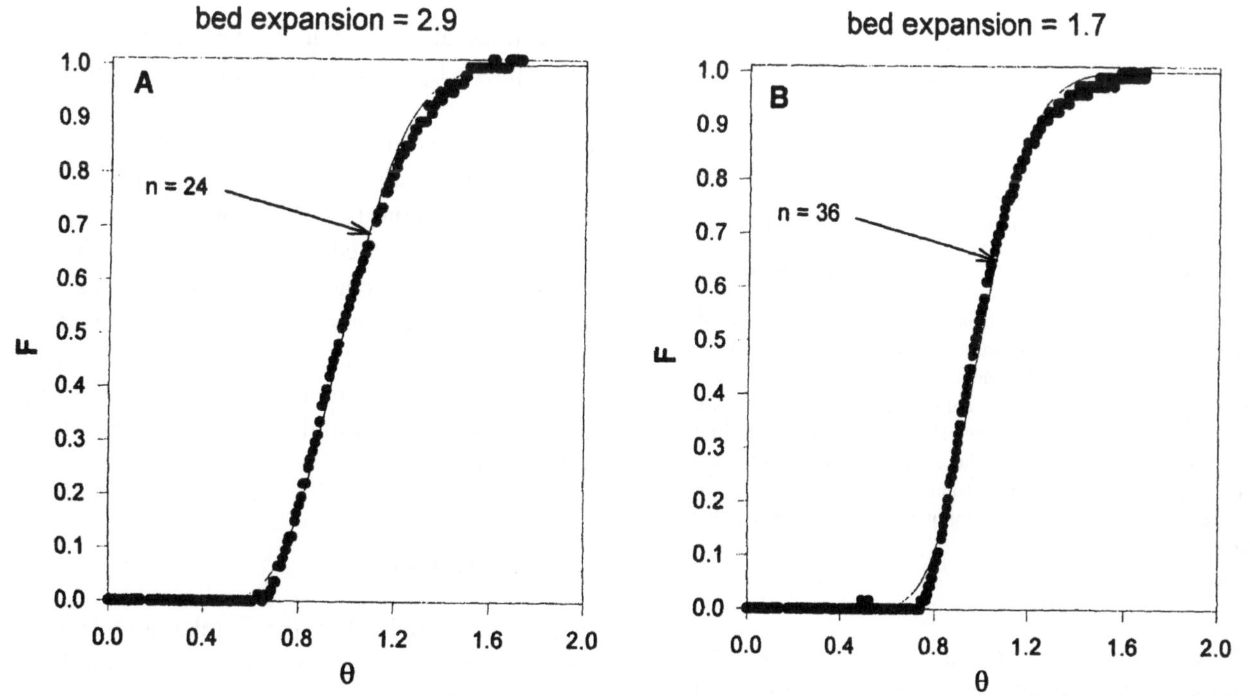

Figure 5. Acetone residence time distribution on a Q zirconia fluidized bed column (particle diameter = 63 μm). Comparison with the tank-in-series model (equation 1). Sedimented bed height 11 cm, column i.d 11 mm. Tracer = 1% v/v acetone in 3 g/l NaCl. A: 2.9 bed expansion, number of plates=24, B: 1.7 bed expansion, number of plates=36.

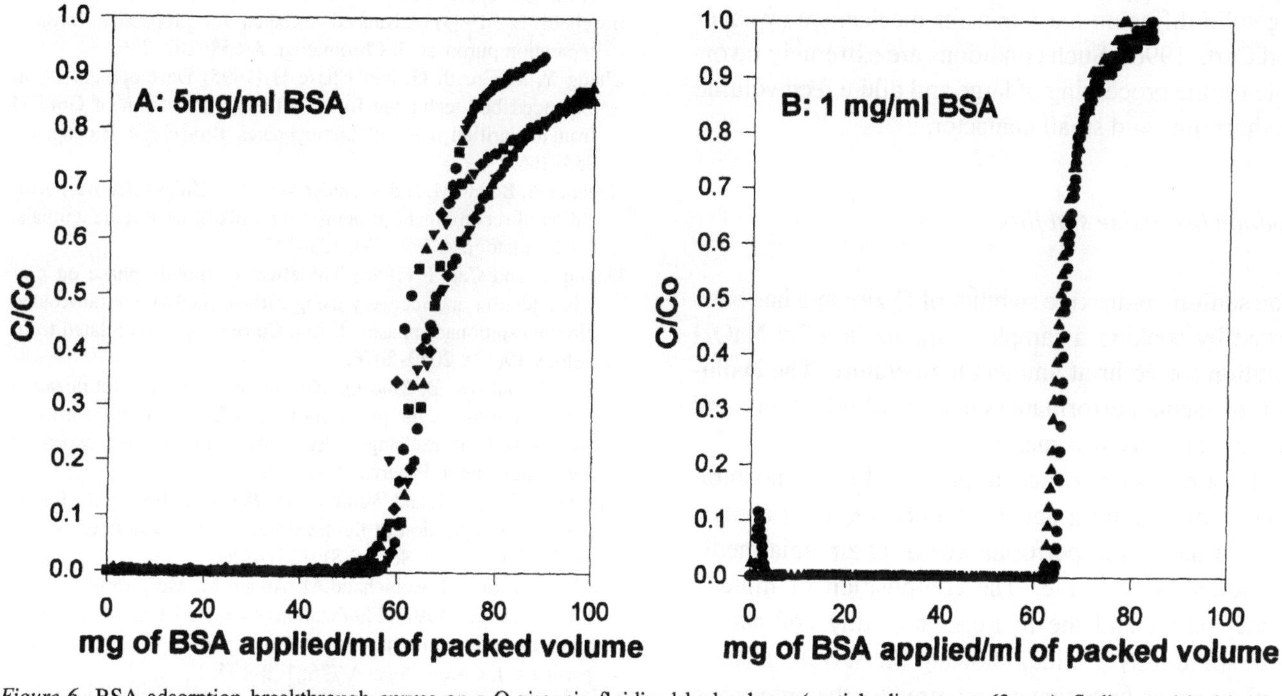

Figure 6. BSA adsorption breakthrough curves on a Q zirconia fluidized bed column (particle diameter = 63 μm). Sedimented bed height 11 cm, column i.d. 11 mm. Equilibration buffer; 50 mM TrisHCl buffer pH 8.6. A: BSA concentration = 5 mg/ml, (●) fluidizing superficial velocity=180 cm/h, (■) fluidizing superficial velocity=320 cm/h, (▲) fluidizing superficial velocity=520 cm/h, B: BSA concentration = 1 mg/ml, (●) fluidizing superficial velocity=180 cm/h , (▲) fluidizing superficial velocity=520 cm/h.

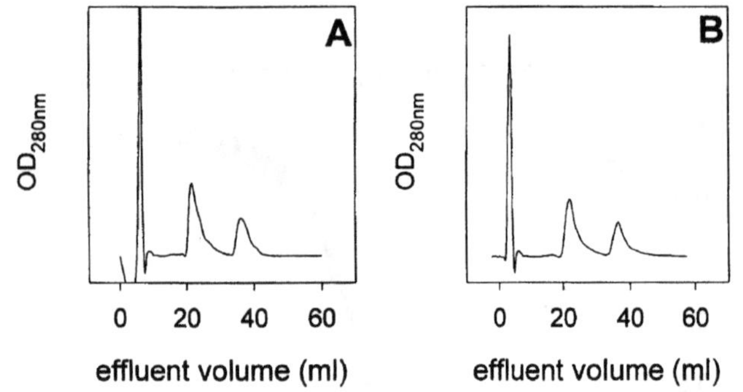

Figure 7. Stability test of Q zirconia to NaOH treatment. Separation of standard proteins on a Q ziconia packed bed column (particle diameter = 63 μm - 7.4 ml column volume - 1 ml load: 5 mg/ml cytochrome *c*, 2 mg/ml BSA, 2 mg/ml transferrin - linear gradient elution 50 mM TrisHCl buffer pH 8.6 / 500mM NaCl, 50mM TrisHCl buffer pH 8.6 in 12.5 column volume. A: Chromatogram obtained on a new media, B: Chromatogram obtained after a 96 hours storage of the media in 200 ml of 5 N NaOH prior to column packing.

Figure 6 indicates that the impact of the linear velocity on the breakthrough behavior is low, especially at the front end of the curves. Tailing is however observed on the rear end of the curves.

Additionally, the feed concentration affects the breakthrough behavior with steeper curves observed for more dilute loading solutions. This behavior has already been reported for chromatographic media showing solid diffusion mass transfer mechanism (Weaver and Carta 1996). Such conditions are extremely favorable for the processing of large and dilute feed volume in short time and small contactor.

Sodium hydroxide stability

The sodium hydroxide stability of Q zirconia has been tested by soaking a sample of media in a 5N NaOH solution for 96 hr at ambient temperature. The evolution of media performances in term of selectivity and protein capacity was measured.

Figure 7 shows that despite the high concentration of the cleaning agent (5N) and the long contact time (4 days), the performances of Q zirconia media are remarkably stable. The composition of mineral oxide matrix and the hydrogel are fully compatible with sodium hydroxide, allowing the severe cleaning in place procedures that are required for the processing of untreated biological materials.

Conclusions

Dense mineral oxide porous beads combined with high capacity hydrogel are suitable adsorbents for use in fluidized bed adsorption. This investigation has shown that performances of fluidized-bed biomolecule adsorption can be improved by using small and dense zirconia porous beads whose pores are filled with cationic anchored hydrogel. High fluidization velocities can be achieved with this media, while preserving a high mass transfer, particularly for dilute solution of biomolecules. Moreover, the excellent chemical stability of zirconia is compatible with harsh cleaning methods required for the processing of crude biological feed stocks.

Acknowledgment

Support was provided by the Eureka organization of the E.E.C and the French ministry of research (Eureka project EU 1647).

References

Batt B, Yabannavar V, and Singh V (1995) Expanded bed adsorption process for protein recovery from whole mammalian cell culture broth. Bioseparation 5: 41–52.

Boschetti E. (1994) Advanced sorbents for preparative protein separation purposes. J. Chromatogr. A 658: 207–236.

Chang Y, McCreath G, and Chase H (1995) Development of an expanded bed technique for an affinity purification of G6PDH from unclarified yeast cell homogenates. Biotechnol. Bioeng. 48: 355–366.

Degener A, Belew M, and Velander W (1998) Zn2+ selective purification of recombinant proteins from milk of transgenic animals. J. Chromatogr. A 799(1-2): 125–137.

Dunlap C and Carr P (1996) The effect of mobile phase on protein retention and recovery using carboxymethyl dextran-coated zirconia stationary phases. J. Liq. Chromatogr. and related technology 19(13): 2059–2076.

Finette G, Baharin B, Mao Q, and Hearn M (1998) Optimization considerations for the purification of alpha(1)- antitrypsin using silica-based ion-exchange adsorbents in packed and expanded beds. Biotechnol. Progress 14(2): 286–236.

Foscolo P, Gibilaro L and Waldram S (1983) A unified model for the particulate expansion of fluidized beds and flow in fixed porous media. Chem. Eng. Sci. 38(8): 1251–1260.

Griffith C, Morris J, Robichaud M, Annen M, McCormick A, and Flickinger M (1997) Fluidization characteristics of and protein adsorption on fluoride-modified porous zirconium oxide particles. J. Chromatogr. A 776: 179–195.

Levenspiel O (1972) Chemical Reaction Engineering, John Wiley & Sons, New York, Chichester, Brisbane, Toronto, Singapore.

Raymond F, Rolland D, Gauthier M and Jolivet M (1998) Purification of a recombinant protein expressed in yeast: optimization of analytical and preparative chromatography. J. Chromatogr. B 706: 113–121.

Richardson J and Zaki W (1954) Sedimentation and fluidization: part I. Trans. Instn. Chem. Engrs. 32: 35–53.

Spence C, Schaffer C, Kessler S and Bailon P (1994) Fluidized-bed receptor-affinity chromatography. Biomedical Chromatography 8: 236–241.

Thommes J, Weiher M, Karau A and Kula M (1995) Hydrodynamics and performances in fluidized bed adsorption. Biotechnol. Bioeng. 48: 367–374.

Walas S (1997) Chemical reactors. Perry's Chemical Engineers' Handbook, R. Perry (ed.), Mc Graw-Hill, New York.

Weaver L and Carta G (1996) Protein adsorption on cation-exchangers: comparison of macroporous and gel-composite media. Biotechnol. Progress 12(3): 342–355.

Author for correspondence: N. Voute, Biosepra S.A., 35 Avenue Jean Jaurès, 92390 Villeneuve la Garenne, France. Tel: 146-859200; Fax: 147-922655; E-mail: nvoute@biosepra.com

Bioseparation **8:** 131–138, 1999.
© 1999 *Kluwer Academic Publishers.*

Expanded bed adsorption on supermacroporous cross-linked cellulose matrix

Anita Pai, Shyamal Gondkar, Sumati Sundaram & Arvind Lali
Chemical Engineering Division, Department of Chemical Technology, University of Mumbai, Matunga, Mumbai 400 019, India
Fax: 91 22 4145614; e-mail: arvind@udct.ernet.in

Key words: column efficiency, expanded bed, lactate dehydrogenase, macroporous adsorbent

Abstract

Rigid spherical macroporous adsorbent beads (CELBEADS) prepared by cross-linking of cellulose were characterised and found eminently suitable for use as expanded bed affinity chromatography matrix. Chromatographic runs were performed on a 10 mm diameter column with three solutes tyrosine, papain and bovine serum albumin under non-retaining conditions on CELBEADS and Streamline^{TM} DEAE, a commercial agarose based expanded bed matrix. Performance of the runs was measured in terms of height equivalent to theoretical plate, HETP. Variation in HETP with velocity on Streamline^{TM}DEAE gave flat profiles in packed bed and increasing trend in expanded bed. On CELBEADS, the HETP curves in both packed and expanded bed modes followed profiles typical of macroporous adsorbents i.e. increasing and levelling with velocity. HETP values obtained for papain and bovine serum albumin on CELBEADS were lower than those obtained on Streamline^{TM} DEAE at all velocities. Lactate dehydrogenase was purified from porcine muscle homogenate using Cibacron blue conjugated to CELBEADS using a protocol reported for supports with surface hydroxyl groups. Elution of the enzyme was investigated both in packed mode as well as in expanded mode at a flow rate of 1 ml min^{-1}. The purification procedure took about 60 minutes and a purification fold of about 14 was achieved in both cases. The adsorbent could be cleaned in place with 5 M urea and used repeatedly without loss of performance.

Introduction

There has been considerable interest in recent years in application of expanded bed systems to biopurifications with the emergence of improved 'designer' sorbents. Packed beds pose some operational difficulties when large volumes of unclarified and/or viscous extract solutions are involved. Even with clarified feeds, the operating flow rates with the available small, soft and gel based adsorbent particles must be highly constrained to avoid high operating pressures. Expanded bed adsorption conceptually offers a number of advantages over packed beds for large scale operations, particularly to isolate and purify bioproducts from large volumes of particulate matter containing crude extracts (Frej et al., 1994). Counter balancing these advantages however, are potential problems whose magnitude remains unknown (Buijs and Wesselingh, 1980). These include possible reduction in bed capacity, unfavourable mixing at high flow rates and particle aggregation in viscous broths; all tending to reduce column efficiencies.

The celebrated van Deemter equation has been accepted to account for all possible effects that contribute to zone broadening in packed beds and thus affect column efficiency which is quantified in terms of height equivalent to theoretical plate, HETP (van Deemter et al., 1956). The equation is

$$\mathrm{HETP}, H = A + B/u_o + C.u_o$$

The three terms of the van Deemter equation account for eddy diffusion (A), molecular diffusion (B/u_o) and intraparticle effects (Cu_o) respectively, where u_o is the superficial bed velocity and A, B and C are system and operation dependent constants. For most practical cases in protein separations the molecular diffusion has negligible contribution to HETP and the applicable equation takes the form

$$H = A + C.u_o$$

In preparative packed bed chromatography, with adsorbents having particle size in the range of 30 to 300 μm, the contribution of the eddy dispersion term, A has been shown to be almost constant (Yamamoto et al., 1987). Thus to improve column efficiency at a given velocity, one is left with the only option of attempting to reduce the intraparticle interaction term, C. There are several ways of reducing the intraparticle contributions to HETP. One way is use of a non-porous support that can significantly reduce the C term, but has the outweighing disadvantage of drastically reducing adsorbent capacity. Another way is reduction of adsorbent particle size which leads to a decrease of the diffusion time constant. This does lead to a healthy reduction in HETP as in HPLC but has the disadvantage of prohibitively high pressure drops in preparative scale columns. A third way has been to increase particle permeability by carefully providing large pores for transport, connected to smaller diffusive pores as in 'perfusion chromatography' (Afeyan et al., 1990).

Macroporous adsorbent supports have been shown to result in smaller HETP values at high flow rates since these adsorbents exhibit a plateauing of the HETP curve with velocity (Rodrigues et al., 1996). In this paper we report use of a new synthesised supermacroporous support (pore size: $\sim$ 3 μm) as an expanded bed adsorbent matrix. The developed adsorbent beads, based on cross-linked cellulose, were found to give not only a stable expanded bed over a large velocity range, but also gave a constant HETP after a threshold liquid velocity. The beads showed negligible non-specific adsorption and a good density of hydroxyl groups for activation and use as an affinity matrix that could be fully regenerated by cleaning-in-place procedure like washing with 1 M NaCl, 1 M NaOH or 5 M urea. Cibacron blue, a pseudo-affinity dye ligand, was immobilized onto the beads using procedure reported for other hydroxyl based supports and was used to purify lactate dehydrogenase from porcine muscle homogenate. There has been considerable interest in attempting expanded bed elution of the adsorbed protein from the matrices. Elution of lactate dehydrogenase was performed at a given flow rate in both packed and expanded bed modes.

Experimental

Bead characterization

Rigid porous cross-linked cellulose beads (called CELBEADS hereafter) were prepared according a procedure for which patent application has been filed. The procedure involved cross-linking of cellulose in a porogen solvent and hardening of the dispersed beads in an aqueous solution. These beads were characterised for their properties requisite of an adsorbent for protein purifications. Shape and size of CELBEADS was analysed on an IP PLUS Image Analyser. Porosity and surface area analysis was carried out by mercury porositimetry. Bead porosity was also measured by gel filtration experiments on a 10 mm dia. Bio-Rad ECONO column using different molecular weight marker molecules. A number of proteins were tested on CELBEADS for non-specific adsorption at different ionic strengths.

Suitability of CELBEADS for expanded bed operation was found by bed expansion experiments on the set-up described below. Bed expansion studies for CELBEADS as well as for Streamline™DEAE (a composite cross-linked agarose gel based anion exchanger adsorbent especially designed for expanded bed operations and marketed by Amersham Pharmacia Biotech, Uppsala, Sweden) were carried out with deionised water by varying inlet flow rate.

Experimental setup and methods

The experimental setup used for column runs consisted of a 10 mm inner dia., 250 mm long glass column provided with two 140 mm long BioRad ECONO adapters fitted with 90 μm stainless steel mesh distributors. These adapters could be moved in and out from the top and bottom ends according to the adsorbent bed height desired in the column. The bottom adapter was connected to an Alitea XV peristaltic pump which fed the column with appropriate solutions. The outlet or the top adapter was connected to an on-line Hitachi UV-Vis 1100 spectrophotometer working at 280 nm and on to an ISCO Retriever II fraction collector. The absorbance readings at the spectrophotometer were recorded on a continuous chart recorder. All experiments were performed in upflow mode. The adapters were always so adjusted as to provide one tube diameter height of 'free board' over the expected height of the expanded bed surface.

Chromatographic experiments under non-retaining conditions were performed on both Streamline™DEAE and CELBEADS. Separate runs were made with each of the three biomolecules, viz. tyrosine (MW 181 Da), papain (MW 23 000 Da) and bovine serum albumin, BSA (MW 66 000 Da). While tyrosine was loaded in 50 mM Tris/HCl buffer at pH 7.0, papain and BSA were loaded in the same buffer containing 1M NaCl to ensure non-retaining conditions on pre-equilibrated adsorbent. Chromatographic peaks were obtained at flow velocities ranging from 0.6 to 6.0 cm min^{-1} in both packed and expanded modes by injecting 4 ml of sample as a small plug in the flow line. The chromatograms obtained on the chart recorder showed that the elution bands were almost symmetrical but for a very slight tailing in the end. Total plate number N for a run was calculated by measuring the retention time and peak width at half peak height. The operating bed height divided by N gave height equivalent to theoretical plate. HETP was determined as a function of superficial velocity for each of the three molecules.

Chromatographic procedure for purification of LDH from porcine muscle homogenate

Porcine muscle extract from locally procured fat free porcine muscle was prepared in 50 mM Tris/HCl buffer pH 8.0 by a procedure similar to that used by Lali et al. (1997). The extract was stored at −20 °C in suitable aliquots and thawed prior to use and had LDH activity and protein content of 110 U ml^{-1} and 6.4 mg ml^{-1} respectively. Cibacron Blue 3GA (SIGMA) was coupled to the hydroxyl groups on the beads using the procedure reported for agarose based gel supports. (Hermanson et al., 1992). The amount of dye immobilized was calculated by estimating the unbound dye. Cibacron conjugated matrix, CB-CELBEADS with a dye loading of 7.4 μmoles ml^{-1} was thus prepared. The experimental setup used was same as mentioned above. Adsorbent beads were poured as a suspension in the working buffer (50 mM Tris/HCl buffer pH 8.0) into the column from the top. After the desired volume of the adsorbent was added, the top adaptor was inserted in such a way as to provide one tube diameter height of 'free board' over the expected height of the expanded bed. An equilibrating buffer wash of five expanded column volumes at 1 ml min^{-1} was given. Desired amount of crude extract was then passed through the column at 1 ml min^{-1}. To remove residual crude, suspended matter as well as substances physically adhering to the adsorbent, working buffer containing 0.15 M KCl was passed through the adsorbent bed at same flow rate till eluate protein content dropped to base-line on spectrophotometer. Elution with 0.5 M KCl in working buffer was performed in both packed bed mode as well as expanded mode.

LDH was assayed according to reported procedure (Worthington, 1982) by spectrophotometrically monitoring (340 nm) the amount of NADH consumed for LDH catalysed reduction of pyruvate at room temperature (35 °C). Protein content was measured using BioRad protein assay Kit-I with bovine serum albumin as standard.

Results and discussion

Characterization of the beads

Image analysis of the beads indicated that CELBEADS were spherical in shape and fell in a size range of 150–200 μm. Porosity and surface area analysis by mercury porositimetry gave average pore radius as 33 500 Å and pore volume was 59.8%. The surface area estimated was 26.4 m^2 g^{-1}. Pore volume estimated by gel filtration experiments with BSA and blue dextran (from BioRad) was also found to be 59%.

Non-specific adsorption of several proteins was attempted on CELBEADS in batch mode and was found to be negligible in all cases. For example, incubation of 1 ml of CELBEADS equilibrated in 5 ml of 130 mg ml^{-1} of BSA at pH 7 resulted in less than 30 μg BSA adsorbed per millilitre of CELBEADS.

Experimental (in 10 mm dia. column) and reported bed expansion plots for Streamline™DEAE are shown in Figure 1. The two plots are not very different and the expansion data could be fitted by the Richardson-Zaki relation. The 10 mm dia. column thus provided a true expanded bed with no wall effects. This is in accordance with the generally accepted rule that the wall effect on bed behaviour is negligible so long as the column diameter to particle diameter ratio is more than 20 (Lali et al., 1989). Figure 1 also includes bed expansion data with CELBEADS as well as with CB-CELBEADS. It was found that CELBEADS gave particulate and stable expansion behaviour that fitted the Richardson-Zaki relation.

HETP analysis for Streamline™DEAE

HETP experiments were performed under non-retaining conditions on both Streamline™DEAE and CELBEADS in packed bed as well as expanded bed

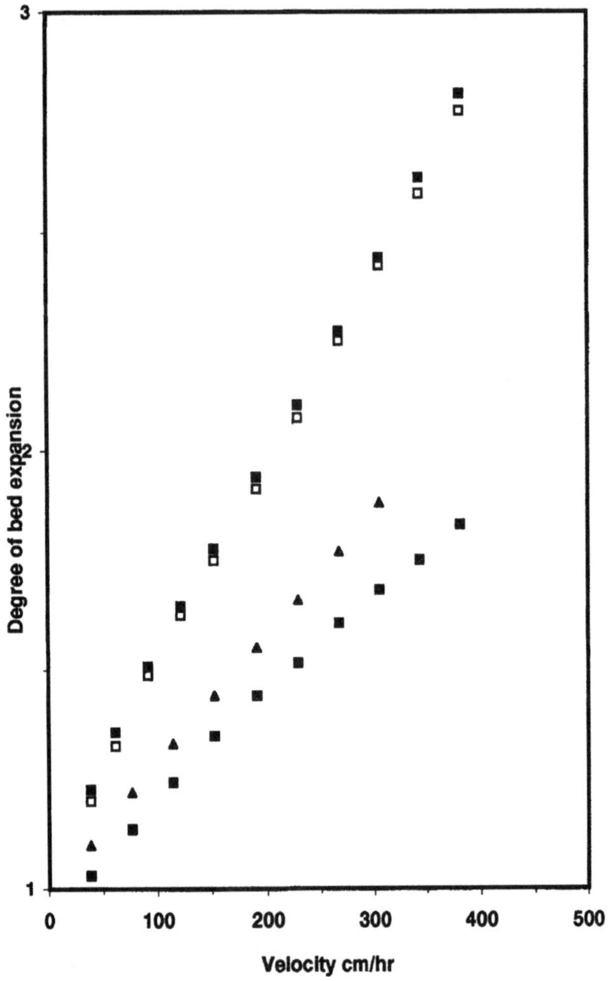

Figure 1. Bed expansion characteristics: ■ Streamline™DEAE (reported); □ Streamline™DEAE (measured in 10 mm glass column); ▲ CELBEADS; ■ CB-CELBEADS.

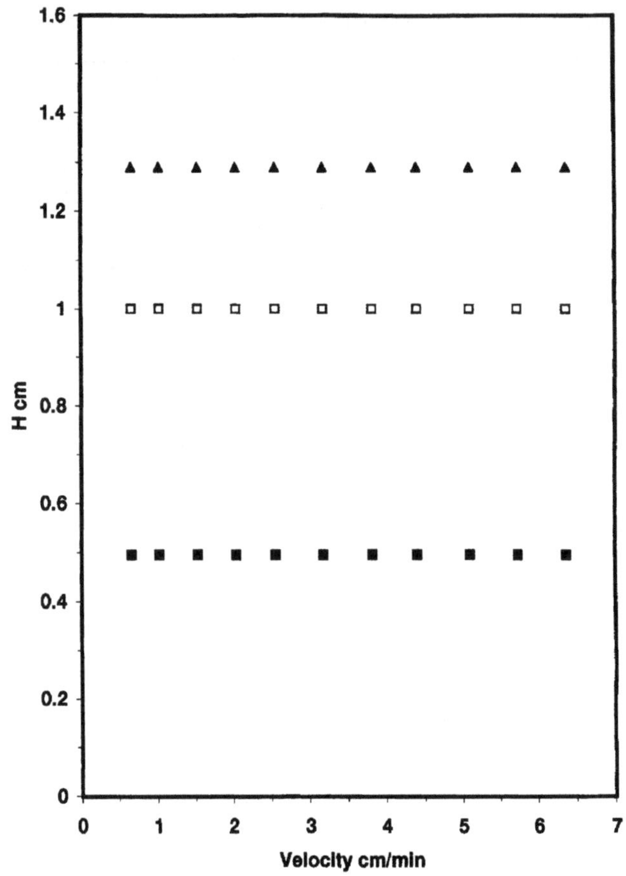

Figure 2. Height equivalent to theoretical plate variation with superficial liquid velocity in packed bed mode on Streamline™DEAE: ■ Tyrosine; □ Papain; ▲ BSA.

operation. Figure 2 shows the observed HETP variation with superficial liquid velocity u_o, in packed bed on Streamline™DEAE with tyrosine, papain and BSA. It can be seen that in the range of velocities tested, HETP was not a function of velocity and remained constant for all three solutes and increased with molecular weight. Constant HETP values are possible under two circumstances: Either on non-retaining non-porous adsorbents, or with macroporous adsorbents. A few chromatographic runs with tyrosine and bovine serum albumin were made on 100 μm spherical non-porous glass beads. The experiments gave constant HETP with both tyrosine and BSA. Constant HETP with velocity has also been shown to be a property of macroporous adsorbent matrices wherein there is a certain convective flow within the pores (Rodrigues et al., 1996). Rodrigues et al. (1992) developed the following relationship for HETP on macroporous adsorbents in the form of a modified van Deemter equation,

$$H = A + C.u_o.f(\lambda)$$

where $f(\lambda)$ is a function of λ, the intraparticle Peclet number related to convection in the macropores. For adsorbents with small pores there is no intraparticle convection possible and $f(\lambda)=1$. Under conditions of significant convection through macropores λ is high and HETP is constant as

$$H_{plateau} = A + C.u_{oc}$$

where u_{oc} is the critical velocity above which HETP is constant. Rodrigues (1997) found that HETP was constant with velocity for rigid gel beads with a large exclusion limit, an observation reportedly supported by other workers on membrane chromatography of proteins. Streamline™DEAE classifies as a rigid gel support with the reported exclusion limit of 4×10^6 for globular proteins and hence shows flat HETP profile.

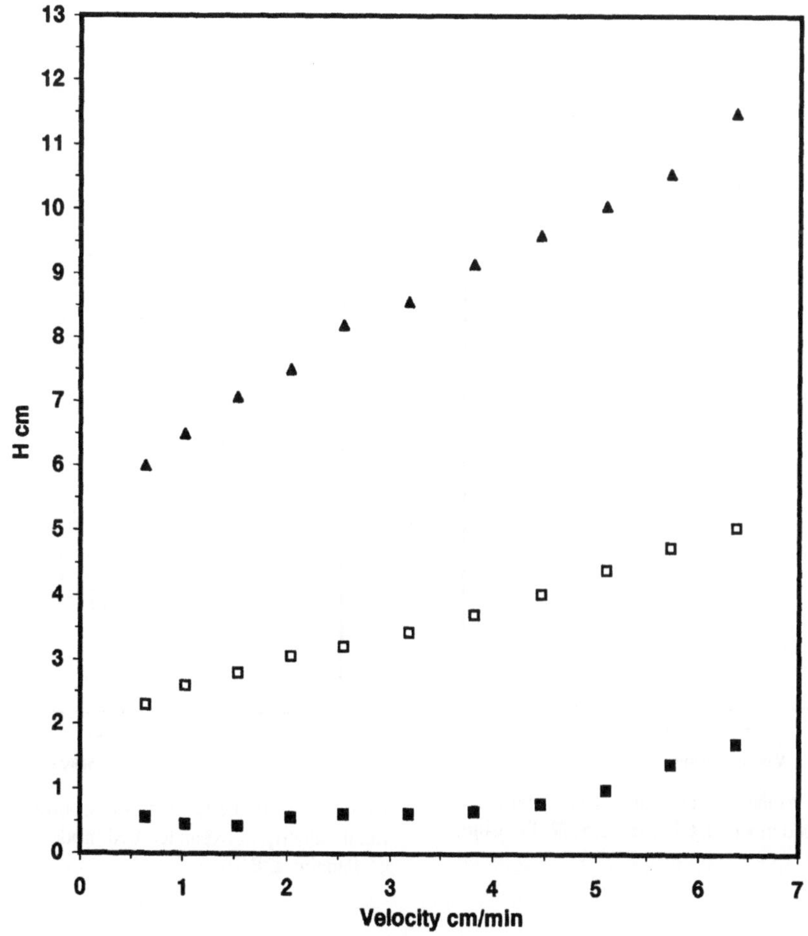

Figure 3. Height equivalent to theoretical plate variation with superficial velocity in expanded bed mode on Streamline^{TM} DEAE: ■ Tyrosine; □ Papain; ▲ BSA.

Further, since a pore size distribution within the gel may be expected to exist, a Streamline gel bead may offer different pore volume fractions to BSA, papain and tyrosine. With the permeabilities of the gel sphere for different solutes, at a given flow rate, also being related to molecular size, it is expected that the HETP value obtained for larger molecule would be higher than one for smaller molecule (Figure 2).

The plot of HETP versus velocity for expanded bed with Streamline^{TM}DEAE was noted to be different. HETP was found to increase with velocity and the rate of increase was higher for larger solute molecule (Figure 3). The term A arising from the van Deemter equation accounts for the eddy diffusion in the bed and is given as

$$A = 2.\Lambda.d_p$$

where d_p is the adsorbent particle diameter and Λ is the parameter related to flow structure in the bed which remains almost constant for packed bed where bed structure does not change with flow. However, Λ would not be constant if bed porosity and particle arrangement changed with flow rate, as happens in an expanded bed. Since flow structure in the bed affects zone broadening, the parameter Λ can be said to be related to liquid phase axial dispersion, Da. Variation in Da with velocity is known to be markedly lesser for packed beds than for fluidized or expanded beds. Chung and Wen (1968) presented a relation for axial dispersion in packed and expanded beds. This relation gives Da being proportional to $u_o^{0.52}$ for packed bed and proportional to $u_o^{1.74}$ in case of fluidized beds for small, low density particles. A number of workers have reported that the eddy diffusion term in van Deemter equation is not constant and is $A.u_o^n$ with the index n ranging from 0.2 to 0.33 for packed bed chromatography (Knox, 1966) which is of same order as the index for Da variation with velocity in packed beds. Variation of Da with velocity however is more pronounced in expanded beds which explains increase in HETP with velocity on expanded beds with Streamline^{TM}DEAE.

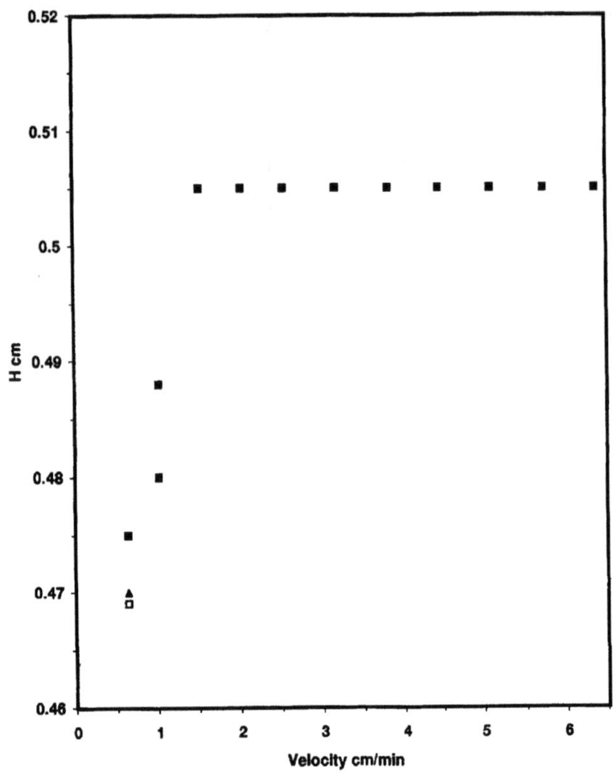

Figure 4. Height equivalent to theoretical plate variation with superficial velocity in packed bed mode on CELBEADS: ■ Tyrosine; □ Papain; ▲ BSA.

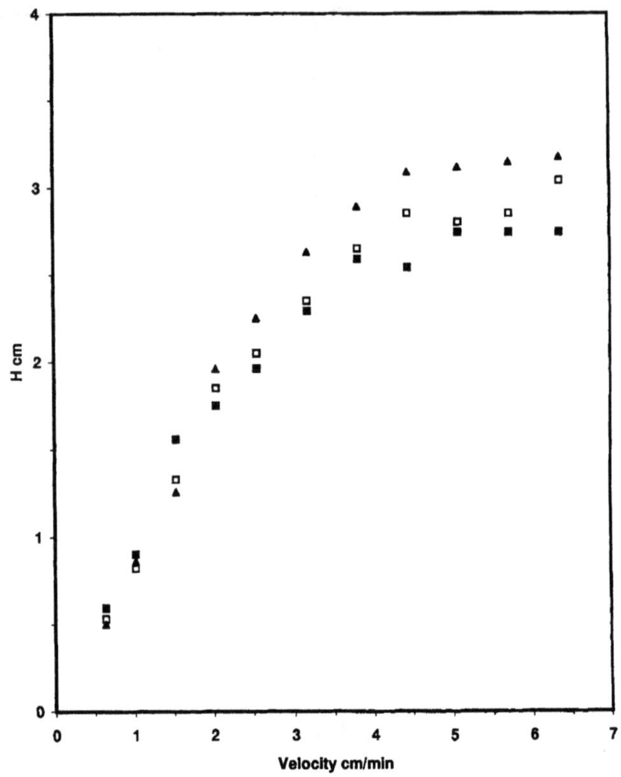

Figure 5. Height equivalent to theoretical plate variation with superficial velocity in expanded bed mode on CELBEADS: ■ Tyrosine; □ Papain; ▲ BSA.

HETP analysis for CELBEADS

Figure 4 shows variation in HETP with velocity for packed bed runs on CELBEADS. It can be seen that the plots for all three solutes, viz. tyrosine, papain and BSA coincided and behaved similar. This is expected since the average pore size of CELBEADS being about 3 μm, most of the pore volume (59%) should be available to the solutes and the particle permeabilities may not be much different. The plot in Figure 4 follows the typical behaviour of macroporous adsorbents (Rodrigues et al., 1996) and the indicated critical velocity, $u_{oc}$for commencement of convective flow in the pores is about 1.5 cm min^{-1}. Reported u_{oc} for POROS Q/M, a macroporous adsorbent form Perseptive Biosystems , UK (0.8 μm average pore size) is about 15 cm min $^{-1}$ (Vega et al., 1998), i.e. ten times higher than obtained with CELBEADS. Using the relations for $HETP_{plateau}$ and from the slope of the HETP plot for $u_o<u_{oc}$, the ratio of convective velocity in the pores to superficial velocity was estimated at different velocities and was also found to be about ten times higher than obtained by Vega et al. (1998) with POROS Q/M at much higher superficial velocities. Thus a supermacroporous matrix like the developed cellulose matrix, CELBEADS, can be expected to give lower HETP values than possible with other adsorbents at comparable flow rates. This fact is brought out by comparison of Figures 2 and 4. The plateau HETP value on CELBEADS for all the three solutes was about 0.5 cm, whereas the plateau HETP value for tyrosine, papain and BSA on Streamline was about 0.5, 1.0 and 1.3 cm, respectively.

Figure 5 shows HETP variation with velocity for expanded bed runs on CELBEADS. Unlike Streamline^{TM}DEAE where HETP increased with velocity in the range tested, CELBEADS showed a behaviour similar to that in packed bed except that u_{oc} shifted to a higher value (3.7 cm min^{-1}). There also was slight variation with respect to the molecular weights of the solutes and the plateau HETP values were also higher ($\sim$ 3 cm). Comparison of Figures 3 and 5 with respect to BSA indicates an HETP three times lesser on CELBEADS than on Streamline^{TM} DEAE in expanded bed mode at a velocity of above 3 cm min^{-1}. The role of extra large pores in a macroporous matrix can be expected to be further dominant under retaining conditions where the effect of C term in van Deemter equation is more significant.

Table 1. A typical run for purification of L- lactate dehydrogenase from porcine muscle extract

Data	Expanded bed elution	Packed bed elution
Settled bed height	77 mm	77 mm
Crude specific activity	17.2 U/mg	17.2U/mg
Loading flow rate	1 ml min^{-1}	1 ml min^{-1}
Expanded bed height	88 mm	88 mm
Total crude loaded	20 ml	20 ml
Total activity loaded	2206 U	2206 U
Total protein loaded	128 mg	128 mg
Activity adsorbed	988 units (44.7%)	998 (45%)
Protein adsorbed	30 mg (23.4%)	30 mg (23.4%)
Elution flow rate	1 ml min^{-1}	1 ml min^{-1}
Eluted activity	1155 U (116%)	1190 U (120%)
Eluted protein	5.0 mg (17%)	4.6 mg (15.4%)
Eluted specific activity	231 U/mg	259 U/mg
Purification factor	13.4	15.0

Purification of LDH on CELBEADS

The muscle extract was found to be unsuitable for application on a packed column but could easily pass through the expanded bed of CELBEADS. Figures 6 and 7 are the chromatograms obtained for expanded bed and packed bed elution cases respectively. Table 1 gives details of the performance of the operations. The band widths in both cases were almost same and the height equivalent to theoretical plate (HETP) values for the eluting bands were determined to be 0.32 cm and 0.30 cm for expanded and packed bed elutions respectively.

An important stage in the operation of an expanded bed is the cleaning-in-place procedure, as working with 'dirty' extracts can significantly 'foul' the matrix. Lali et al. (1997) attempted purification of LDH from muscle extract on modified Streamline and found that the regenerated capacity of the adsorbent fell successively even after rigorous cleaning. In the present work with CELBEADS it was found that 5 M urea passed upward through the expanded bed at 0.6 ml min^{-1} for 3 h cleaned the CELBEADS and there was no loss of adsorption capacity or bed expansion characteristics over 5 cycles of operations tested.

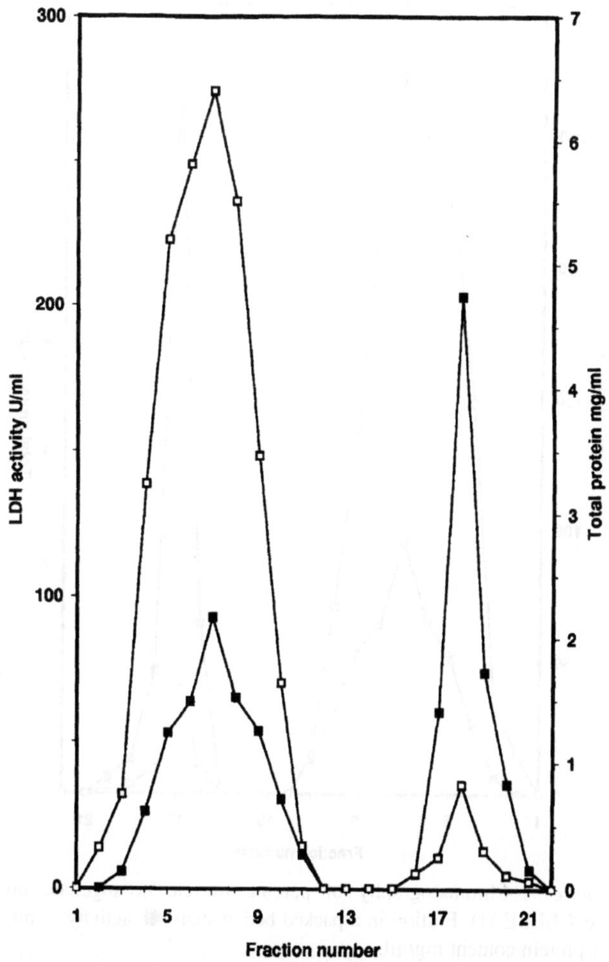

Figure 6. Chromatography of porcine muscle homogenate on CB-CELBEAD. Elution in an expanded bed system: ■ activity U/ml; □ protein content mg/ml.

Conclusions

The following conclusions are drawn from the work presented:

1. A macroporous adsorbent that permits intraparticle convection when used as an expanded bed matrix exhibits similar plateauing of the HETP with velocity as observed when used in packed bed. The threshold velocity for plateauing is higher in expanded bed operation than in packed bed mode.
2. The adsorbent that is macroporous offers higher chromatographic efficiency at all flow rates than adsorbents that are not macroporous, both in packed bed and expanded bed operations.
3. On account of the above two points, a macroporous adsorbent offers possibility of effective expanded bed elution of adsorbed molecules.
4. The supermacroporous beaded adsorbent matrix, CELBEADS prepared as rigid cross-linked cellu-

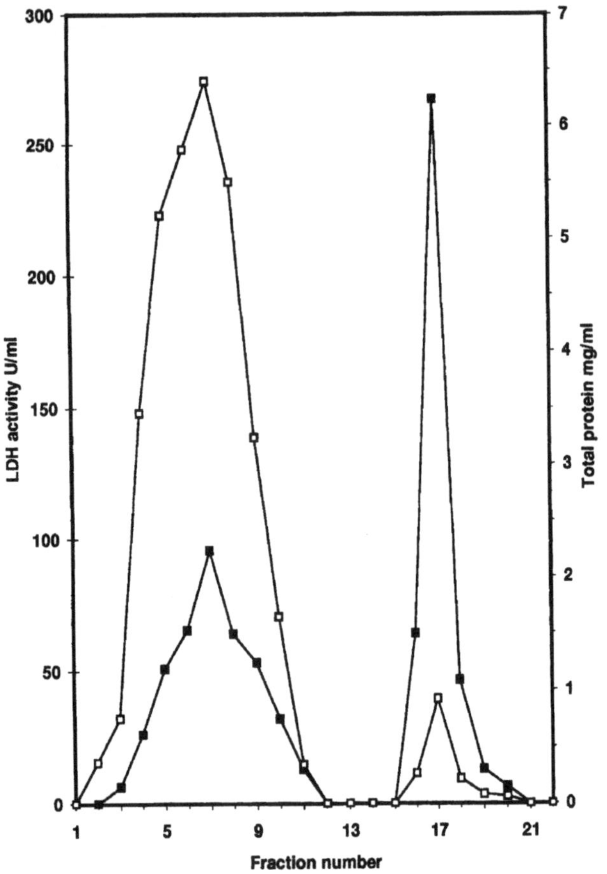

Figure 7. Chromatography of porcine muscle homogenate on CB-CELBEAD. Elution in a packed bed system: ■ activity U/ml; □ protein content mg/ml.

lose beads, was characterised and found suitable for use as a protein purification matrix. The matrix beads were spherical and suitable for expanded bed adsorption/chromatography. CELBEADS when used in packed bed mode gave performance similar to macroporous adsorbents.

5. The new matrix could be used for dye affinity purification of lactate dehydrogenase from porcine muscle extract by conjugating Cibacron blue. The expanded bed purification procedure used on a 10 mm dia. column resulted in purification that was same for both packed bed and expanded bed elution modes. The matrix could be cleaned in place using 5 M urea without any loss of capacity and performance for 5 purification cycles.

Acknowledgement

This work was carried out under the project sponsored by Department of Biotechnology, Government of India (Project no BT/R&D/15/20/95).

References

Afeyan NB, Gordon NF, Mazsaroff I, Varady L, Fulton SP, Yang YB and Regnier FE (1990) Flow through particles for the high performance liquid chromatographic separation of biomolecules: Perfusion Chromatography. J. Chromatogr. 519: 1–29.

Barnfield Frej AK, Hjorth R and Hammarström A (1994) Pilot scale recovery of recombinant Annexin V from unclarified *Escherichia coli* homogenate using expanded bed adsorption. Biotechn. Bioeng. 44: 922–929.

Buijs A and Wesselingh JA (1980) Batch fluidised ion-exchange column for streams containing suspended particles. J. Chromatogr. 201: 319–327.

Chung SF and Wen CY (1968) Longitudinal dispersion of liquid flowing through fixed and fluidized beds. AIChE J. 14: 857–865.

Hermanson GT, Mallia AK and Smith KP (1992) Immobilised affinity ligand techniques. Academic Press, California.

Knox JH (1966) Evidence for turbulence and coupling in chromatographic column. Anal. Chem. 38: 253–261.

Lali AM, Khare AS and Joshi JB (1989) Behaviour of solid particle in viscous non Newtonian solutions: Settling velocity wall effects and bed expansion in solid liquid fluidised beds. Powder Tech. 57: 39–50.

Lali A, Kaul R, Galaev IY and Mattiasson B (1997) Purification of L-lactate dehydrogenase form crude homogenate of porcine muscle by expanded bed affinity chromatography. Iso. Purif. 2: 289–300.

Rodrigues AE, Lopes JC, Dias MM, Loureiro JM and Lu ZP (1992) Importance of intraparticle convection in the performance of chromatographic processes. J. Chromatogr. 590: 93–100.

Rodrigues AE, Chenou C and Rendueles de la Vega (1996) Protein separation by liquid chromatography using permeable POROS Q/M particles. Chem. Eng. J. 61: 191–201.

Rodrigues AE (1997) Permeable packings and perfusion chromatography in protein separation. J. Chromatogr. 699: 47–61.

Vega de la RM, Chenou C, Loureiro JM and Rodrigues AE (1998) Mass transfer mechanisms in Hyper D media for chromatographic protein separation. Biochem. Eng. J. 1: 11–23.

Yamamoto S, Nomura M and Sano Y (1987) Factors affecting the relationship between the plate height and the linear mobile phase velocity in gel filtration chromatography of proteins. J. Chromatogr. 394: 363–367.

van Deemter JJ, Zuiderweg FJ and Klinkenberg A (1956) Longitudinal diffusion and resistance to mass transfer as causes of non ideality in chromatography. Chem. Eng. Sc. 5: 271–289.

Worthington Enzymes and Related Biochemicals, Worthington Biochemical, Freehold, NJ (1982) 109.

Address for correspondence: Arvind Lali, Chemical Engineering Division, Department of Chemical Technology, University of Mumbai, Matunga, Mumbai 400 019, India.
Fax: 91-22-4145614; E-mail: arvind@udct.ernet.in

Bioseparation **8:** 139–144, 1999.
© 1999 *Kluwer Academic Publishers.*

Characterisation of STREAMLINE™ Phenyl

Johan Färenmark, Jan Gustavsson, Inger Lagerlund & Lena Sandberg
Amersham Pharmacia Biotech AB, 751 84 Uppsala, Sweden

Received 8 September 1998; accepted in revised form 7 April 1999

Key words: capture, expanded bed, hydrophobic interaction chromatography, stability

Abstract

STREAMLINE Phenyl is a new hydrophobic interaction chromatography support designed for use in expanded bed adsorption. The phenyl groups are linked to STREAMLINE matrix via highly stable ether linkages. Within this development project the chemical and chromatographic stability as well as the breakthrough capacity for human IgG has been studied. The chemical stability was monitored as the carbon leakage from the matrix to the storage solution, pH 1–14 at 20 and 40 °C. The carbon content in the supernatant was determined with Total Organic Carbon (TOC) technique. In the chromatographic stability study STREAMLINE Phenyl was stored in eight different storage solutions under ambient conditions for 12 weeks and then tested in a chromatographic function test. The results show that the adsorbent is chemically stable and that the chromatographic properties are retained under the tested conditions. The breakthrough capacity study demonstrates the importance of the bed height for obtaining maximal dynamic capacity. Further, there is a good correlation between breakthrough data generated from packed bed and expanded bed runs.

Introduction

STREAMLINE columns and adsorbents are products specially designed for purifying proteins using expanded bed adsorption. The properties of expanded bed adsorption makes it the ultimate capture step for initial recovery of target proteins from crude feed stocks. The basic principles (Chase, 1994; Thömmes, 1997) as well as industrial applications (Hjorth, 1997) of expanded bed adsorption has recently been reviewed. Until now, only adsorbents for use in ion exchange and affinity chromatography have been available.

Hydrophobic Interaction Chromatography (HIC) is a well known technique (Jansson and Rydén, 1998) that has come to increasing use during recent years. STREAMLINE Phenyl is a new chromatographic support designed to combine the strength of HIC and expanded bed adsorption. The adsorbent has phenyl groups coupled to the matrix via uncharged, chemically stable ether linkages, which results in a hydrophobic medium with minimal ionic properties.

In this paper we present the results from a stability study and some determinations of dynamic breakthrough capacity for a model protein (human IgG) when using different bed heights of STREAMLINE Phenyl in packed and expanded bed modes.

Materials and methods

Equipment

The chromatographic studies were performed on FPLC™ system. Detectors used were either UV-1 or UV-M (280 nm) monitors. Columns used for the chromatography were HR 10/10, STREAMLINE 25 or STREAMLINE 50 columns. All chromatographic equipment was obtained from Amersham Pharmacia Biotech AB, Uppsala, Sweden.

For the measurement of the carbon leakage a TOC analyser (TOC-5050A, Shimadzu, Kyoto, Japan) was used.

Chemicals

Model protein used in the tests were cytochrome C, ribonuclease A, lysozyme (Sigma-Aldrich, Stock-

Table 1. Characteristics of STREAMLINE Phenyl

Bead structure:	Macroporous cross-linked 6% agarose containing crystalline quartz core material
Mean particle size:	100–300 μm
Mean particle density:	Approximately 1.2 g ml^{-1}
Flow velocities:	200–400 cm h^{-1}
Degree of expansion (H/H_0) at 300 cm h^{-1}:	2—3
Recommended pH range for use:	
–working range	3–13
–cleaning in place	2–14
Temperature stability:	4–40 °C

holm, Sweden) and human IgG (Gammanorm, Pharmacia AB, Stockholm, Sweden). All buffer chemicals used were analytical grade.

STREAMLINE Phenyl and Phenyl Sepharose™ Fast Flow (high sub) (Amersham Pharmacia Biotech, Uppsala, Sweden) were used as chromatographic supports. General characteristics for STREAMLINE Phenyl are summarised in Table 1.

Chromatographic function test

STREAMLINE Phenyl was packed in a HR 10/10 with adsorption buffer (0.1 M NaH_2PO_4, 1.2 M $(NH_4)_2SO_4$, pH 7.0) using a flow rate of 2.0 ml min^{-1} to a bed height of approximately 10 cm corresponding to a column volume of 8 ml. A sample solution was prepared dissolving the model proteins cytochrome C, ribonuclease A, and lysozyme (2:8:1) in adsorption buffer to a total protein concentration of 18.33 mg ml^{-1}. The column was connected to a FPLC system with a UV-M monitor attached to the outlet of the column. It was then equilibrated with 30 ml of adsorption buffer before 100 μl of the sample solution was injected. The proteins were eluted with a linear gradient, 3.6 column volumes, of desorption buffer (0.1 M NaH_2PO_4, pH 7.0 to which 111.1 g ethylene glycol per litre buffer was added) at a flow of 0.5 ml min^{-1}. The retention time for the respective protein is defined as the time from gradient start until the UV monitor records a peak maximum.

Degree of expansion

The degree of expansion was determined in a STREAMLINE 50 column in which 9–11 cm settled adsorbent was expanded with 0.50 M NaCl solution at a liquid velocity of 275 cm h^{-1}. The degree of expansion is expressed as the ratio of expanded bed height over settled bed height.

Stability studies

Storage of STREAMLINE Phenyl in different solutions under ambient conditions

STREAMLINE Phenyl was rinsed with distilled water and then mixed with the storage chemicals and distilled water needed to give the desired concentrations of 20% ethanol, 70% ethanol, 80% ethylene glycol, 8 M urea, 30% isopropanol, 8 M guanidine hydrochloride, 1 M sodium hydroxide, and 1 M acetic acid respectively. The gel slurries were filled in plastic bottles and stored on a shelf in the laboratory, i.e. normal daylight and temperature (on average 22 °C). Gel samples were removed after 4 and 12 weeks, washed thoroughly and tested for chromatographic function and degree of expansion.

Determination of carbon release from STREAMLINE Phenyl

Approximately 200 ml STREAMLINE Phenyl was rinsed thoroughly on a glass filter funnel with 10×200 ml Milli-Q water. The gel was sucked dry and transferred to a beaker together with Milli-Q water where it was kept overnight. Then it was rinsed with 5×200 ml 0.1 M NaCl and 5×200 ml Milli-Q water. The gel was sucked dry and portions of 10.00 g were weighed into glass filter funnels. Each sample was equilibrated with the storage solution of different pH (2×25 ml). The sample was sucked dry and the gel together with 50 ml storage solution were transferred to a flask which were stored at 20 or 40 °C for one week. 10 ml of the supernatant was collected from each of the stored flask. The samples were acidified with a known volume of concentrated HCl and purged with oxygen

Table 2. The degree of expansion of STREAMLINE Phenyl, at a liquid velocity of 275 cm h^{-1}, after 12 weeks storage under ambient conditions in different solutions

Storage solution	Weeks	0	12
	Degree of expansion		
20% Ethanol		2.5	2.3
70% Ethanol		2.5	2.4
30% Isopropanol		2.5	2.5
80% Ethylene glycol		2.5	2.7
1 M Sodium hydroxide		2.5	2.5
1 M Acetic acid		2.5	2.4
8 M Urea		2.5	2.4
8 M Guanidine hydrochloride		2.5	2.4

Table 3. Chromatographic function test results for STREAMLINE Phenyl after 4 and 12 weeks storage under ambient conditions in different solutions. Peak I–Cytochrome C, Peak II–Ribonuclease A, Peak III–Lysozyme

Storage solution	Weeks	0			4			12		
	Retention time (minutes) Peak I–III	I	II	III	I	II	III	I	II	III
20% Ethanol		17	38	63	17	37	61	18	38	63
70% Ethanol		17	38	63	19	40	64	19	40	66
30% Isopropanol		17	38	63	18	38	63	19	39	64
80% Ethylene glycol		17	38	63	18	39	63	18	38	63
1 M Sodium hydroxide		17	38	63	18	40	62	18	40	64
1 M Acetic acid		17	38	63	19	39	63	19	35	58
8 M Urea		17	38	63	17	37	61	17	38	63
8 M Guanidine hydrochloride		17	38	63	18	39	63	18	38	63

for three minutes. The non-purgeable organic carbon was detected on a TOC analyser.

Breakthrough capacity

The breakthrough capacity was determined for human IgG (1 mg ml^{-1} in adsorption buffer 1.0 M $(NH_4)_2SO_4$, 20 mM NaH_2PO_4, pH 7.0). All studies were performed in a STREAMLINE 25 column. In the expanded bed experiments STREAMLINE Phenyl was added until the settled bed heights were 15 and 28 cm. Adsorption buffer was pumped through the column at 300 cm h^{-1} until the expanded bed was stable then the IgG solution was pumped through the column at 300 cm h^{-1}. IgG solution was applied until the outlet UV reading was at least 50% of the UV reading of the initial IgG solution. The adaptor was then lowered to the gel surface and the column was rinsed with adsorption buffer at 100 cm h^{-1} and finally the IgG was eluted with desorption buffer (20 mM NaH_2PO_4, pH 7.0) at 100 cm h^{-1}. All fractions were collected and a mass balance was calculated. The breakthrough capacity is calculated as the amount of IgG applied per ml chromatographic support, corrected for the dead volume of the column, at a certain ratio of inlet and outlet concentration of the feed.

In the packed bed experiment STREAMLINE Phenyl was slightly compressed with the top adaptor to give a bed height of 15 cm. The column was equilibrated with adsorption buffer and sample was loaded and eluted in the same manner as above.

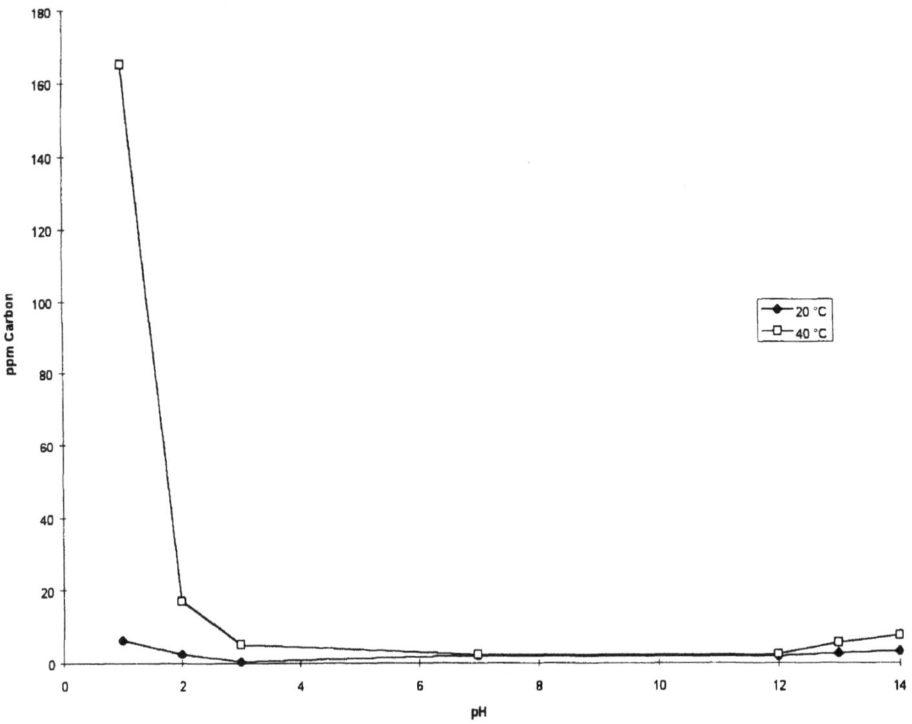

Figure 1. The influence of pH and temperature on the release of carbon from STREAMLINE Phenyl. Copyright Amersham Pharmacia Biotech. Reproduced with permission.

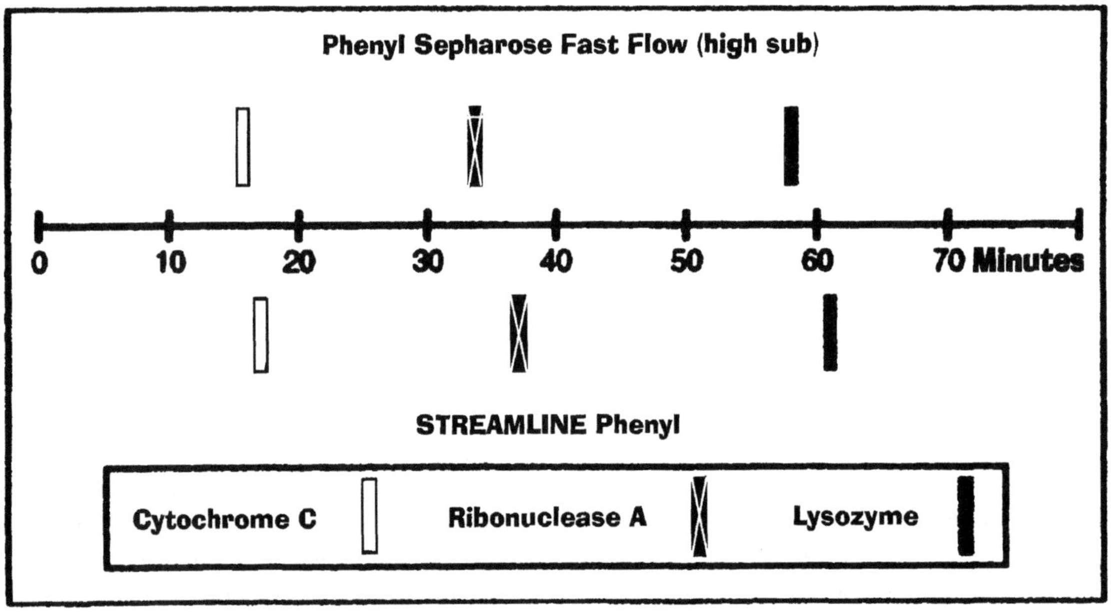

Figure 2. Typical retention times of model proteins on Phenyl Sepharose Fast Flow (high sub) and STREAMLINE Phenyl. Copyright Amersham Pharmacia Biotech. Reproduced with permission.

Results and discussion

Stability studies

Chromatographic stability

The results from the stability study are summarised in Tables 2 and 3. As can be seen 12 weeks of storage in different solutions has not significantly affected the chromatographic performance of STREAMLINE Phenyl. There might be an effect from storage in 1 M acetic acid. This effect is probably caused by base matrix breakdown that take place at acidic pH.

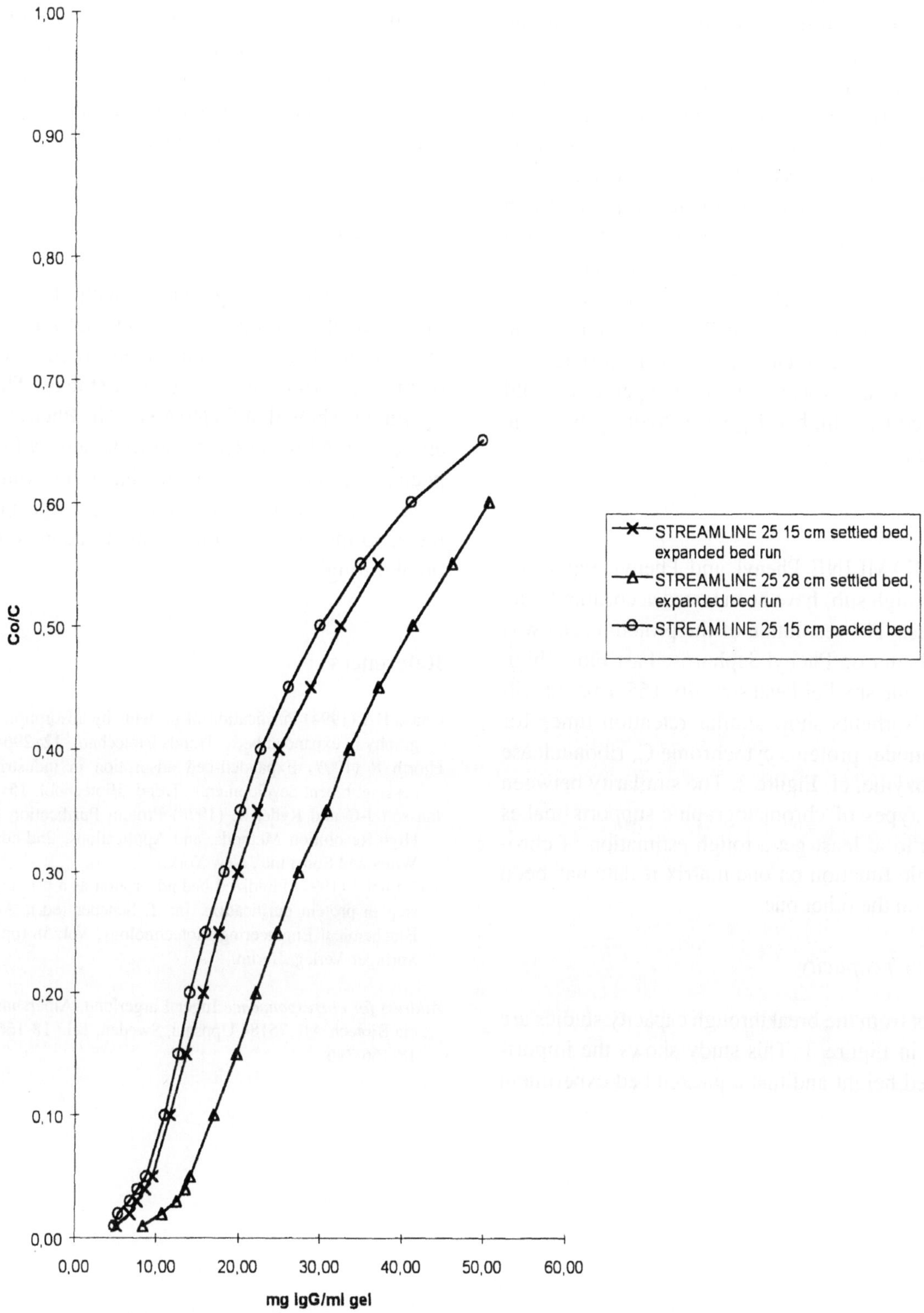

Figure 3. Breakthrough capacity curves for human IgG on STREAMLINE Phenyl in a STREAMLINE 25 column at a liquid velocity of 300 cm h^{-1}. Copyright Amersham Pharmacia Biotech. Reproduced with permission.

Chemical stability

The TOC technique is a very suitable test method for leakage of STREAMLINE Phenyl, since the main component of the product is carbon. However, it has the limitation that it is impossible to study leakage in any carbon-containing solvent (e.g. ethanol, urea, guanidine-HCl). The carbon leakage is extremely low all over the tested pH range, cf. Figure 1. The highest carbon leakage after 1 week at 20 °C was obtained for the pH 1 sample where approximately 6 ppm carbon were detected in the supernatant. This leakage level indicates a loss of approximately 0.1% of the carbon content of the gel. At pH 1 it can be seen that when increasing the temperature from 20 to 40 °C the carbon leakage increases substantially. The increased decomposition of the adsorbent at higher temperature could be explained by a higher degree of hydrolysis of the base matrix.

Selectivity

Both STREAMLINE Phenyl and Phenyl Sepharose Fast Flow (high sub) have been tested according to the function test. The resolution of the protein peaks was of course better on Phenyl Sepharose Fast Flow (high sub) due to the smaller bead size, 46–165 μm, but still the two adsorbents show similar retention times for the three model proteins cytochrome C, ribonuclease A and lysozyme, cf. Figure 2. The similarity between these two types of chromatographic supports makes it possible to at least get a rough estimation of chromatographic function on one matrix if data has been generated on the other one.

Breakthrough capacity

The results from the breakthrough capacity studies are presented in Figure 3. This study shows the importance of bed height and that a packed bed experiment can give useful data for prediction of the behaviour in expanded bed. Recovery from the cycle (load, wash and elution) was more than 90% in all experiments. The breakthrough capacity is dependent on what system and which test protein that has been used, hence one has to be careful to draw any comparative conclusions between investigations made using different equipment or different proteins.

Conclusions

STREAMLINE Phenyl has selectivity similar to Phenyl Sepharose Fast Flow (high sub) and is an alternative to other expanded bed adsorbents for capture of proteins from unclarified feed stocks. The stability studies show that STREAMLINE Phenyl is robust and can stand harsh chemical treatment, which is especially important for media that is in contact with crude feed. Bed height is a very important parameter to investigate if one wants to achieve maximal breakthrough capacity.

References

Chase HA (1994) Purification of proteins by adsorption chromatography in expanded beds. Trends Biotechnol. 12: 296–303.

Hjorth R (1997) Expanded-bed adsorption in industrial bioprocessing: recent developments. Trends Biotechnol. 15: 230–235.

Jansson J-C and Rydén L (1998) Protein Purification Principles, High Resolution Methods, and Applications, 2nd edition. John Wiley and Sons, Inc., New York.

Thömmes J (1997) Fluidized bed adsorption as a primary recovery step in protein purification. In: T. Scheper (ed.), Advances in Biochemical Engineering/Biotechnology, Vol. 58 (pp.185–230). Springer-Verlag, Berlin.

Address for correspondence: Inger Largerlund, Amersham Pharmacia Biotech AB, 75184 Uppsala, Sweden. Tel.: 18-166-298; Fax: 18-166-396

Bioseparation **8:** 145–151, 1999.
© 1999 *Kluwer Academic Publishers.*

Use of Streamline chelating for capture and purification of poly-His-tagged recombinant proteins

Santosh Noronha, Jeanne Kaufman & Joseph Shiloach*
Biotechnology Unit, LCDB, NIDDK, NIH, Bethesda, MD 20892

Received 7 July 1998; accepted 11 November 1998

Key words: EBA, IMAC, 6×His, TBV25H, CRM H21G

Abstract

Expression of recombinant proteins with poly-histidine tags enables their convenient capture and purification using immobilized metal affinity chromatography (IMAC). The 6×His-tagged protein binds to a chelating resin charged with metal ions such as Ni^{2+}, Cu^{2+} or Zn^{2+}, and can therefore be separated from proteins which have lower, or no, affinity for the resin. Two recombinant proteins, a malaria transmission-blocking vaccine candidate secreted extracellularly by *S. cerevisiae* and a modified diphtheria toxin produced intracellularly by *E. coli*, were expressed with 6×His tags and could therefore be purified using IMAC. In an effort to further simplify the initial capture of these proteins, an expanded bed adsorption technique using a chelating resin (Streamline Chelating) was introduced. It was possible to capture the intracellular diphtheria protein from *E. coli* directly after cell lysis, without prior centrifugation or filtration. The extracellular malaria vaccine candidate was also directly captured from a high cell density yeast culture. Detailed information on the experimental work performed, and the capture processes developed, is provided.

Abbreviations: BCIP/NBT – 5-bromo-4-chloro-3-indolyl phosphate/nitro blue tetrazolium; CV – column volume; EBA – expanded bed adsorption; IMAC – immobilized metal affinity chromatography; MWCO – molecular weight cut off; NiNTA – nickel nitrilo triacetic acid; PBS – phosphate-buffered saline; SDS-PAGE – sodium dodecyl sulphate polyacrylamide gel electrophoresis

Introduction

Expression of recombinant proteins with poly-histidine tags (Hochuli et al., 1988) enables their convenient capture and purification using immobilized metal affinity chromatography (IMAC; Porath et al., 1975). The 6×His-tagged protein binds to a chelating resin charged with metal ions such as Ni^{2+}, Cu^{2+}, or Zn^{2+}, and can therefore be separated from other proteins that have lower or no affinity for the resin. As a result, this expression method has become increasingly popular, with a large number of recombinant proteins being expressed with histidine tags. In addition, several IMAC resins are now commercially available. In most cases, some pretreatment is required before the protein mixture containing the 6×His-tagged protein is applied to the IMAC resin. Expanded bed adsorption (EBA) is a convenient method of capturing a desired protein from an unclarified feed stock, such as a bacterial culture or disrupted bacterial cells, eliminating the need for centrifugation and filtration (Hjorth, 1997). Combining EBA and IMAC could simplify the capture and purification process even further by taking advantage of the specific binding of the histidine tagged protein to the chelating resin.

Two recombinant proteins needed for clinical trials were expressed with 6×His tags and could therefore be purified by IMAC. CRM H21G (Johnson and Nicholls, 1994), a modified diphtheria toxin, is produced intracellularly by *E. coli*, while TBV25H (Kaslow and Shiloach, 1994), a malaria

* Corresponding author
Joseph Shiloach, Bldg. 6, Rm. B1-33, NIH, Bethesda, MD 20892, USA. Tel.: +1301-496-9719; E-mail: ljs@helix.nih.gov

transmission-blocking vaccine candidate, is secreted extracellularly by *S. cerevisiae*.

The CRM H21G diphtheria toxin was needed for the preparation of conjugated vaccines (Shelly et al., 1997) and immunotoxins (Neville et al., 1996). Diphtheria toxins are secreted extracellularly by their native producer (*Corynebacterium diphtheriae*), and therefore purification is relatively simple (Carroll et al., 1988). Production of these proteins depends on the iron concentration in the media and expression levels are low. Therefore, there has been some effort to produce the desired modified toxin in *E. coli*. Recombinant proteins in *E. coli* are often retained intracellularly and therefore after disruption of the microorganism, the desired protein needs to be separated from other intracellular proteins. A modified diphtheria toxin which maintained the conformation of the wild type, but had greatly reduced enzymatic activity, was constructed by replacing the histidine in position 21 with glycine (Johnson and Nicholls, 1994). The toxin, with a histidine tag, was expressed intracellularly and was purified using NiNTA agarose. This purification process could not be easily scaled up; therefore the EBA approach was investigated. In the following work, the use of an IMAC resin in expanded bed operation is described and toxin production by native and recombinant bacteria compared.

The production of TBV25H, a malaria transmission-blocking vaccine candidate in *S. cerevisiae*, has been described elsewhere (Noronha et al., 1998). TBV25H was constructed by modifying Pfs25, a *Plasmodium falciparum* surface protein, with the addition of yeast α-factor for folding in the yeast secretory pathway. A 6×His tag was also added to facilitate purification using IMAC. The current approach to protein recovery involves centrifugation and ultrafiltration of high cell density yeast cultures, followed by IMAC using NiNTA agarose. While this procedure captures the recombinant protein present in the culture supernatant, it is difficult to scale up. The use of an expanded bed with an IMAC resin was investigated as an alternative approach to purification, and the relative merits of the two procedures are discussed in this work.

Materials and methods

CRM H21G

Frozen *E. coli* cell paste was suspended in ice cold binding buffer (20 mM Tris, 500 mM NaCl, 5 mM imidazole, pH 8.0) at 0.25 (g cells wet weight)(ml buffer)$^{-1}$. The resultant suspension was passed twice through a homogenizer (Manton Gaulin, Wilmington, MA) at 62 MPa. The broken cell extract was adjusted to 0.13 (g cells)(ml buffer)$^{-1}$.

Streamline Chelating high sub, a prototype resin with 10% higher capacity than the commercially available Streamline Chelating (Pharmacia Biotech, Uppsala, Sweden) was saturated with 100 mM $NiSO_4$. The resin was then washed with water and equilibrated with binding buffer. After the sample had been loaded in expanded mode, the column was packed, and the resin washed with 20 mM Tris, 500 mM NaCl, 30 mM imidazole, pH 7.2. Bound protein was eluted in downward mode with 20 mM Tris, 500 mM NaCl, 250 mM imidazole, pH 7.2.

SDS–PAGE gels (8–16% gradient, Novex, San Diego, CA) were run in reducing conditions. Gels were scanned and band areas calculated using NIH Image 1.61. Proteins from SDS–PAGE gels were transferred onto nitrocellulose TransBlot transfer media (BioRAD, Richmond, CA) for 1 h at 30 V. The resulting blot was blocked with 10% goat serum in 1× PBS (Jackson Immunoresearch, West Grove, PA), and incubated with the primary antibody, horse anti-DT (CDC, Atlanta GA) for 1 h and washed with 1× PBS. The blot was then incubated with the phosphatase-labeled secondary antibody, goat anti-horse (Kirkegaard and Perry, Gaithersburg, MD) for 1 h. The blot was washed with 1× PBS and developed with the BCIP/NBT substrate system (Kirkegaard and Perry). Total protein content was assayed using the BCA method (Pierce, Rockford, IL).

TBV25H

All expanded bed experiments were performed on a 2.5-cm column (Streamline 25, Pharmacia Biotech, Uppsala, Sweden) with 60 ml of Streamline Chelating resin. The resin was equilibrated with 3 column volumes (CV) of 100 mM $NiSO_4$ and 1 × PBS, pH 7.4. After the sample was loaded, the resin was washed with 5 CV each of 2 × PBS, pH 7.4, and 2 × PBS, pH 6.4, in succession. Protein bound to the resin was eluted using 0.25 mM sodium acetate, pH 4.5.

The yield of TBV25H in the yeast fermentation and expanded bed runthrough fractions was determined by purifying the 6×His-tagged protein from pretreated samples using NiNTA agarose. Sample pretreatment was performed by centrifuging culture broth at 9000 rpm for 30 min and ultrafiltering the super-

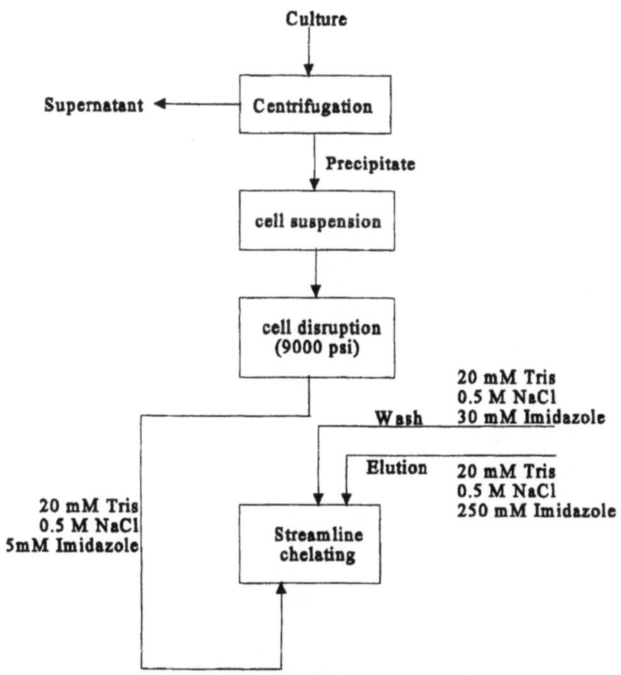

Fig. 1. Flow diagram of the expanded bed capture process for CRM H21G.

natant through a 5000 MWCO filter (Prep/Scale-TFF, Millipore, Bedford, MA) to 20% of the original volume. One hundred ml of the ultrafiltered supernatant were contacted with 6 ml of NiNTA agarose (Qiagen, Valencia, CA) overnight at 4°C in batch mode. This material was then loaded onto a small column, the resin was washed and bound protein eluted using the same buffers as with Streamline Chelating.

Eluted fractions were ultrafiltered through a 10000 MWCO filter (Macrosep 10k, Filtron Technology Corporation, Northborough, MA). Gel filtration of protein eluted from both metal affinity resins (NiNTA agarose and Streamline Chelating) was performed on a Superdex HR 16/60 column (Pharmacia). A flow rate of 0.5 cm min^{-1} of 1 × PBS, pH 7.4, was maintained and 6 ml of sample was loaded onto the column. SDS–PAGE analysis was performed on 4–20% gradient gels (Novex). Total protein was estimated using the BCA method.

Results

CRM H21G

Binding capacity of CRM diphtheria toxin to Streamline Chelating

CRM toxin, or similar material, appears immediately in the runthrough of an expanded bed process. Since detection was done by SDS–PAGE, it was not possible to evaluate the binding capacity of Streamline Chelating for the CRM diphtheria toxin by performing a classical breakthrough curve analysis. Therefore, the binding capacity was evaluated by analyzing the eluant, rather than the runthrough. Different amounts of starting material were loaded onto the same volume of resin. The results are summarized in Table 1. The relative amount of the CRM toxin present in the eluant was independent of the loading ratio and was between 40 and 50% of the total protein eluted. The lowest loading ratio (0.19 g cells per ml resin) yielded 0.8 mg per ml resin compared to 1.0 and 1.3 mg protein from 2- and 4-fold higher loading ratios.

Table 1. Effect of loading ratio on CRM H21G capture. A 13% cell suspension was passed through the column. The eluant was analyzed by SDS-PAGE and the percentage of CRM toxin calculated after scanning, as described in Materials and Methods.

Sample[1] g ml^{-1}	Elution (mg protein)(ml resin)$^{-1}$	% CRM in total protein eluted
0.78	1.3	45
0.39	1.0	38
0.19	0.8	53

[1]g cells (wet weight) per ml of resin.

Expanded bed operation for capture of CRM diphtheria toxin

Expanded bed adsorption was performed using a ratio of 0.19 g of disrupted cells per ml resin, based on the results of the binding capacity experiment. The cell extract was diluted 4-fold and pumped upward at a flow rate of 225 cm h^{-1} onto 300 ml of Streamline Chelating resin in a 5-cm column (Streamline 50, Pharmacia). The overall process is summarized in Fig. 1. Gel electrophoresis of the various fractions is shown in Fig. 2a. Expanded bed operation was identical to the small scale experiment described in Table 1. Three hundred mg of total protein (containing 50% CRM toxin) were obtained from 60 g of cells (wet weight). SDS–PAGE analysis indicated the several other proteins had bound to the resin. These other proteins could be degradation products of CRM H21G, or intracellular native proteins with affinity to the IMAC resin. Immunoblot analysis revealed that CRM toxin and its degradation products were present in the runthrough (Fig. 2b).

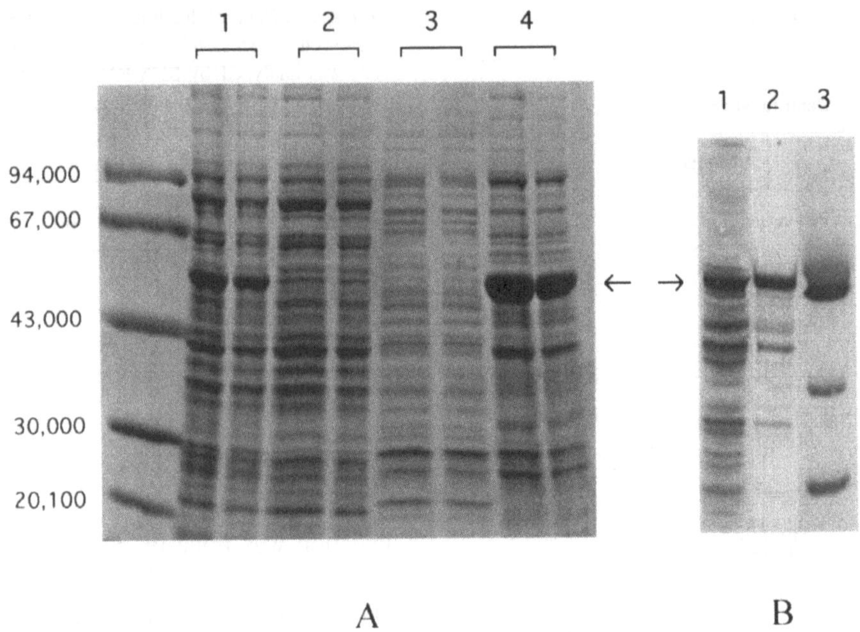

Fig. 2. (A) Capture of CRM H21G on Streamline Chelating. Fractions were analyzed on an 8–16% SDS–PAGE gel. (1) Bacterial extract; (2) column runthrough; (3) wash; and (4) elution. (B) Immunoblot analysis of expanded bed fractions. (1) Runthrough; (2) Elution; and (3) CRM H21G standard. Toxin subunits are visible in the immunoblot as the two lower bands of the standard.

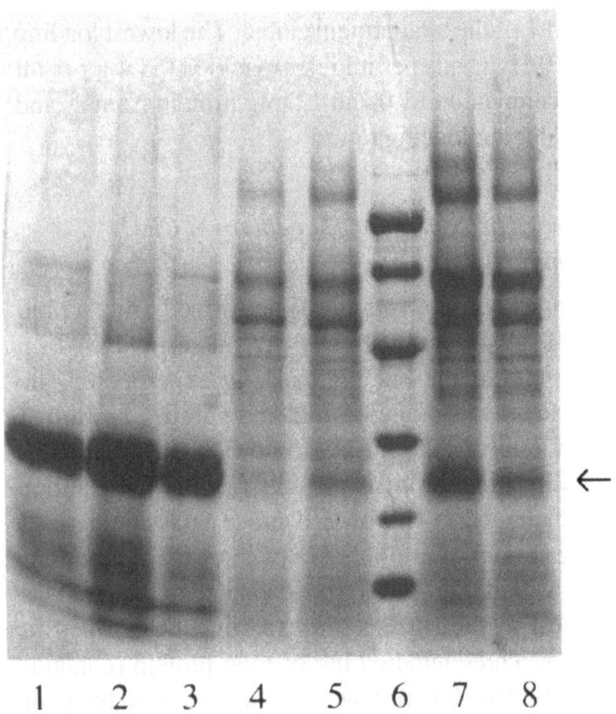

Fig. 3. Capture of TBV25H on IMAC resins. Fractions were analyzed on a 4–20% SDS–PAGE gel. (1) Streamline Chelating elution; (2) chelating high-sub elution; (3) NiNTA agarose elution; (4) Streamline Chelating runthrough; (5) NiNTA agarose runthrough; (6) marker; (7) culture broth; (8) inoculum.

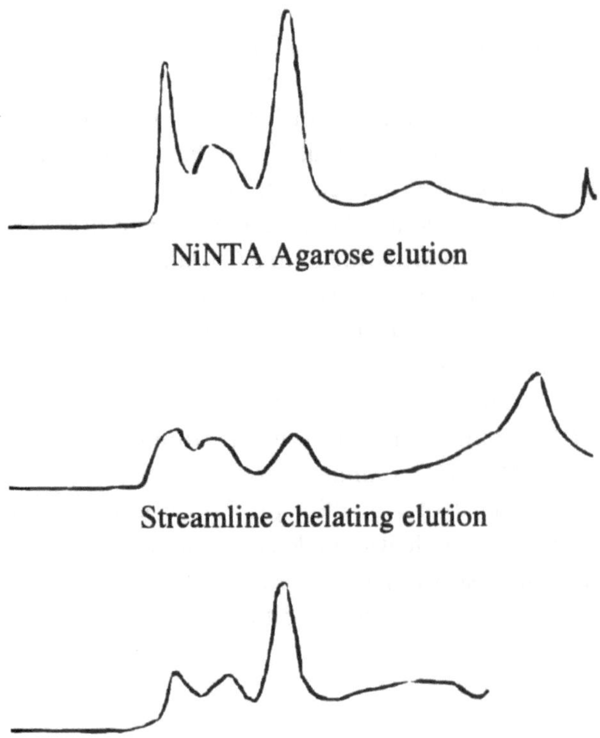

Fig. 4. Gel filtration of IMAC resin elutions.

TBV25H

NiNTA elutions from pretreated samples were previously determined by gel filtration to contain negligi-

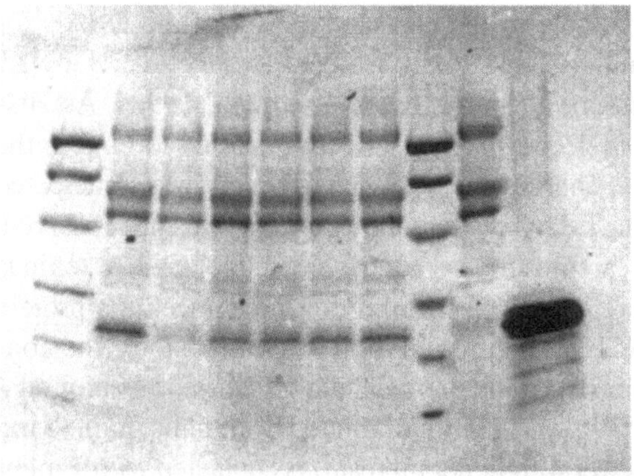

Fig. 5. Determination of specific capacity of Streamline chelating for TBV25H. Runthrough fractions were analyzed on a 4–20% SDS–PAGE gel. Lanes: 1, 8, marker; 2, culture supernatant; 3–7, EBA runthrough fractions; 9, inoculum; 10, EBA elution.

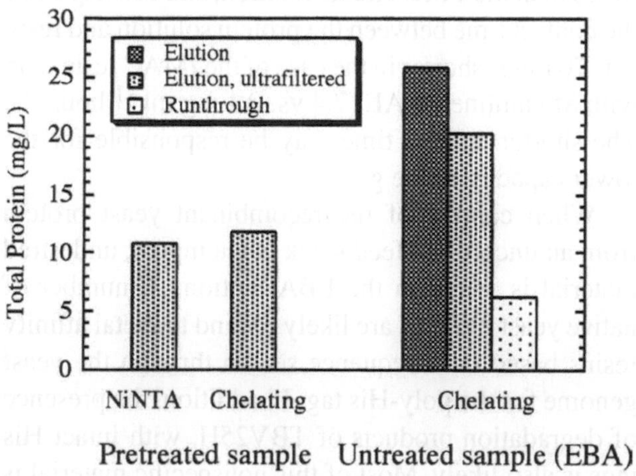

Fig. 6. Effect of sample pretreatment on TBV25H capture.

ble levels of contaminants. The protein present was confirmed to be TBV25H by sequencing, immunoactivity and bioactivity studies (Kaslow and Shiloach, 1994). This approach was therefore used to determine the yield of TBV25H in the yeast fermentation and various expanded bed fractions.

Adsorption of TBV25H to Streamline Chelating

The product obtained from loading 1 *l* of untreated culture broth onto Streamline Chelating was analyzed using SDS–PAGE and gel filtration. One litre of culture was also processed and loaded onto NiNTA agarose as a control. SDS–PAGE (Fig. 3) indicated that both resins had captured TBV25H. However, the total protein content of the EBA elution was higher than that of NiNTA agarose (36 vs 12 mg). On ultrafiltration of the EBA elution, a significant drop in total protein retained was observed (19 mg), indicating the presence of contaminants. This was clearly illustrated by gel filtration of the EBA elution prior to, and after, ultrafiltration, and in comparison to a NiNTA elution used as a control (Fig. 4). The first three peaks on the gel filtration chromatograms represent immunogenic protein. The broad peak which follows represents the contaminants, which presumably are degradation products or other impurities with affinity to IMAC resins.

Table 2. Binding of TBV25H as a function of loading ratio.

Sample volume (l)	EBA elution (mg)	Ultrafiltered elution (mg)	Loss in run through (mg)
1	38	22	6
2	42	28	15
4	48	38	30

Binding as a function of loading ratio

In order to determine conditions for EBA operation, the effect of increasing the loading ratio (sample volume to resin volume) on the binding of protein to Streamline Chelating was investigated. Sixty ml of resin were contacted with 1, 2 and 4 l of culture (Table 2). The amount of protein bound to the resin did not increase in proportion to the loading ratio (38, 42 and 48 mg for 1, 2 and 4 l of sample, respectively). Therefore, the binding capacity of the resin for total protein was approximately 1 mg per ml resin. The elutions were ultrafiltered and the retentates analyzed for total protein (22, 28 and 38 mg).

The fraction of protein retained after ultrafiltration increased with the loading ratio, indicating that contaminants were being steadily displaced by TBV25H as larger volumes of sample were passed through the column. Analysis of the runthrough (using NiNTA agarose-based purification, and SDS–PAGE) indicated the presence of unbound TBV25H (6, 15 and 30 mg). The amount of material not captured was in direct proportion to the loading ratio.

To determine the specific binding capacity of Streamline Chelating for TBV25H, a breakthrough experiment was performed. 1.5 l of culture was diluted 1:1 with water and loaded onto a 2.5-cm column with 60 ml of resin. One hundred and fifty-ml fractions of the runthrough were collected. These fractions were analyzed for the reappearance of the TBV25H band using SDS–PAGE (Fig. 5). TBV25H was observed to reappear within 300 ml of the runthrough and was present thereafter. To achieve total capture, the specific capacity of Streamline Chelating to TBV25H in untreated culture broth was approximately 0.05 mg TBV25H per ml resin. Forty-six mg of total protein were present in the elution, of which 84% was retained upon ultrafiltration.

Effect of supernatant pretreatment on TBV25H binding to Streamline Chelating

When pretreated supernatant was contacted with NiNTA agarose and Streamline Chelating in batch mode, a comparable amount of protein was eluted from both resins (11–12 mg l^{-1} sample, Fig. 6). Expanded bed operation using untreated material resulted in a higher amount of total protein in the elution (26 mg l^{-1}), and, as before, unbound TBV25H was detected in the runthrough. Clearly, an ultrafiltration step before metal affinity purification is responsible for the removal of contaminants which would interfere with TBV25H binding in untreated supernatant.

The presence of biomass as a possible cause for interference to TBV25H binding to the affinity resin was ruled out by an expanded bed experiment performed with and without biomass present in the sample. The extent and nature of proteins (TBV25H and total protein) bound to the resin was found to be the same in both cases (results not shown).

Discussion

Expanded bed adsorption is a simple method for capturing proteins directly from cell extracts or from culture supernatant. Preliminary steps such as centrifugation, microfiltration or diafiltration are usually not required for EBA operation. This system has been successful in the capture of proteins onto ion exchangers, and onto an affinity matrix containing protein A. In the present work, immobilized metal affinity chromatography (IMAC) is combined with expanded bed technology. Two protein preparations were analyzed: an intracellular protein produced in *E. coli*, and an extracellular protein produced by *S. cerevisiae*. Both proteins were expressed with a histidine tag, enabling binding to a matrix containing metals such as nickel or copper.

The binding capacity of the diphtheria toxin to Streamline Chelating (0.4 mg toxin per ml resin) is comparable to the capacity of Streamline DEAE for a similar CRM toxin secreted extracellularly by *C. diphtheriae* (Shiloach and Kaufman, 1998). As His-rich molecules are selectively expected to bind to the nickel charged resin, higher recovery was anticipated for the IMAC process; however, this was not observed. The capture strategy for the extracellular CRM toxin is different from that utilized for the intracellular protein CRM H21G. After dilution of the feedstock, the concentration of the extracellular CRM toxin is around 4 $\mu g\ ml^{-1}$, and 150 ml of diluted bacterial culture (5 mg cells wet weight per ml resin) is contacted with 1 ml of resin. With CRM H21G, only 2 ml of cell extract containing 32.5 mg (wet cell weight) per ml were passed through 1 ml of resin, with the CRM concentration in the feedstock estimated to be 0.5 mg ml^{-1}. Clearly, the intracellular protein preparation was more concentrated than the extracellular solution, and consequently the contact time between the protein solution and resin was 70 times shorter in the case of the IMAC resin than with Streamline DEAE (7.4 vs. 0.1 mg ml^{-1} $hour^{-1}$). The shorter contact time may be responsible for the lower capacity of the gel.

When capture of the recombinant yeast protein from an unclarified feed stock is attempted, undesired material is found in the EBA elution. A number of native yeast proteins are likely to bind to metal affinity resins based on a sequence search through the yeast genome for the poly-His tag. In addition, the presence of degradation products of TBV25H, with intact His tags is also likely. Most of this nonspecific material is smaller than 10000 MWCO and is removed by ultrafiltration. The presence of this material in the feed stock could possibly prevent the intact protein from binding to the metal affinity matrix. The fraction of protein in an expanded bed elution retained after ultrafiltration increases with loading ratio, indicating that TBV25H displaces other material bound to the resin. Total capture of TBV25H is observed only when a very small volume of sample is contacted with the resin (2.5 ml untreated sample per ml resin).

A significant observation for both protein preparations is the presence of unbound material in the runthrough. In the case of TBV25H, this material is lost at a constant rate during the course of sample loading. Attempts to reduce this loss by increasing the contact time between the resin and sample by either recirculation (up to four times) or dilution of the feed stock (up to 8-fold) did not succeed (results not shown).

When nonspecific material is removed by pretreatment of the protein preparation, Streamline Chelating and NiNTA agarose perform similarly, presumably because of reduced interference to binding of TBV25H. The estimated loading ratio for capture of TBV25H (0.05 mg per ml resin) is 20-fold lower than the capacity of the resin for total protein in expanded bed operation.

Adsorption onto Streamline Chelating was found to be not very specific. While it was possible to capture most of the diphtheria toxin from cell extracts using the expanded bed approach, the extracellular malaria vaccine candidate (TBV25H) could not be successfully captured without preliminary removal of certain contaminants from the culture supernatant. The operation of an expanded bed involves flow rates (in the range of 300–400 cm h^{-1}) which are significantly higher than those used with packed beds. This could restrict the time available for affinity binding of specific proteins. In addition, since the metal is weakly bound to the IDA backbone, this could also result in some of the metal leaching off the column. When a purified CRM H21G preparation was contacted with three packed columns in series using gravity flow, the fraction of protein present in each runthrough to that bound to the resin was observed to be constant (results not shown). This would indicate that protein ligand binding kinetics are limiting. The presence of other molecules in the sample which could competitively bind to the metal would further reduce the extent of binding of the desired protein. Further studies on the binding kinetics of 6×His-tagged proteins in EBA processes are needed.

References

Carroll SF, Barbieri JT & Collier RJ (1988) Diphtheria toxin: purification and properties. Methods Enzymol. 165: 68–76.

Hjorth R (1997) Expanded bed adsorption in industrial bioprocessing: recent developments. Trends Biotechnol. 15: 230.

Hochuli E, Bannwarth W, Döbeli H, Gentz R & Stüber D (1988) Genetic approach to facilitate purification of recombinant proteins with a novel metal chelate adsorbent. Bio/Technology 6: 1321–1325.

Johnson VG & Nicholls P (1994) Histidine 21 does not play a major role in Diphtheria toxin catalysis. J. Biol. Chem. 269: 4349–4354.

Kaslow DC & Shiloach J (1994) Production, purification and immunogenicity of a malaria transmission-blocking vaccine candidate: TBV25H expressed in yeast and purified Using nickel-NTA agarose. Bio/Technology 12: 494–499.

Neville DM, Scharff J, Hu HZ, Rigaut K, Shiloach J, Slingerland W & Jonker M (1996) A new reagent for the induction of T-cell depletion, anti-CD3-CRM9. J. Immunotherapy 19: 85–92.

Noronha SB, Kaslow DC & Shiloach J (1998) Transition phase in the production of recombinant proteins in yeast under the *ADH2* promoter: An important step for reproducible manufacturing of a malaria transmission blocking vaccine candidate. J. Ind. Microbiol. Biotechnol., 20: 192–199.

Porath J, Carlsson J, Olsson I & Belfrage G (1975) Metal chelate affinity chromatography, a new approach to protein fractionation. Nature 258: 598–599.

Shelly MA, Jacoby H, Riley GJ, Graves BT, Pichichero M & Treanor JJ (1997) Comparison of pneumococcal polysaccharide and CRM_{197}-conjugated pneumococcal oligosaccharide vaccines in young and elderly adults. Infect. Immun. 65: 242–247.

Shiloach J & Kaufman J (1998) The combined use of expanded bed adsorption and gradient elution for capture and partial purification of mutant diphtheria toxin (CRM9) from *Corynebacteria diphtheriae*. Sep. Sci. Technol., in press.

Bioseparation **8:** 153–158, 1999.
© 1999 *Kluwer Academic Publishers.*

A simplified purification procedure of α-lactalbumin from milk using Ca^{2+}-dependent adsorption in hydrophobic expanded bed chromatography

Wim Noppe, Petra Haezebrouck, Ignace Hanssens & Marcel De Cuyper
Interdisciplinary Research Centre, Katholieke Universiteit Leuven, Campus Kortrijk, B-8500 Kortrijk, Belgium

Received 14 August 1998; accepted in revised form 9 December 1998

Key words: expanded bed adsorption, α-lactalbumin, milk proteins

Abstract

The technique of expanded bed adsorption is originally designed for a direct recovery of proteins from fermentor feedstocks. In this article we describe the use of expanded bed adsorption for the recovery of α-lactalbumins from defatted milk using the hydrophobic Streamline Phenyl gel. α-Lactalbumins are Ca^{2+}- binding proteins. Upon Ca^{2+} removal, they undergo a significant conformation change rendering them more hydrophobic. Based on this unique property we develop a protocol for fast and efficient purification of α-lactalbumin from milk. The use of this technique results in a reduction of the number of chromatographic purification steps.

Abbreviations: BLA – bovine α-lactalbumin; CIP – cleaning in place; EBA – expanded bed adsorption; GLA – goat α-lactalbumin

Introduction

α-Lactalbumin is a small 14.4 kDa protein, produced in the lactating mammary gland. It plays an important role in the lactose synthesis. Combined with galactosyltransferase it catalyses the reaction of UDP-galactose and glucose to form lactose (Holpert et al., 1990). α-Lactalbumin has also been found to have strong calcium binding ability (Segawa et al., 1983; Van Ceunebroeck et al., 1985; Desmet et al., 1987). At room temperature and in the absence of Ca^{2+}, the tertiary structure of α-lactalbumin is partially unfolded by which hydrophobic clusters from the protein interior are accessible for the external medium. In the presence of Ca^{2+}, the native protein state is regained and the hydrophobic interior is screened. For our studies on the partially folded protein state (Vanderheeren et al., 1996; Haezebrouck et al., 1998) we are searching for an efficient, simple and fast purification method. Several techniques have been reported for purification of α-lactalbumin from different species. The purification set-ups mostly consist of combined classical chromatographic techniques such as ion exchange and size exclusion chromatography. (Aschaffenburg et al., 1956; Conti et al., 1989; Shaw et al.,1993; Stevenson et al., 1994). More specific techniques make use of the strong binding affinity of α-lactalbumin for metal ions and of the metal-ion-dependent interaction of the protein with hydrophobic agents. Such specific techniques are Ca^{2+} dependent hydrophobic interaction chromatography (Lindahl et al., 1984), immobilized metal affinity chromatography (Blomkalns et al., 1997), membrane adsorber chromatography (Splitt et al., 1996). All these techniques, however, are applied for the purification of α-lactalbumin from whey and include casein precipitation in advance. Formerly, by means of expanded bed chromatography, we purified lysozyme from equine milk (Noppe et al., 1996). In this paper we report the purification of α-lactalbumin from cow and goat milk by the use of expanded bed chromatography based on Ca^{2+}-dependent hydrophobic interactions.

Materials and methods

Materials

All materials and equipment are from Amersham Pharmacia Biotech (Uppsala, Sweden) unless otherwise stated. The experiments are performed at room temperature (20–22 °C).

BLA in the binding capacity experiments is from Sigma (St. Louis, USA). The protein is decalcified on a Sephadex G-25 column in 10 mM HCl , freezed dried and stored at −20 °C until further use. GLA used in these experiments is prepared from fresh goat milk by the classical procedure. It is decalcified and stored as described for BLA.

Serva W tablets (Serva, Heidelberg, Germany) contains 100 mg Serva Blue W pro tablet. Phastgel Blue R tablets contain 675 mg Coomassie Blue R-350 pro tablet.

Methods

Expanded bed adsorption set-up

Scouting experiments are performed on a XK16/20 column filled with 25 ml of Streamline Phenyl gel (protoype gel lot nr LJB55094 97-05-06LS). The column is fed by a P-1 peristaltic pump.

Expanded bed experiments are performed on a Streamline C-50 column filled with 300 ml Streamline Phenyl gel (prototype gel lot nr. LJB55094 97-05-06LS). In these experiments the flow- and gel bed configurations are processed by two Masterflex peristaltic pumps, equipped with 7017-20 pumpheads and Pharmed tubing (Cole Palmer, USA).

A single path UV-1 dectector (filter kit 280 nm) equipped with an industrial flow cell (6 mm internal diameter, 5 mm optical length) is connected to the Streamline C-50 system. A REC2 recorder registers the flow through. The complete system is operated manually.

Expanded bed adsorption chromatography

The Streamline C-50 column is filled with 300 ml Streamline Phenyl to a bed height of approximately 15 cm. The movable adaptor is set at 50 cm height. The column is equilibrated in upward mode at 200 cm/h with a buffer containing 50 mM Tris–1 mM EDTA which is adjusted to pH 7.5 with HCl (buffer A). Column expansion is about 1.9.

The sample of skim milk is mixed with water and 500 mM Tris–250 mM EDTA buffer to reach a final concentration of 50 mM Tris–25 mM EDTA. The pH is adjusted to 7.5 with NH_4OH.

Loading of the diluted sample is done also in upward mode at 200 cm/h. The column finally expands about 2.1 times due to sample viscosity. After loading, the column is washed in upward mode at 200 cm/h with buffer A. In the meantime the adaptor is lowered to about 2 cm above the expanded bed surface. When pure eluent flows through the column and the UV signal reaches its baseline, the pump is stopped and the gel is allowed to settle. The adaptor is lowered to the gel surface and elution of the bound proteins is started. The elution is done downward in packed mode at 100 cm/h. The elution buffer contains 50 mM Tris and 25 mM Ca^{2+} at pH 7.5. (Buffer B). The eluted protein fraction is collected for further polishing steps. CIP is done by consecutive washing steps with water and 0.5 M NaOH–1 M NaCl. After CIP the column is re-equilibrated with buffer A.

Size Exclusion Chromatography

Size exclusion chromatography is performed on a Superdex 75HR XK16/60 column for polishing if necessary. The column is connected to a FLPC system. The column is equilibrated and run in 50 mM Tris-HCl–0.2 M NaCl–0.02% NaN_3,, pH 8.0 at a flow rate of 1 ml/min.

Purity Controls

Electrophoresis (gradient SDS-PAGE) is performed on ExcelGel SDS gradient 8–18%. LMW Kit (molecular weight range 14–94 kDa) is used for molecular weight markers. Isoelectric focusing is performed on Servalyt Precotes® 3-10 (Serva, Heidelberg, Germany). Serva protein test mixture 3-10 (Serva, Heidelberg, Germany) is used for pI markers.

Phast Blue R tablets and Serva Blue R tablets are used to visualize the protein bands of the electrophoresis and isoelectric focusing, respectively. Phastgel Blue T tablets contain 675 mg Coomassie Blue R-350 pro tablet. For staining one tablet is dissolved in 400 ml water. Serva Blue tablets contain 100 mg Serva Blue W pro tablet. For staining four tablets are dissolved in 250 ml 0.4% trichloroacetic acid in water.

Circular dichroism spectroscopy

Circular dichroism spectra are measured on a Jasco J-600 spectropolarimeter (Jasco, Japan). The spectra are taken in the near-UV range (250–320 nm) with a 10 mm quartz cuvet. The different spectra are obtained from α-lactalbumins dissolved in 10 mM Tris–2 mM Ca^{2+}, pH 7.5 at a protein concentration of 0.4 mg/ml.

Results and discussion

Binding capacity of Streamline Phenyl gel for α-Lactalbumin

At first we will determine the binding capacity of the specific gel lot that will be used for further purification. Therefore apo-BLA or apo-GLA is dissolved in 50 mM Tris–25 mM EDTA, pH 7.5 at a concentration of 1 mg/ml. The protein solution is loaded on the scouting column at 200 cm/h until a breakthrough of α-lactalbumin is observed. The column is extensively washed with buffer A to remove unbound protein. When the baseline is reached, the adsorbed α-lactalbumin is eluted with buffer B. The protein fraction is collected and the amount of α-lactalbumin is calculated from the volume of the protein fraction and its light absorption at 280 nm. The results indicate that the column capacity for both α-lactalbumins is about 0.7 mg/ml. In the next purification set-up we intend to obtain a maximal output of α-lactalbumin at a limited amount of gel, therefore it will be arranged that the volume of milk used for the protein recovery is sufficiently large to saturate the column.

Purification of α-lactalbumin from milk

Defatting

In order to prevent column saturation due to fat globules present in the milk, defatting is most necessary. Fresh milk, from different origin, is defatted by centrifugation at 7500 *g* during 10 minutes at 4 °C. The fat layer is removed and the supernant is filtered through four layers of cheese cloth to remove remaining fat particles. The defatted milk is stored in the freezer till further use.

Expanded bed chromatography

To avoid the problems encountered in packed bed with casein containing milk samples, we isolate BLA and GLA from milk by expanded bed chromatography. This technique, originally developed to purify proteins from crude fermentor feedstocks, is used for the purification of α-lactalbumin from milk. Column clogging by coagulation of casein micelles is no longer a problem in the expanded mode. The casein micelles flow through the expanded gel bed without any clogging of the gel. To 500 ml of defatted milk, 200 ml of 500 mM Tris–250 mM EDTA and 1300 ml of water are added. The final buffer concentration in the diluted milk is 50 mM Tris and 25 mM EDTA. The pH of the diluted milk is adjusted to 7.5 with dilute NH_4OH. The high amount of EDTA in the sample buffer is necessary to complex the Ca^{2+} present in the milk (Lindahl et al., 1984).

The diluted milk sample is then applied on the Streamline Phenyl column, previously equilibrated with buffer A, at 200 cm/h in upward flow. All non-binding proteins, including the casein micelles, flow through. In contrast, α-lactalbumin exposes highly apolar surfaces in the absence of Ca^{2+} and binds on the hydrophobic gel. The other milk proteins do not show this property and therefore do not have affinity for the hydrophobic gel. After extensively washing with buffer A, the bound α-lactalbumin fraction is eluted from the column with 50 mM Tris–25 mM Ca^{2+} buffer at pH 7.5, downward in packed mode at 100 cm/h. The elution pattern is shown in Figure 1. The α-lactalbumin fraction is dialyzed against water

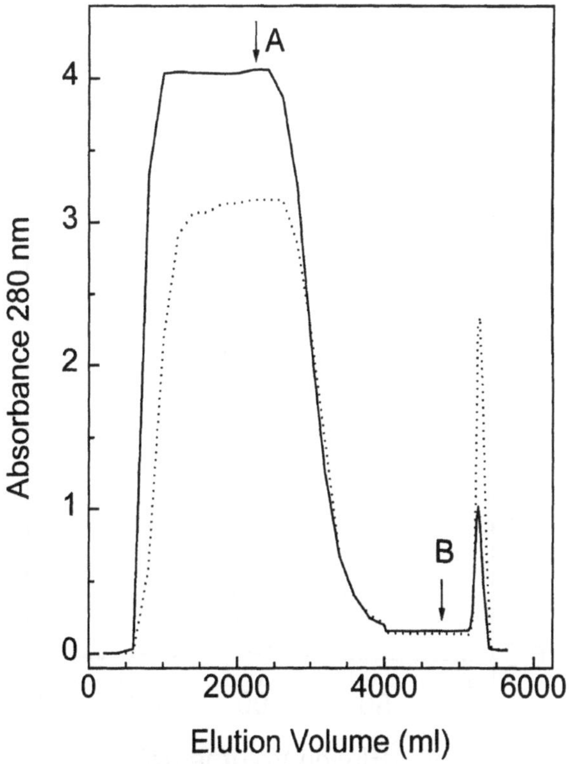

Figure 1. Elution profiles of Streamline Phenyl chromatography of cow milk (—) and goat milk (. . .). After application of the skimmed milk in 50 mM Tris–25 mM EDTA, the column is washed with 50 mM Tris–1 mM EDTA from point A and eluted with 50 mM Tris–25 mM Ca^{2+} from point B.

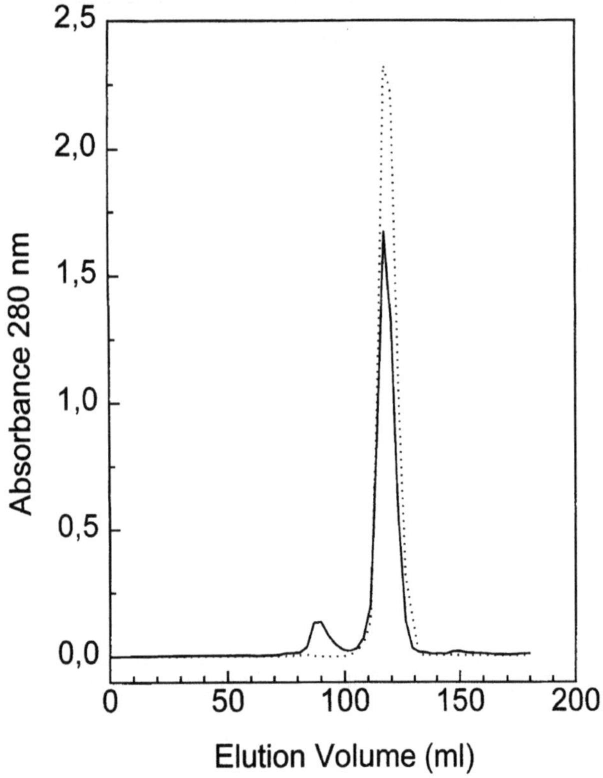

Figure 2. Elution profiles of size exclusion chromatography on Superdex 75 HR of the α-lactalbumin fraction from Streamline phenyl. Cow milk (—) and goat milk (. . .). The column is run at 1 ml/min in 50 mM Tris 0.2 mM NaCl – 0.02% NaN_3, pH 8.5.

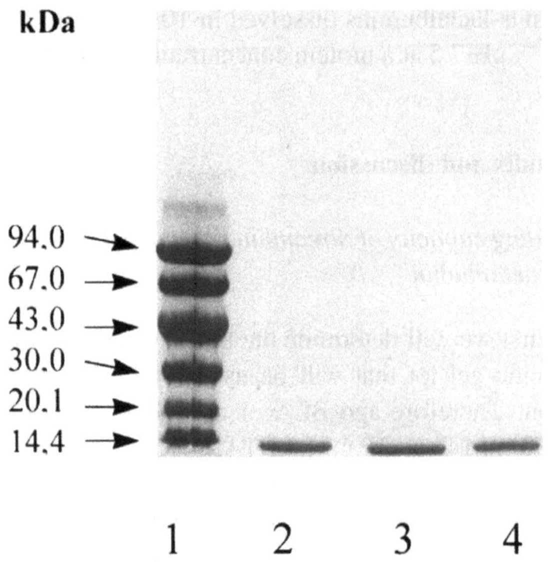

Figure 3. SDS-PAGE of α-lactalbumin fraction after size exclusion chromatography. SDS-PAGE is performed on ExcelGel SDS 8–18% gradient gel. Lane 1, Marker proteins; lane 2, Commercial BLA (Sigma, USA) lane 3, BLA fraction from milk; lane 4, GLA fraction from milk.

and freeze dried. The protein is stored in the freezer at −20 °C until further purification if necessary.

Size exclusion chromatography

If necessary, further polishing of the previous α-lactalbumin fraction can be performed on a Superdex 75 HR column. The column is equilibrated in 50 mM Tris–0.2 M NaCl–0.02% NaN_3, pH 8.0, at 1 ml/min. Crude α-lactalbumin (15–20 mg) is dissolved in the same buffer (3 ml) and is applied to the column. The elution pattern is shown in Figure 2. The purified α-lactalbumin fraction is dialyzed against water and freeze dried. As seen on Figure 2 it is not always necessary to run the polishing step. With goat milk we immediately obtained pure α-lactalbumin by Streamline Phenyl chromatography. BLA on the other hand needed the second purification step. By SDS-PAGE (data not shown) it has been observed that the impurity consists of a single protein with a molecular weight of 30–31 kDa. This refers to γ-casein (MW= 30.6 kDa). It is worth to notice that Quiles (1994) did not detect γ-casein in goat milk. Our result seems to confirm the lack of γ-casein in goat milk.

Purity controls

The purity of α-lactalbumin fractions after size exclusion chromatography is checked by SDS-PAGE (Figure 3) and isoelectric focusing (Figure 4). The result shows a pure protein with a molecular weight of 14.4 kDa and isoelectric points of approximately 5.30 for BLA and 5.40 for GLA. The small difference for the isoelectric points is due to a slightly different composition and distribution of acid and basic amino acids in both types of α-lactalbumin. Moreover, at neutral pH, the calculated net charge for BLA is −6 and for GLA −5, by which a lower isoelectric point for BLA is reasonable. The structural integrity of the proteins is checked by circular dichroism spectroscopy. The spectra of the Ca^{2+} forms are typical for the folded state of α-lactalbumin in the near UV range (250–320 nm) (Figure 5), indicating that a native protein is obtained.

Conclusions

Expanded bed chromatography was proven to be a successful technique for the purification of proteins from milk. The problems that caseins cause in packed bed chromatography were evaded by the use of the expanded bed technique. A comparison of the classical procedure and the new one (Figure 6) demonstrates that by using Streamline Phenyl gel we are able to avoid the casein precipitation step, simplifying the

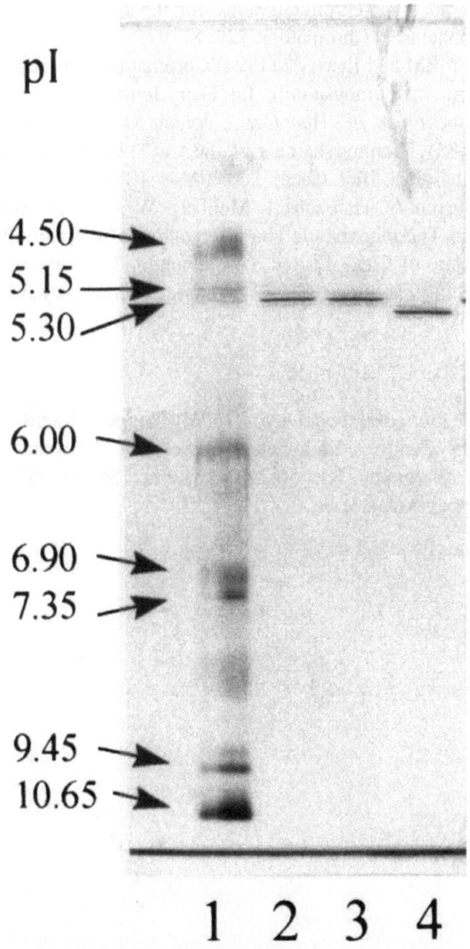

Figure 4. Isoelectric focusing of α-lactalbumin fraction after size exclusion chromatography. The isoelectric focusing is performed on Servalyt Precotes 3–10. Lane 1, pI markers; lane 2, Commercial BLA (Sigma); lane 3 , BLA fraction from milk; lane 4, GLA fraction from milk.

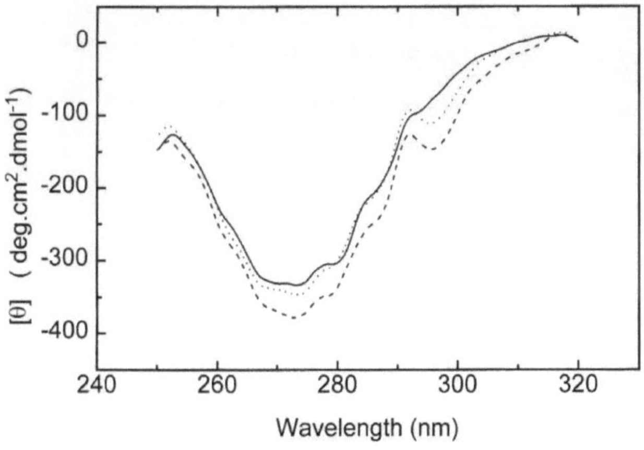

Figure 5. CD spectra of Ca^{2+}-α-lactalbumin in the near UV range 250–320 nm α-Lactalbumin is dissolved in 10 mM Tris-2 mM Ca^{2+} pH 7.5 at 0.4 mg/ml. Commercial BLA-Sigma (- - -); BLA from milk (—); GLA from milk (. . .).

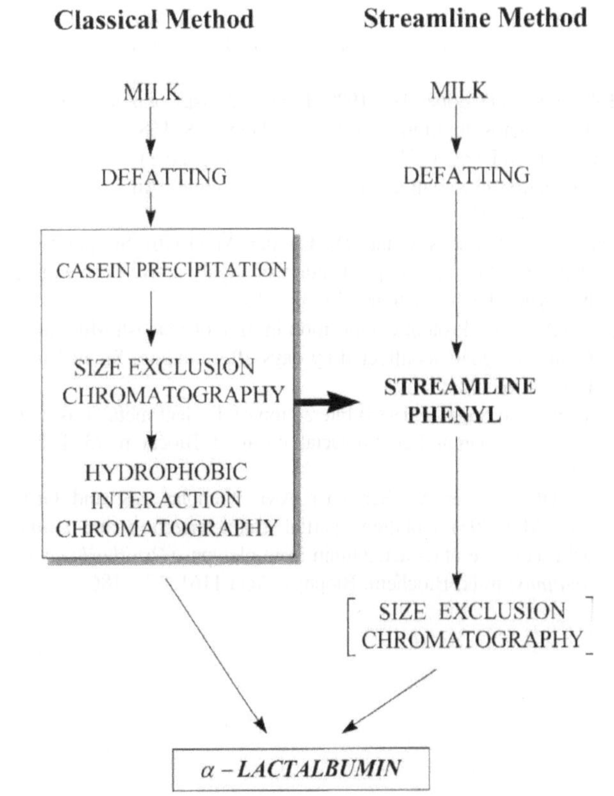

Figure 6. Comparison between classical and Streamline method for purification of α-lactalbumin from milk.

purification of milk α-lactalbumin. The protein is obtained with a high purity and it conserves its native conformation in the prescence of Ca^{2+}.

Acknowledgements

Dr François Geeraerts (Amersham Pharmacia Biotech Benelux) is greatly acknowledged for providing us with the Streamline Phenyl prototype gel.

References

Aschaffenburg R and Drewry J (1956) Improved method for the preparation of crystalline β-lactoglobulin and α-lactalbumin from cow's milk. Biochem. J. 65: 273–277.

Blomkalns AL and Gomez MR (1997) Purification of bovine α-lactalbumin by immobilized metal ion affinity chromatography. Prep. Biochem. & Biotechn. 27: 219–226

Conti A, Napolitano L, Lai P, Pinna W and Godovac-Zimmerman J (1989) Isolation of donkey whey proteins and N-terminal amino acid sequence of α-lactalbumins A and B ,β- lactoglobulins I and II and lysozyme. Milchwissenschaft 44: 138–141.

Desmet J, Hanssens I and Van Cauwelaert F (1987) Comparison of the binding of Na^+ and Ca^{2+} to bovine α-lactalbumin. Biochim. Biophys. Acta 912: 211–219.

Haezebrouck P, Noyelle K and Van Dael H (1998) Equilibrium and kinetic folding of pigeon lysozyme. Biochemistry 37: 6772–6780.

Holpert M and Cooper TG (1990) Improved assay of α-lactalbumin and galactosyltranferase. Anal. Biochem. 188: 168–175.

Lindahl L and Vogel HJ (1984) Metal-ion-dependent hydrophobic interaction chromatography of α-lactalbumins. Anal. Biochem. 140: 394–402.

Noppe W, Hanssens I and De Cuyper M (1996) Simple two-step procedure for the purification of highly active pure equine lysozyme. J. Chromatogr. 719: 327–331.

Quiles A (1994) Protein composition in milk of Spanish Murciano-Granadina goats as affected by days after kidding. Small Rum. Res. 14: 67–92.

Segawa T and Sugai S (1983) Interaction of divalent metal ions with bovine, human and goat α-lactalbumins. J. Biochem. 93: 1321–1328.

Shaw DC, Messer M, Scrivener AM, Nicholas KR and Griffiths M (1993) Isolation, partial characterization and amino acid sequence of α-lactalbumin from platypus (*Ornithorhynchus anatinus*) milk. Biochem. Biophys. Acta 1161: 177–186.

Splitt D, Mackenstedt I and Freitag R (1996) Preparative membrane adsorber chromatography for the isolation of cow milk components. J. Chromatogr. 729: 87–97.

Stevenson EM and Leaver J (1994) Chromatographic separation of the proteins of mouse milk. Int. Dairy Journal 4: 205–220.

Van Ceunebroeck JC, Hanssens I, Joniau M and Van Cauwelaert F (1985) Thermodynamics of the Ca^{2+} binding to bovine α-lactalbumin J. Biol. Chem. 260: 10944–10947.

Vanderheeren G, Hanssens I, Meijberg W and Van Aerschot A (1996) Thermodynamic characterization of the partially unfolded state of Ca^{2+} bovine α-lactalbumin: Evidence that partial unfolding can precede Ca^{2+} release. Biochemistry 35: 16753–16759.

Address for correspondence: W. Noppe, Interdisciplinary Research Centre, Katholieke Universiteit Leuven, Campus Kortrijk, Belgium. Tel: 56-246-418; Fax: 56-256-997; E-mail: wim.noppe@kulak.ac.be

Bioseparation **8:** 159–168, 1999.
© 1999 *Kluwer Academic Publishers.*

Expanded bed adsorption as a unique unit operation for the isolation of bacteriocins from fermentation media

Raf Callewaert & Luc De Vuyst
Division of Industrial Microbiology, Fermentation Technology and Downstream Processing (IMDO), Department of Biotechnology, Faculty of Sciences, Vrije Universiteit Brussel (VUB), Pleinlaan 2, B-1050 Brussels, Belgium

Received 7 July 1998; accepted in revised form 4 November 1998

Key words: amylovorin L471, bacteriocin, expanded bed adsorption, lactic acid bacteria, *Lactobacillus amylovorus*

Abstract

Expanded bed adsorption using a strong cation exchanger allowed the direct isolation of amylovorin L471, a bacteriocin from *Lactobacillus amylovorus* DCE 471, from the fermentation medium. The pH of the loading and elution buffer were optimised in a packed bed with cell-free culture supernatant. Bound bacteriocin was eluted with 1.0 M NaCl. The highest recovery (30%) was obtained at the lowest pH (3.6). At higher pH values the recovery was lower, namely 12%, 15% and 7% at pH 4.5, 6.5 and 8.0, respectively. In expanded bed mode, direct isolation of the bacteriocin from the fermentation medium at pH 3.6 (loading and elution) initially resulted in a recovery of 12%. After optimisation of the pH (loading and elution at pH 3.6 and 6.5, respectively), the recovery for amylovorin L471 increased up to 30% and higher. Recovery of enterocin A from *Enterococcus faecium* CTC 492 fermentation medium averaged 15% (loading and elution at pH 3.6 and 6.0, respectively). With pediocin, produced by *Pediococcus acidilactici* ATCC 8042, 26% recovery was obtained at a pH of 6.5 during loading and elution. Low recoveries can be ascribed to non-optimal operation conditions (pH of loading and elution buffer), inactivation of the bacteriocin on a cationic resin, and the formation of more insoluble and less active, strongly hydrophobic bacteriocin aggregates upon further purification.

Introduction

Bacteriocins are ribosomally synthesised, extracellularly released peptides, proteins or protein complexes with a bactericidal activity against closely related species (Tagg et al., 1976; Klaenhammer, 1988; De Vuyst and Vandamme, 1994; Jack et al., 1995; Nes et al., 1996). Large numbers of bacteriocins have been detected among lactic acid bacteria (Piard and Desmazeaud, 1992; Hoover and Steenson, 1993; De Vuyst and Vandamme, 1994; Dodd and Gasson, 1994; Jack et al., 1995). These bacteriocins can be divided into four major classes depending on their genetic and physicochemical properties (Klaenhammer, 1993). The first class (I) consists of small, heat-stable antimicrobial peptides, which contain the unusual amino acids lanthionine, methyl-lanthionine and dehydrated residues (Sahl et al., 1995; Konings and Hilbers, 1996; Jack et al., 1998). The second class (II) are lanthionine non-containing, small, strongly hydrophobic and heat-stable bacteriocins (Nes et al., 1996), which can further be divided into three classes: (IIa) pediocin-like or *Listeria*-active peptides; (IIb) bacteriocin complexes; (IIc) *sec*-dependent peptides. The third class (III) consists of large, hydrophilic and heat-labile proteins (Joerger and Klaenhammer, 1986; Joerger and Klaenhammer, 1990). The fourth class (IV) represents complex bacteriocins composed of a protein and other chemical moieties. However, the latter class of bacteriocins has not yet been characterised adequately at biochemical level.

Bacteriocins produced by lactic acid bacteria are of interest for potential application in the food industry as future food biopreservatives to improve food quality, naturalness and safety, thanks to their antimicrobial activity and their technologically favourable properties (Ray and Daeschel, 1992; De Vuyst and Vandamme, 1994; Stiles, 1996). Only nisin (Nisaplin) and pediocin (Alta) are produced commercially on a large scale; details about their production process and downstream processing are not known (De Vuyst and Vandamme, 1994; Anonymous, 1995).

Development of appropriate fermentation and isolation techniques is necessary for validation of research from new bacteriocins produced by lactic acid bacteria. Generally, bacteriocins are isolated from culture supernatant by salt precipitation and solvent extraction (De Vuyst et al., 1996a; Carolissen-MacKay et al., 1997). Recently, several authors reported quick and easy large-scale isolation methods by acid extraction of the bacteriocin from the producing cells (Yang et al., 1992; Daba et al., 1994; Lyon et al., 1995; Elegado et al., 1997; Tahara and Kanatani, 1997). Batch adsorption to an appropriate adsorbent, based on hydrophobic or electrostatic interactions, was also used to isolate different bacteriocins from culture supernatant (Stoffels et al., 1993; Coventry et al., 1996). Mostly the adsorbent was added batch-wise to the cell-free culture supernatant. Bacteriocins were then isolated by centrifugation and dissolving in an appropriate elution buffer. This paper deals with a new bacteriocin isolation method through expanded bed adsorption. It is a unique unit operation in that it does not require centrifugation to remove cells from the fermentation broth or cell debris upon further clarification, and no further concentration is required, allowing process integration of fermentation with the first steps of downstream processing (Chase, 1994; Hjorth, 1997). Protein isolation or purification by the use of expanded bed adsorption has been largely studied and already proved to result in high recoveries, activities and fold purification (Hansson et al., 1994; Hjorth et al., 1995; Johansson et al., 1996; Noppe et al., 1996).

Materials and methods

Bacterial strains and media

The bacteriocin-producing strains and the bacteriocin-sensitive indicator organisms used for detection of bacteriocin activity are listed in Table 1. Stock cultures stored at −80 °C in cultivation broth (cf. infra) plus 25% (v/v) glycerol, were propagated twice in the appropriate broth prior to inoculum preparation for experiments. All lactic acid bacteria were grown at 37 °C in MRS broth (de Man et al., 1960) adjusted to pH 6.5. The growth medium used for *Listeria* was composed of (g l^{-1}) tryptone (Oxoid), 15.0; bacteriological peptone (Oxoid), 2.5; Lab Lemco (Oxoid), 2.5; NaCl, 5.0; yeast extract (Merck), 6.0; and adjusted to pH 7.0. Agar media were prepared by addition of 1.5% granulated agar (Oxoid) to MRS broth or 0.7% granulated agar for overlay assays.

Bacteriocin production

Bacteriocin-producing strains were cultivated in a 6.4 l Biostat®E fermentor (B. Braun Biotech International, Melsungen, Germany) containing 5.5 l of MRS broth adjusted to pH 6.5. The bioreactor was sterilised *in situ* at 121 °C. The carbon source, glucose, was sterilised separately. The fermentor was inoculated with 50 ml of an exponentially growing culture obtained by propagating a fresh culture twice at 37 °C during 12 h. The fermentations were run at 37 °C, without aeration and with slow agitation (50 rpm) to keep the fermentation broth homogenous. Crude culture supernatant was harvested after 10 h of fermentation.

Bacteriocin isolation by expanded bed adsorption

Large-scale bacteriocin isolation was done via expanded bed adsorption (Pharmacia™, Uppsala, Sweden) with a strong cation exchanger (500 ml; Streamline SP™, Pharmacia) consisting of sulphopropyl functional groups bound on a Sepharose matrix as adsorbent (100–300 μm particles). A column with a diameter of 5 cm and a height of 1 m was connected to a peristaltic pump (Watson Marlow 505U, Falmouth, Cornwall, England). Protein elution was monitored at 280 nm using an UV-1 detector (Pharmacia) connected to a REC 101 recorder (Pharmacia). A 50 mM sodium phosphate buffer (pH 6.5 and 8.0) or a 100 mM sodium acetate buffer (pH 3.6 and 4.5) was used for equilibration of the adsorbent. The adsorbent was expanded and equilibrated by a stepwise increase of an upward flow rate from 2 l h^{-1} to 4 l h^{-1} and finally 5 l h^{-1}. When the expanded bed was stable, the adsorbent bed was expanded 2.5 times from an initial height of 28.5 cm. Subsequently, 5.0 l crude fermentation medium (still containing the cells, approximately 1.6 g cell dry mass l^{-1}) was applied with an upward flow rate of 5 l h^{-1}. Broth components passed through the column while protein molecules were captured on the adsorbent. The adsorbent bed rose during loading and was finally nearly 3.0 times expanded. The adsorbent was washed with the same buffer in expanded bed mode with an upward flow rate of 5 l h^{-1}. After the washing step, the bed was packed again. Subsequently

Table 1. Strains used in this study

Bacteriocin-producing strain	Bacteriocin	Bacteriocin-sensitive indicator strain	Reference
Lactobacillus amylovorus DCE 471	amylovorin L471	*Lactobacillus delbrueckii* subsp. *bulgaricus* LMG 6901^T	De Vuyst et al. (1996a)
Enterococcus faecium CCM 4231	enterocin 4231	*Listeria innocua* RZS	Lauková et al. (1993)
Enterococcus faecium CTC 492	enterocin A	*Listeria innocua* CTC 1014	Aymerich et al. (1996)
Enterococcus faecium RZS C13	enterocin C13	*Listeria innocua* RZS	Vlaemynck et al. (1994)
Lactococcus lactis subsp. *lactis* NIZO 22186	nisin A	*Lactococcus* sp.	De Vuyst and Vandamme (1992)
Pediococcus sp. RZS C9	pediocin C9	*Pediococcus acidilactici* ATCC 8042	De Vuyst et al. (1996a)

the flow rate was reversed and lowered to 2 l h^{-1}. An upper adapter was moved down to the top of the adsorbent. Finally, the captured bacteriocin was eluted in packed bed mode with 0.5 or 1.0 M NaCl added to the appropriate elution buffer.

The pH of elution and loading buffer was optimised in packed bed mode with 1.5 l cell-free culture supernatant adjusted to different pH values (pH 3.6, 4.5, 6.5 or 8.0) and using the appropriate buffers mentioned above. A downward flow rate of 2 l h^{-1} was applied for loading, washing and elution.

Quantitative determination of bacteriocin activity

Bacteriocin activity, free in the liquid, was determined by an adaptation of the critical dilution method currently used for the assay of bacteriocins (De Vuyst et al., 1996a). Briefly, serial twofold dilutions of filter-sterilised cell-free culture supernatant containing bacteriocin were spotted (10 μl) onto agar plates containing fresh indicator lawns of a sensitive indicator strain. These lawns were prepared by propagating fresh cultures to an optical density at 600 nm of 0.45 and adding 100 μl of the cell suspension to 3.5 ml of overlay agar. Overlaid agar plates were incubated for at least 24 h. The bacteriocin activity was defined as the reciprocal of the highest dilution which showed complete inhibition of the indicator lawn and was expressed in activity units (AU) per millilitre of culture medium.

Protein determination

Protein concentrations were determined using a Protein Assay Kit from BioRad Laboratories (California, USA).

Stability of amylovorin L471 in cell-free solution and adsorption to the cell surface of the bacteriocin-producing cells as function of pH

The influence of pH on the stability of amylovorin L471 in cell-free solution was checked by addition of partially purified bacteriocin to different citric acid – disodium hydrogen phosphate buffers (100 mM) with pH values ranging from 2.2, 3.0, 4.0, 5.0, 6.0, 7.0 to 8.0. Partially purified amylovorin L471 was obtained via ammonium sulphate precipitation and chloroform/methanol extraction (De Vuyst et al., 1996a). To 3 ml of buffer, 2 ml of partially purified bacteriocin solution were added. Bioactivity was checked after 0, 3, 6, 12 and 24 h of incubation at room temperature.

The adsorption of the bacteriocin to the cell surface was investigated too in the pH range mentioned before. Cells of *Lb. amylovorus* DCE 471 were grown in MRS broth for 12 h. The cells were separated from the culture medium by centrifugation (10 min, 20,000 *g*) and washed twice with 0.85% NaCl. Cells originating from 10 ml medium were suspended in 5 ml buffer of the different pH values. The tubes were shaken (400 rpm) at room temperature. Samples were withdrawn after 3, 12 and 24 h of incubation. Cells were removed by microcentrifugation (10 min, 13,000 *g*) before bioactivity was measured.

Dialysis and lyophilisation

The most active fractions eluted from the expanded bed were subjected to dialysis followed by lyophilisation. Dialysis was done with a MMCO membrane of 3000 Da (Spectra®/Por, Spectrum, Houston, Texas). The retentate was finally subjected to lyophilisation (Secfroid, Lausanne, Switzerland).

Results and discussion

Amylovorin L471 isolation via expanded bed adsorption

Expanded bed adsorption is a promising, large-scale method for protein isolation and purification directly from crude culture broth. Since it eliminates a number of other unit operations such as removal of the cells by centrifugation and concentration of the molecules through ammonium sulphate precipitation, it has potential to be low cost too. This method was used for bacteriocin isolation from lactic acid bacteria bacteriocin-containing culture broths. Amylovorin L471 is a bacteriocin produced by *Lactobacillus amylovorus* DCE 471. This strain was isolated from fresh corn steep liquor (De Vuyst et al., 1996a). Amylovorin L471 was characterised as a small, heat-stable and extremely hydrophobic bacteriocin with a narrow inhibitory spectrum, including *Lactobacillus delbrueckii* subsp. *bulgaricus*, *Lb. delbrueckii* subsp. *delbrueckii* and *Lb. helveticus*. It belongs to the bacteriocin class II (De Vuyst et al., 1996a). Its production is growth-associated and is maximal under controlled conditions of temperature and pH at 37 °C and pH 6.5, respectively (De Vuyst et al., 1996b). *Lb. amylovorus* DCE 471 was grown in 5.5 litres of MRS broth at 37 °C. The pH decreased from 6.5 to 4.2 within 10 h of fermentation. A maximal bacteriocin activity of 3200 AU ml^{-1} was obtained. The cell density at this time was 1.6 g cell dry mass l^{-1}. The pH of crude fermentation broth (5.0 l) was readjusted to 6.5 with 10 N NaOH. The bacteriocin was isolated directly from the crude fermentation medium via expanded bed adsorption (De Vuyst et al., 1996a). Therefore, the adsorbent was expanded and equilibrated with a 50 mM sodium phosphate buffer, pH 6.5. Five litres crude fermentation medium were pumped through the column. Subsequently sodium phosphate buffer was used to rinse the adsorbent. Proteins were eluted from the adsorbent in packed bed mode with a reversed flow of firstly 0.5 M NaCl and secondly 1.0 M NaCl in 50 mM sodium phosphate buffer. In total 19.5% of the total bacteriocin activity present in the crude medium was recovered during elution with 0.5 M NaCl. Further elution with 1.0 M NaCl did not result in a higher recovery. During loading of the adsorbent 80.5% bacteriocin activity was lost in the flow through the column. Consequently, isolation of amylovorin L471 from the crude fermentation broth of *Lb. amylovorus* DCE 471 through expanded bed adsorption using a strong cation exchanger resulted in a very low recovery. Optimisation of the pH was necessary to improve bacteriocin isolation.

Optimisation of the pH in packed bed mode for the isolation of amylovorin L471

The pH of loading and elution was optimised in packed bed mode with pH-adjusted cell-free culture supernatant of *Lb. amylovorus* DCE 471. The pH of 1.5 l cell-free supernatant was adjusted to 3.6, 4.5, 6.5 or 8.0, respectively. The activity of the supernatant (1600 AU ml^{-1}) was independent of the pH. Each time the adsorbent was equilibrated and washed with loading buffer of the corresponding pH. During loading at low pH (3.6), no amylovorin L471 could be detected in the effluent whereas at higher pH values for loading (4.5, 6.5 and 8.0), increasing amounts of bacteriocin were found in the effluent (Figure 1). At low pH values the net charge of amylovorin L471 was more positive than at high pH values resulting in better adsorption during loading. At pH 8.0 nearly all bacteriocin activity (94.1%) was present in the effluent. Lower affinity of amylovorin L471 for the adsorbent at higher pH (8.0) is due to a less positive or even negative charge at this pH. The pI-value of amylovorin L471 can be estimated around pH 8.0 (unpubl.). High pI values are characteristic for most bacteriocins produced by lactic acid bacteria (Jack et al., 1995). At a low loading and corresponding elution pH, a higher total bacteriocin activity was recovered after elution with 1.0 M NaCl. The recoveries were 29.9%, 12.3%, 15.4% and 6.9% for respectively pH 3.6, 4.5, 6.5 and 8.0 (Figure 1). At pH 6.5 and 8.0 all bacteriocin bound to the adsorbent could be recovered by the elution with 1 M NaCl. In contrast, an apparent inactivation or loss of total bacteriocin activity of 70.1% and 65.3% was observed at pH 3.6 and 4.5, respectively. Rather low recoveries of bacteriocin eluting from the column at low pH can be due to the too high affinity of amylovorin L471 for the adsorbent due to its higher positive charge resulting in an improper elution. In addition, the strongly hydrophobic character of amylovorin L471 may lead to the formation of more insoluble and less active bacteriocin complexes ensuing in an apparent lower bacteriocin recovery (De Vuyst et al., 1996a). This aggregation may be dependent on pH and takes place especially in more concentrated solutions. Furthermore, inactivation of the bacteriocin on a cationic resin may occur as described for other bacteriocins (Hastings et al., 1991). In contrast to the higher bacteriocin

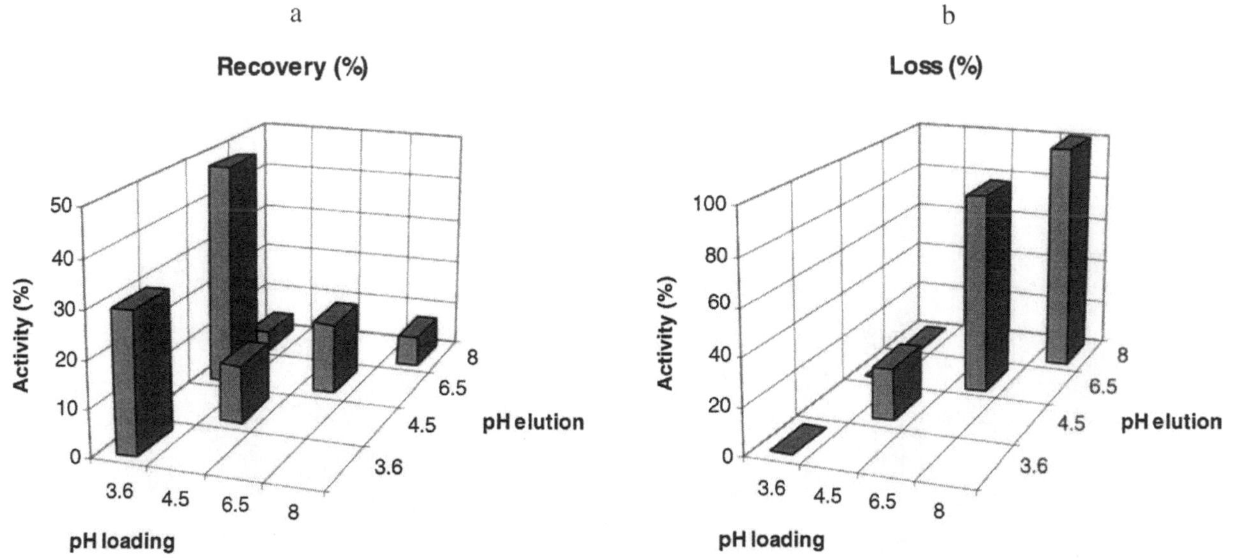

Figure 1. Optimisation of the recovery of amylovorin L471 via expanded bed adsorption in packed bed mode. Fermentation medium adjusted to different pH values (3.6, 4.5, 6.5, 8.0) was loaded; elution took place at these different pH values in the presence of 1.0 M NaCl. Recovery after elution (a) and loss during loading (b) are represented as percentage from the originally applied total activity (1600 AU ml^{-1}; 1.5 l).

Table 2. Specific bacteriocin activity of the eluted fractions after loading and elution at different pH values compared to the original cell-free culture supernatant (1600 AU ml^{-1})

Loading and eluting pH	Protein concentration mg ml^{-1}	Bacteriocin activity AU ml^{-1}	Specific bacteriocin activity AU mg^{-1}	Fold purification
Cell free supernatant	11.0	1600	145	1.0
pH 3.6	7.8	3200	410	2.8
pH 6.5	1.6	1600	1000	6.9
pH 8.0	0.8	800	1000	6.9

recovery, the specific activities of the eluting fractions at pH 3.6 were much lower than at pH 6.5 and 8.0 (Table 2). Loading at pH 3.6 increased the adsorption of all proteins. However, at higher pH-values, adsorption is more selective for the bacteriocin.

Loading at pH 3.6 resulted in a complete adsorption of amylovorin L471 on the adsorbent. The higher positive charge of amylovorin L471 at this pH resulted in a high affinity for the adsorbent. For these loading conditions, subsequent elution with 1.0 M NaCl at pH 3.6, 6.5 and 8.0 resulted in a loss of the total bacteriocin activity of 70.1%, 52.0% and 95.8%, respectively. This can be due to either inactivation or improper elution of the bacteriocin. Elution at pH 6.5 resulted in the highest recovery of 48.0% (Figure 1). This pH improved the elution capacity of the elution buffer by decreasing the net positive charge of amylovorin L471. Elution with buffer at pH 8.0 resulted in a much lower recovery of 4.2%. This high pH most likely resulted in an inactivation of the bacteriocin during elution from the column since no activity was lost in the flow through the column during loading, and elution with pH 3.6 and 6.5 resulted in higher recoveries of bacteriocin activity. This inactivation was confirmed by experiments of bacteriocin stability as function of the pH (see further). On the other hand, at pH 8.0 (pI of bacteriocin) the hydrophobic interactions are favoured.

Stability of amylovorin L471 in cell-free solution as function of the pH

Partially purified amylovorin L471 was obtained via ammonium sulphate precipitation and chloroform/methanol extraction (De Vuyst et al., 1996a). The partially purified amylovorin solution had an activity of 25600 AU ml^{-1}. This was added to different buffers and resulted in a start activity of 19200 AU ml^{-1}. The activities of cell-free amylovorin so-

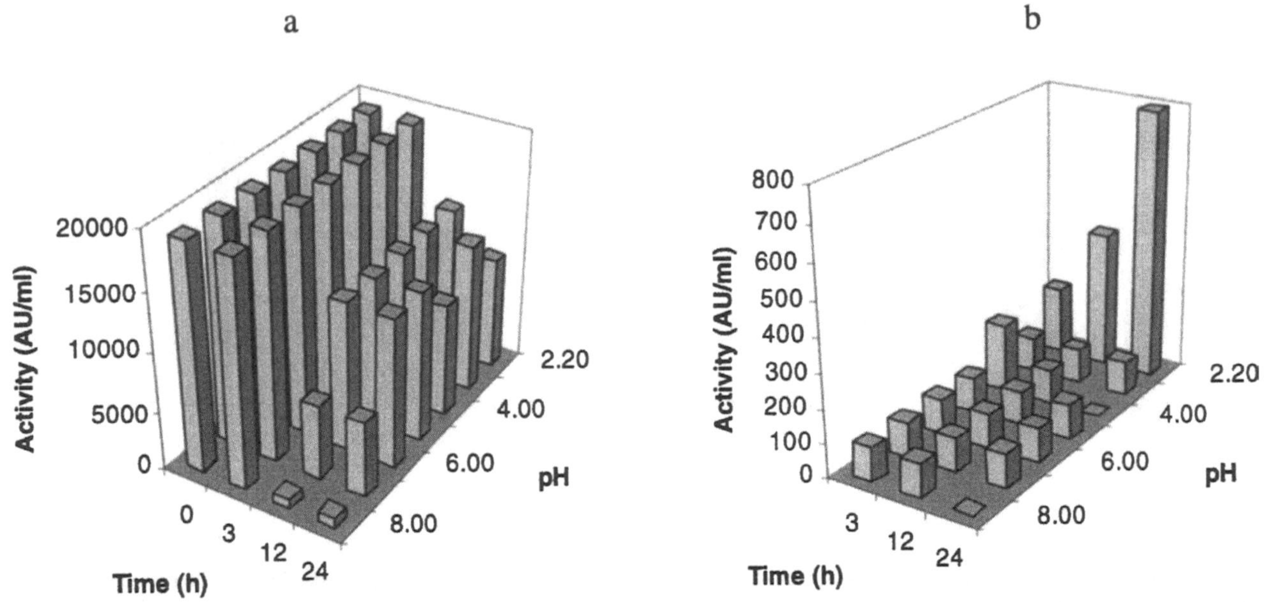

Figure 2. (a) Stability of amylovorin L471 in cell-free solution in function of the pH: partially purified bacteriocin (25600 AU ml^{-1}) was added to buffers at different pH; (b) adsorption of amylovorin L471 to the cell surface of the bacteriocin-producing cells in function of the pH: washed cells dissolved in buffer at different pH values.

lutions as function of pH and time are presented in Figure 2. Amylovorin L471 was stable in a pH range from pH 2.2 to pH 6.0. At pH 7.0, 77.0% of the activity was lost after 12 h of incubation, while at pH 8.0, almost all activity (96.0%) was lost.

Adsorption of amylovorin L471 to the cell surface of the bacteriocin-producing cells as function of the pH

Often a pH-dependent bacteriocin adsorption on the cell surface of the bacteriocin-producing cells is observed (Yang et al., 1992). Based on this phenomenon, acid treatment has been used as first isolation step for the purification of bacteriocins (Daba et al., 1994; Elegado et al., 1997; Tahara and Kanatani, 1997). Also the apparent bacteriocin inactivation upon prolonged fermentation at a controlled pH near neutrality is ascribed to adsorption of the bacteriocin molecules on the cell surface of the producing cells (De Vuyst et al., 1996b). To confirm this hypothesis, washed cells of *Lb. amylovorus* DCE 471 were suspended in buffers with different pH values. These results are presented in Figure 2. Cell-free supernatant still displayed low inhibitory activity. The highest activity was found at pH 2.2. Moreover, the activity at pH 2.2 increased from 200 AU ml^{-1}, after 3 h, to 800 AU ml^{-1}, after 24 h of incubation. At the other pH values, no increase was observed. This indicated that bacteriocin was adsorbed to the cell surface at pH values higher than 3.0 and released at lower pH values. At high pH values, the activity decreased as was also seen in the stability experiments. Decreasing the pH and hence releasing the bacteriocin from the cell surface, before applying crude fermentation broth to the expanded bed, should hence result in higher recoveries.

Isolation of amylovorin L471 in expanded bed mode under optimised conditions

Taking into account the results of the experiments in packed bed and the stability and adsorption experiments, isolation of amylovorin L471 was improved (Figure 3). Elution at pH 6.5 after loading at pH 3.6 resulted in a total recovery of 43.0% (Table 3; Figure 3). No activity was found in the flow through during loading of the column. The results in expanded bed mode did not differ significantly from the results obtained in packed bed mode. Specific bacteriocin activity increased 7.5 times, from 150 AU mg^{-1} (initial value) to 1120 AU mg^{-1} (after elution).

Applying crude culture medium adjusted to pH 3.6 to an expanded bed resulted in a total capture of the bacteriocin on the adsorbent and in a total recovery of only 11.6% after elution at pH 3.6 with 1.0 M NaCl. Compared to loading and elution at pH 6.5 this recovery was much lower. Loss of bacteriocin at pH 3.6 could be due to inactivation by the cationic matrix, since at pH 6.5 adsorbed bacteriocin was completely

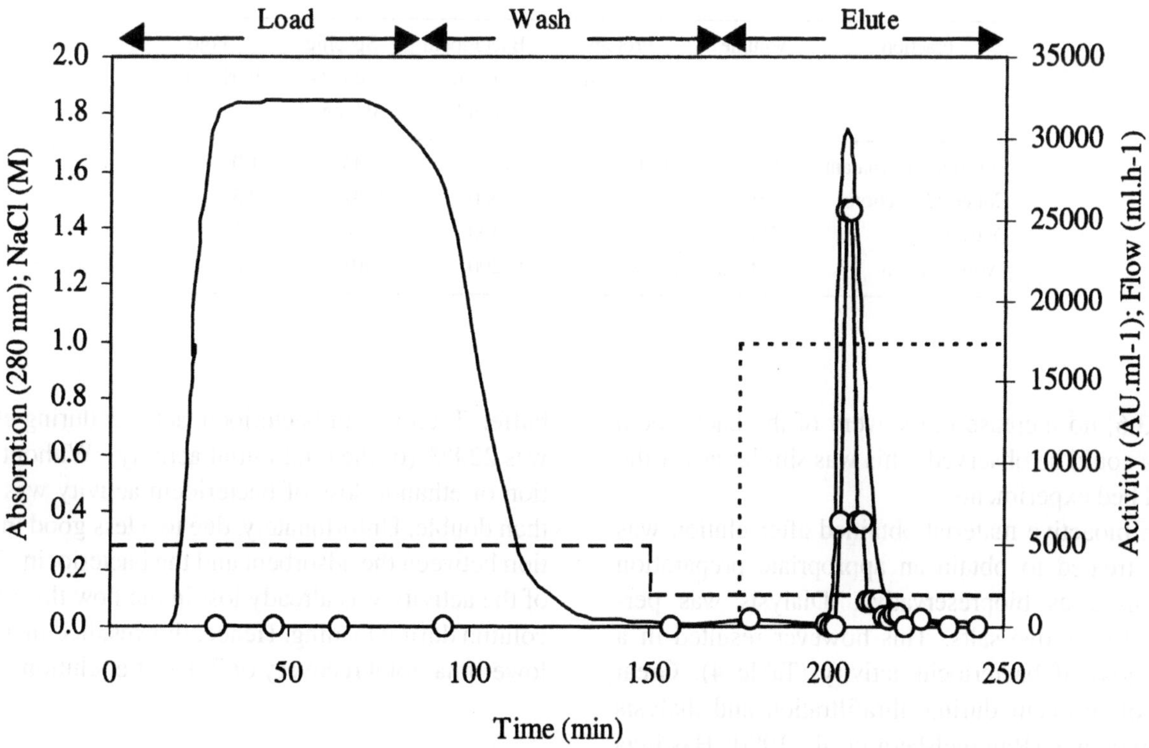

Figure 3. Amylovorin L471 isolation from 5.0 l crude fermentation liquor by expanded bed adsorption with a strong cation exchanger (500 ml; Streamline SP^{TM}, Pharmacia) consisting of sulphopropyl functional groups bound on a Sepharose matrix (100–300 μm particles). A column with a diameter of 5 cm and a height of 1 m (Streamline 50^{TM}, Pharmacia, Uppsala, Sweden) was used. The pH of the liquor was adjusted to pH 3.6. The pH of loading and elution buffer was 3.6 and 6.5, respectively. Flow rate: — —; volumetric bacteriocin activity: —O—; optical density at 600 nm: —; NaCl-concentration: - - - -.

Table 3. Bacteriocin isolation via expanded bed adsorption

Strain	Bacteriocin	Loading pH	Elution pH	Flow through	Recovery	Loss/ inactivation	Reference
Lb. amylovorus DCE 471	amylovorin L471	6.5	6.5	80.5	19.5	0.0	De Vuyst et al. (1996a)
Lb. amylovorus DCE 471	amylovorin L471	3.6	3.6	0.0	11.6	88.4	this work
Lb. amylovorus DCE 471	amylovorin L471	3.6, ethanol	3.6, ethanol	14.0	7.0	79.0	this work
Lb. amylovorus DCE 471	amylovorin L471	3.6	6.5	0.0	43.0	57.0	this work
E. faecium RZS C13	enterocin C13	6.5	6.5	3.7	6.9	89.4	this work
E. faecium CCM 4231	enterocin 4231	6.5	6.5	61.6	1.9	36.5	this work
E. faecium CTC 492	enterocin A	3.6	6.0	54.2	15.6	30.2	Aymerich et al. (1996) this work
Pediococcus sp. RZS C9	pediocin C9	6.5	6.5	74.0	26.0	0.0	De Vuyst et al. (1996a)
Lc. lactis subsp. *lactis* NIZO 22186	nisin A	6.5	6.5	98.6	1.4	0.0	this work

recovered from the column (when loaded at pH 6.5). However, due to too strong interaction of the bacteriocin with the adsorbent after loading at pH 3.6 it can still be sitting on the matrix. Elution with pH 3.6 may not be strong enough for desorption of the bacteriocin. However, even after elution with pH 6.5, when loaded

Table 4. Bacteriocin activity after dialysis and lyophilisation of fractions collected after isolation by expanded bed adsorption with loading and elution at pH 3.6

Fraction	Volume ml	Protein concentration mg ml^{-1}	Bacteriocin activity AU ml^{-1}	Specific activity AU mg^{-1}	Fold Purification
Cell free supernatant	5000	11.0	1600	145	1.0
Collected fractions	150	1.3	1600	1231	8.5
Dialysis	100	2.6	400	154	1.1
Lyophilisation	0.05	127.7	51200	401	2.8

at pH 3.6, no increase in recovery of the bacteriocin activity could be observed. This was similar as for the packed bed experiments.

The bioactive material obtained after elution was further treated to obtain an appropriate preparation to be used as biopreservative. Dialysis was performed to remove salts. This however resulted in a drastic loss of bacteriocin activity (Table 4). Great losses often occur during ultrafiltration and dialysis of bacteriocins (Rammelsberg et al., 1990; Hastings et al., 1991). Subsequent lyophilisation of dialysed amylovorin L471 resulted in an active powder with a specific amylovorin activity that was two times higher than before (Table 4). Nevertheless, expanded bed adsorption is an ideal step to obtain bacteriocin preparations to be used as biopreservative. However, further optimisation of loading and elution conditions or exploration of other adsorbents might improve the rather low recoveries.

Stabilisation of amylovorin L471 activity during isolation

At loading pH 3.6 all bacteriocin was captured by the adsorbent. However, total bacteriocin activity decreased during the isolation procedure. This loss of bacteriocin activity could be ascribed to the strong hydrophobic properties of amylovorin L471. Aggregation leads to less soluble bacteriocin complexes (De Vuyst et al., 1996a). Possible secondary hydrophobic interactions with the matrix of the adsorbent, besides ionic interaction, could also take place; both phenomena are pH dependent. Too strong interaction with the matrix of the adsorbent can lead to difficult elution of the bacteriocin and may explain the improper elution characteristics. In addition, strong interaction may result in inactivation. In an attempt to stabilise the bacteriocin and to avoid interaction with the matrix, 5% (v/v) ethanol was added to the loading and elution buffer. Total loss of bacteriocin activity during elution was 22.8% (of the total initial activity). Without addition of ethanol, loss of bacteriocin activity was more than double. Unfortunately, due to a less good interaction between the adsorbent and the bacteriocin, 70.2% of the activity was already lost in the flow through the column during loading. Hence, this resulted in a much lower final total recovery of 7.0% after elution.

Isolation of other bacteriocins

Expanded bed adsorption was tested for other lactic acid bacteria bacteriocins too (Table 3). Similar results as for amylovorin L471 were obtained. Isolation of enterocin RZS C13 and enterocin CCM 4231 from culture media with activities of 3200 AU ml^{-1} and 25600 AU ml^{-1}, respectively, was performed at non-optimal conditions of loading pH resulting in very low recoveries (Table 3). After optimisation of the pH for enterocin A, a higher recovery compared to the two other enterocins could be obtained (Table 3). From a 5.0 l culture with an activity of 25600 AU ml^{-1}, 15.6% of the total activity could be recovered with a specific activity that was 3.7 times higher after expanded bed adsorption (6000 AU mg^{-1}). Isolation of nisin A from a culture supernatant with an activity of 800 AU ml^{-1} resulted in a recovery of only 1.4%, which is extremely low. In contrast, a recovery of 26.0% was obtained for pediocin (De Vuyst et al., 1996a). To conclude, bacteriocin molecules were often lost in the flow through the column during loading of the adsorbent indicating non-optimised process conditions. Generally no inactivation of the bacteriocins was observed by loading at high pH (6.5). However by adjusting the pH before loading, isolation can be improved as was seen for, for instance, enterocin A.

Conclusions

The pH is an important factor for optimisation of isolation of lactic acid bacterium bacteriocins via expanded bed adsorption based on ion exchange. In addition, for bacteriocin isolation the pH seems to be very critical for conserving bacteriocin activity. At low pH the bacteriocin has a more positive charge and hence affinity for the negatively charged adsorbent increases. At this pH a higher loss of activity is observed, probably due to a strong interaction with the adsorbent. At higher pH, no inactivation is observed. However, bacteriocin is lost in the flow through during loading of the adsorbent at higher pH. Too high pH values during elution, after loading at low pH, result in an inactivation of the bacteriocin due to the alkaline conditions. Elution of the bacteriocin at a pH of 6.5 after loading the bacteriocin to the adsorbent at pH 3.6 resulted in the highest recovery. A loss of activity of approximately 50% could not be prevented generally. Further optimisation of the process conditions or the availability of other commercial adsorbents could overcome this problem. Expanded bed adsorption is hence a promising and timesaving method and may be applied for the large-scale isolation of bacteriocins — which are to be applied commercially (eluted bacteriocin can be further dialysed and/or freeze-dried) — or as an ideal first step in the analytical purification of bacteriocins.

Acknowledgements

Part of this research was financially supported by the Research Council of the Vrije Universiteit Brussel, the Belgian National Fund for Scientific Research, and the Biotechnology Program of the European Commission (grants BIO2-CT943055 and ERB-CIPACT-940160).

References

Anonymous (1995) Bactériocines, une autre approche de la conservation. Process 55: 1103.

Aymerich T, Holo H, Håvarstein LS, Hugas M, Garriga M and Nes IF (1996) Biochemical and genetic characterization of enterocin A from *Enterococcus faecium*, a new antilisterial bacteriocin in the pediocin family of bacteriocins. Appl. Environ. Microbiol. 62: 1676–1682.

Carolissen-Mackay V, Arendse G and Hastings JW (1997) Purification of bacteriocins of lactic acid bacteria: problems and pointers. Int. J. Food Microbiol. 34: 1–16.

Chase HA (1994) Purification of proteins by adsorption chromatography in expanded beds. Trends Biotechnol. 12: 296–303.

Coventry MJ, Gordon JB, Alexander M, Hickey MW and Wan J (1996) A food-grade process for isolation and partial purification of bacteriocins of lactic acid bacteria that uses diatomite calcium silicate. Appl. Environ. Microbiol. 62: 1764–1769.

Daba H, Lacroix C, Huang RE and Lemieux L (1994) Simple method of purification and sequencing of a bacteriocin produced by *Pediococcus acidilactici* UL5. J. Appl. Bacteriol. 77: 682–688.

de Man JC, Rogosa M and Sharpe ME (1960) A medium for the cultivation of Lactobacilli. J. Appl. Bacteriol. 23: 130–135.

De Vuyst L, Callewaert R and Pot B (1996*a*) Characterization of the antagonistic activity of *Lactobacillus amylovorus* DCE 471 and large scale isolation of its bacteriocin amylovorin L471. System Appl. Microbiol. 19: 9–20.

De Vuyst L, Callewaert R and Crabbé K (1996b) Primary metabolite kinetics of bacteriocin biosynthesis by *Lactobacillus amylovorus* and evidence for stimulation of bacteriocin production under unfavourable growth conditions. Microbiol. 142: 817–827.

De Vuyst L and Vandamme EJ (1994) Bacteriocins of lactic acid bacteria: microbiology, genetics and applications. Blackie Academic & Professional, London.

Dodd HM and Gasson MJ (1994) Bacteriocins of lactic acid bacteria. In: Gasson MJ and De Vos WM (eds) Genetics and biotechnology of lactic acid bacteria. (pp 211–251). Blackie Academic & Professional, London.

Elegado FB, Kim WJ and Kwon DY (1997) Rapid purification, partial characterization, and antimicrobial spectrum of the bacteriocin, pediocin AcM, from *Pediococcus acidilactici* M. Int. J. Food Microbiol. 37: 1–11.

Hansson M, Stahl S, Hjorth R, Uhlen M and Moks T (1994) Single-step recovery of a secreted recombinant protein by expanded bed adsorption. Biotechnol. 12: 285–288.

Hastings JM, Sailer M, Johnson KL, Ray KL, Vederas JC & Stiles ME (1991) Characterization of leucocin A-UAL 187 and cloning of the bacteriocin gene from *Leuconostoc gelidum*. J. Bacteriol. 173: 7491–7500.

Hjorth R (1997) Expanded-bed adsorption in industrial bioprocessing: recent developments. Trends Biotechnol. 15: 230–235.

Hjorth R, Kampe S and Carlsson M (1995) Analysis of some operating parameters of novel adsorbents for recovery of proteins in expanded beds. Bioseparation 5: 217–223.

Hoover D and Steenson L (1993) Bacteriocins of lactic acid bacteria. Academic Press, New York.

Jack RW, Bierbaum G and Sahl H-G (1998) Lantibiotics and related peptides. Springer-Verlag, Berlin.

Jack RW, Tagg JR and Ray B (1995) Bacteriocins of Gram-positive bacteria. Microbiol. Rev. 59: 171–200.

Joerger MC and Klaenhammer TR (1986) Characterization and purification of helveticin J and evidence for a chromosomally determined bacteriocin produced by *Lactobacillus helveticus* 481. J. Bacteriol. 167: 439–446.

Joerger MC and Klaenhammer TR (1990) Cloning, expression, and nucleotide sequence of the *Lactobacillus helveticus* 481 gene encoding the bacteriocin helveticin J. J. Bacteriol. 172: 6339–6347.

Johansson HJ, Jagersten C and Shiloach J (1996) Large scale recovery and purification of periplasmic recombinant protein from *E. coli* using expanded bed adsorption chromatography followed by new ion exchange media. J. Biotechnol. 48: 9–14.

Klaenhammer TR (1988) Bacteriocins of lactic acid bacteria. Biochimie 70: 337–349.

Klaenhammer TR (1993) Genetics of bacteriocins produced by lactic acid bacteria. FEMS Microbiol. Rev. 12: 39–85.

Konings RNH and Hilbers CW (1996) Lantibiotics: A unique group of antibiotic peptides. Anton. Leeuw. Int. J. G. (Special issue) 69: 87–202.

Lauková A, Mareková M and Javorsky P (1993) Detection and antimicrobial activity spectrum of a bacteriocin-like substance produced by *Enterococcus faecium* CCM 4231. Lett. Appl. Microbiol. 16: 257–260.

Lyon WJ, Olson DG and Murano EA (1995) Isolation and purification of enterocin EL1, a bacteriocin produced by a strain of *Enterococcus faecium*. J. Food Prot. 58: 890–898.

Nes IF, Diep DB, Håvarstein IS, Brurberg MB, Eijsink V and Holo H (1996) Biosynthesis of bacteriocins in lactic acid bacteria. Anton. Leeuw. Int. J. G. 70: 113–128.

Noppe W, Hanssens I and De Cuyper M (1996) Simple two-step procedure for the preparation of highly active pure equine milk lysozyme. J. Chromatogr. 719: 327–331.

Piard JC and Desmazeaud MJ (1992) Inhibiting factors produced by lactic acid bacteria. 2: Antibacterial substances and bacteriocins. Lait 72: 113–142.

Rammelsberg M, Muller E and Radler F (1990) Caseicin 80: purification and characterization of a new bacteriocin from *Lactobacillus casei*. Arch. Microbiol. 154: 249–252.

Ray B and Daeschel M (1992) Food Biopreservatives of Microbial Origin. CRC Press, Florida.

Sahl H-G, Jack RW and Bierbaum G (1995) Biosynthesis and biological activities of lantibiotics with unique post-translational modifications. Eur. J. Biochem. 230: 827–853.

Stiles ME (1996) Biopreservation by lactic acid bacteria. Anton. Leeuw. Int. J. G. 70: 331–345.

Stoffels G, Sahl H-G and Gudmundsdóttir Á (1993) Carnocin UI49, a potential biopreservative produced by *Carnobacterium piscicola*: large scale purification and activity against various Gram-positive bacteria including *Listeria* sp. Int. J. Food Microbiol. 20: 199–210.

Tagg JR, Dajani AS and Wannamaker LW (1976) Bacteriocins of Gram-positive bacteria. Bacteriol. Rev. 40: 722–756.

Tahara T and Kanatani K (1997) Isolation and partial characterization of crispacin A, a cell-associated bacteriocin produced by *Lactobacillus crispatus* JCM 2009. FEMS Microbiol. Lett. 147: 287–290.

Vlaemynck G, Herman L and Coudijzer K (1994) Isolation and characterization of two bacteriocins produced by *Enterococcus faecium* strains inhibitory to *Listeria monocytogenes*. Int. J. Food Microbiol. 24: 211–225.

Yang R, Johnson MC and Ray B (1992) Novel method to extract large amounts of bacteriocins from lactic acid bacteria. Appl. Environ. Microbiol. 58: 3355–3359.

Address for correspondence: L. De Vuyst, Division of Industrial Microbiology, Fermentation Technology and Downstream Processing (IMDO), Department of Biotechnology, Faculty of Sciences, Vrije Universiteit Brussel (VUB), Pleinlaan 2, B-1050 Brussels, Belgium (Fax: 32-2-6292720; e-mail: ldvuyst@vub.ac.be)

Bioseparation **8:** 169–188, 1999.
© 1999 *Kluwer Academic Publishers.*

Pilot scale recovery of monoclonal antibodies by expanded bed ion exchange adsorption

Nicole Ameskamp, Christoph Priesner, Jürgen Lehmann & Dirk Lütkemeyer
Institute of Cell Culture Technology, Faculty of Technology, University of Bielefeld, P.O. Box 100131, 33501 Bielefeld, Germany

Received 8 September 1998; accepted in revised form 4 March 1999

Key words: cation exchange chromatography, dynamic binding capacity, expanded bed adsorption, feed viscosity, monoclonal antibody, pilot scale

Abstract

The aim of the investigations was to estimate the scale up properties of an efficient chromatographic first capture step for the recovery of murine IgG_1 from undiluted and unclarified hybridoma cell culture broth using an ion exchange matrix in expanded bed mode. The tested new sulfopropyl-based ion exchange matrix (Streamline™ SP XL, Amersham Pharmacia Biotech) stands out due to its enhanced capacity compared to its precursor (Streamline™ SP). Defining the working pH in preliminary electrophoretic analyses (titration curve, SDS-PAGE) and small-scaled chromatographic binding studies showed, that the optimal value for the IgG purification was pH 4.6, where a co-chromatography of the medium supplement albumin (500 mg l^{-1}, pI = 4.8) could not be avoided. Further scouting experiments dealt with the dynamic capacity of the matrix, which was evaluated by frontal adsorption analysis. In packed bed mode no break-through of the target protein was achieved even after 6.5 mg IgG per ml matrix were applied. These results could not be reproduced in expanded bed mode with cell-free supernatant, where the dynamic capacity was found to be only 1.5 mg IgG/ml SP XL. Processing cell-containing broth resulted in an additional decrease of the value down to 0.5 mg ml^{-1}, presumably caused by the remarkable biomass adsorption to the matrix. The search for the reasons led to the examination of the hydrodynamic conditions. Buffer experiments with a tracer substance (acetone) pointed out, that the flow in expanded bed was significantly more influenced by back-mixing effects and channel formations than in packed bed. These effects could be compensated with an enhanced viscosity of the liquid phase, which was achieved by the addition of glucose. As a result of the improved hydrodynamic conditions in the expanded bed, the dynamic capacity could be increased from 0.5 to more than 4.5 mg IgG/ml matrix for the processing of cell culture broth with 400 mM glucose. Finally, the scale up from a Streamline™ 25 to a Streamline™ 200 column was performed under conditions, which proved to be optimal: 100 L of unclarified hybridoma broth were concentrated with a binding rate of 95% in less than 3.5 hours. Loading the column no break-through of the target protein was achieved. However, the eluate still contained debris and cells, which points out the major disadvantage of the method: the biomass attachment to the matrix.

1. Introduction

Today, animal cell culture technology is widely used to produce biologically active proteins for diagnostic and therapeutic applications. The progresses in purposeful recombination of DNA yield a large amount of biomolecules with desired structure and function. Powerful fermentation systems were developed, so that nowadays the cell culture technology is well-established even in industrial large-scale productions (Werner, 1990).

Eucaryonts are able to express complex proteins with correct post-translational modifications, like glycosylation (Goochee et al., 1992) and formation of the right tertiary structure by disulfide bonds. Due to these benefits animal cells are often the expression system of choice, although they are much more delicate to handle than, for example, bacteria or yeast cells.

One important product are monoclonal antibodies, which are synthesised by hybridoma cells (Köhler & Milstein, 1975). Monoclonal antibodies are economically used for fundamental research as well as for diagnostic and therapeutic applications. Since the binding characteristic of immunoglobulins is very specific, they served for instance as affinity ligands in chromatographic purification processes (Eketorp, 1992) or for the investigation of protein structures (Jockusch, 1986) and tissue compositions. Additionally, antibodies are widely used in sensitive assays for the detection and quantification of soluble substances. Thereby monoclonal antibodies are nowadays powerful tools in environmental and diagnostic analyses (Nowinski et al., 1983). Since 1986 they are even permitted for therapeutical approaches such as the immunological treatment of cancerous tumours.

In order to maintain the biological function of a cell culture product, the fermentation strategy as well as the following downstream process must guarantee gentle conditions. From the economic point of view the latter is more relevant and decides on the final product quality.

Cell culture harvest usually consists of a complex mixture of substances. Besides the biomass, it contains unconsumed medium components, products of the cell metabolism, and host substances, like cytosolic enzymes or DNA, which are released by damaged cells. All these contaminants have to be removed, without influencing the stability of the product.

A conventional downstream process for an extracellular target molecule is divided into a number of successive steps, including the biomass separation, the concentration and the fractionation. During the removal of the biomass the integrity of the fragile cells has to be guaranteed in order to avoid a contact of the product with intracellular enzymes, which might cause a loss of the biological activity (Lind et al., 1991). In practice the centrifugation and the microfiltration are established methods. The latter is characterised by the problems of fouling and clogging processes (Le and Gollan, 1989), which cause a short life expectancy of the filtration membranes. As a negative consequence, the required frequency of their replacement enhances the overall production expenses. The recent development of the centrifugation technique on the other hand, led to high-throughput systems, which are meanwhile suitable for the use in cell culture technology. Special disc stack separators for example were modified by an improved feed device, which considers the shear sensitivity of cells (Tebbe et al., 1996). However, in most cases the supernatant has to be additionally filtrated for the subsequent downstream steps, which demand absolute particle-free solutions.

After the removal of the biomass the target protein is usually concentrated by means of ultrafiltration before the final fractionation takes place. The latter is carried out by a broad range of chromatographic methods, which differ from each other by their separation mechanisms.

With the exception of some special materials, like membrane adsorbers (Lütkemeyer et al., 1993), the common chromatographic matrices are based on stable agarose, dextran or cellulose beads (Arshady, 1991), which are derivated with various ligands. They fractionate protein solutions on the principles of ion exchange, hydrophobic interactions, difference in molecular size, metal chelate affinity or by the specificity of antibody/antigen and enzyme/substrate interactions. The suitability of a matrix depends on the product properties and the required purification degree. A combination of the chromatographic methods most likely fulfils the purposes.

This short summary indicates that the conventional multi-step downstream process can become very costly and time-consuming. During the last years a new technology has been introduced (Chase, 1994), which integrates the cell removal, the concentration and the first product capture in a single step procedure. The alternative method, called expanded bed adsorption EBA, is based on the enhancement of the bed voidage of a chromatographic column, which enables the load with particle- and even cell-containing solutions (Draeger and Chase, 1991). The fluidization of the matrix is achieved by special stirring devices (UpFront Chromatography A/S) or by applying the liquid in upward flow direction (Amersham Pharmacia Biotech). Choosing a suitable resin and the right conditions the target protein will be retained on the matrix, whereas unspecific soluble substances as well as the biomass pass the column unhindered. Ideally, an expanded bed adsorption process results in a particle-free product concentrate, rendering unnecessary any preliminary centrifugation or filtration steps.

In comparison with the conventional chromatographic materials, the EBA matrices are especially modified in order to achieve a stable fluidized bed. High through-put capacities without rinsing the gel out of the column are enabled by the use of denser matrix particles (1.2–1.3 g ml^{-1} Amersham Pharmacia Biotech, 1.4–1.6 g ml^{-1} UpFront). Defined distributions in weight and diameter lead to a classification of the expanded bed, which reduces circulation and back mixing effects. Karau (Karau et al., 1997) has examined the influence of the particle size distribution on the adsorption performance. The basic principles for the characterisation of a fluidized bed were summarised by Thömmes for example (Thömmes, 1997).

Nowadays, quite a large number of different matrices with a wide range of functional groups are available for the use in expanded bed adsorption. Although the EBA technology is still under development many applications are already established (Hjorth, 1997). The purification of recombinant human albumin from the yeast *Pichia pastoris* in production scale was introduced by Noda et al. (1996). Other examples deal with the capture of proteins from *Escherichia coli* fermentation broth (Barnfield Frej et al., 1994; Johansson et al., 1996).

This paper describes the investigation of the scale up properties of an expanded bed procedure, carried out with the cation exchange matrix Streamline™ SP XL (Amersham Pharmacia Biotech). Preparing the 100 L pilot scale runs, the main point of interest besides the definition of the chromatographic parameters was to optimise the binding capacity of the Streamline™ SP XL matrix. Within these scouting experiments the feed viscosity was modified by the addition of glucose in order to stabilise the fluidized bed. Previous investigations showed that the dynamic binding capacity decreases in the presence of increasing concentrations of glycerol (De Luca et al., 1994; Chang and Chase, 1996). The reasons for this decrease were related to the resulting enlargement of the bed voidage, which had drastically increased the axial dispersion of the liquid phase in the expanded bed. However, since up to 32% (v/v) glycerol were added, the feed viscosity was more than three times higher than in the initial aqueous buffer. In contrast to these studies, our aim was to examine the influence of an only moderately enhanced feed viscosity on the EBA process, not least because of the necessary maintenance of physiological conditions.

2. Materials and methods

2.1. Cell cultivation and product

The studies were carried out with a mouse/mouse hybridoma cell line, which produced a monoclonal antibody of the subclass IgG_1. The serum-free culture medium was based on a 1:1 mixture of Dulbecco's Modified Eagle Medium (DMEM) and Ham's F12 (Gibco BRL), which was supplemented with some essential amino acids and the protein additives human serum albumin (HSA, 500 mg L^{-1}), human transferrin (HTF, 10 mg L^{-1}) and bovine insulin (BI, 10 mg L^{-1}).

For the chromatographic experiments in bench scale the cultivation was extended from a 1 L-*Superspinner* system (Heidemann et al., 1994) to a 5 L- (B. Braun Biotech International) and 20 L-bioreactor (Diessel) respectively, depending on the required amount of biological material. The cells for the final expanded bed adsorption scale up were grown in a 100 L-bioreactor (Biostat UD100, B. Braun), which was bubble-free aerated through a silicon tube basket (Lehmann et al., 1987). In order to avoid the aggregation of cells, the cultivation was done without the usual pH control by automated injection of 2 M NaOH. As an alternative, the buffer capacity of the medium was increased.

All fermentations were run in split batch mode, which resulted in a harvest cycle of approx. three days. At the end of each batch the number of viable cells reached up to $3 \cdot 10^6$ cells/ml at a high viability of more than 80%. The product concentration amounted to values between 30 and 70 mg IgG_1/L.

2.2. Chromatographic equipment

2.2.1. Matrix

The Streamline™ SP XL matrix (Amersham Pharmacia Biotech) used is a strong cation exchanger, which provides a higher capacity than its precursor Streamline™ SP. The sulphonate groups are attached via long dextran chains to the highly cross-linked 6% agarose base resin. In adaptation to the operation mode in expanded bed, the beads are size distributed between 100 and 300 μm and weighted down by inert quartz cores to approx. 1.3 g ml^{-1} to build up a stable, classified fluidized bed at high throughput capacities.

2.2.2. Method scouting in bench scale

The optimisations of the chromatographic binding conditions took place in packed bed mode using an FPLC™ system and a C10/10 column (d = 1 cm, l = 10 cm) with 4 ml Streamline™ SP XL (H_0= 5.1 cm). Working in packed bed chromatography requires a preliminary clarification step in the downstream process. Therefore, the cell culture broth had to be centrifuged and filtrated (0.2 μm) before it could be applied to the C10/10 column.

The expanded bed runs in bench scale were performed with a modified FPLC™ or standard GradiFrac™ system and a Streamline™ 25 column (d = 2.5 cm, l = 100 cm) containing 75 ml of adsorbent (H_0= 15.3 cm). The same equipment was used for the examination of the dynamic binding capacity under different feed conditions and for related buffer experiments concerning the residence time behaviour of a tracer substance.

The storage vessel for the processing of cell containing harvest included a stirring device, which maintained a homogeneous load during the whole application.

2.2.3. Pilot scale

The EBA pilot system consisted of a Streamline™ 200 column (d = 20 cm, l = 100 cm) – filled with 5 L (7.5 L) Streamline™ SP XL matrix; a peristaltic pump (604U, Watson Marlow); an air-trap, and manual valves. In addition the system included a UV (Amersham Pharmacia Biotech) and a conductivity detector (WTW), which were connected with a PC for the documentation of the chromatography. The flow velocity could be permanently controlled via a scale (QS64B, Sartorius). Figure 1 illustrates a scheme of the whole pilot scale plant. A tempered 100 L stainless steel vessel with a stirring device was used for harvesting the cell culture broth and adjusting the chromatographic conditions.

2.3. Buffer system, chromatographic procedure and matrix regeneration

The buffer system was based on 20 mM acetate adjusted to pH 4.6. The equilibration and washing buffer A contained 150 mM sodium chloride to achieve a conductivity comparable to the feed (approx. 15 mS cm^{-1}). The elution was carried out with a high salt concentration step by adding 1 M NaCl to the stock buffer. In order to increase the viscosity, up to 400 mM glucose were added to the solutions. The buffer consumption amounted to 10 and 5 sedimented bed volumes for the equilibration and washing fraction respectively, and about 5 matrix volumes for the desorption of bound substances.

The flow velocity for all chromatographic runs was 300 cm h^{-1} for the equilibration, loading, and washing of the matrix and 100 cm h^{-1} for the elution. Each step – including the elution – was performed in upward flow direction to prevent a clogging of the matrix beads by the remaining biomass.

After each run with biomass-containing solutions the matrix was regenerated in a cleaning-in-place CIP procedure, which is summarised in Table 1. The success of the cleaning was controlled by examinations of the expansion behaviour.

Table 1. Cleaning in place procedure CIP. Due to the remaining biomass, the regeneration of the matrix required a time- and material-consuming CIP procedure. It took 9 hours to clean the SP XL material for the next run

CIP solution	Flow velocity [cm/h]	Sedimented bed volumes	Volume (SL200) [l]	Time [min]
1. 0.5 M NaOH, 1.0 M NaCl	30	8*	40	240
2. Distilled water	100	10	50	90
3. 30% (v/v) Isopropanol	100	10	50	90
4. 25% (v/v) Acetic acid	100	10	50	90
5. 20% (v/v) EtOH in buffer A	100	3	15	30

*Contact time: 4 h.

2.4. Determination of the dynamic binding capacity

The dynamic binding capacity was determined by frontal adsorption analysis using the bench scaled columns. The purpose of the examinations was to compare the influence of different operating modes and feed conditions on the chromatographic capacity of the Streamline™ SP XL matrix. Therefore, several IgG_1 break-through curves were recorded in packed and expanded bed, applying up to 10 L of hybridoma cell culture broth to the Streamline™ column. The feed modifications referred to the viscosity of the liquid phase, which was increased by the addition of various amounts of glucose (200; 400 mM) in order to obtain a more stable fluidized bed and thereby an enhanced dynamic binding capacity.

The IgG's break-through was recorded by the determination of the rising antibody content in the flow-through. It was related to the concentration in the initial feedstock and plotted versus the applied IgG amount per millilitre of matrix. Within regular intervals small aliquots were taken from the flow-through and measured by means of immunoassays. The dynamic binding capacity was defined as mg IgG_1 applied per ml matrix at a break-through of 10% of the concentration in the feed. The flow velocity was kept constant at 300 cm h^{-1} for all analyses.

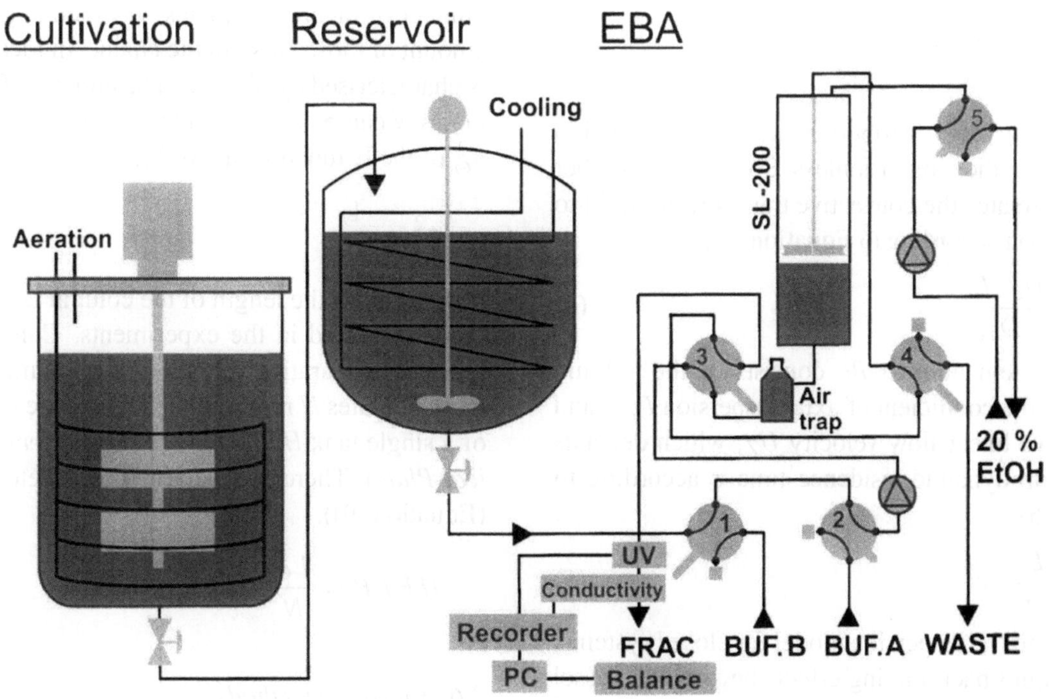

Figure 1. Scheme of the pilot scale equipment, which consists of a 100 L bioreactor, a storage vessel and the chromatographic unit with a Streamline™ 200 column.

2.5. Determination of the residence time distribution RTD

The residence time behaviour was examined in order to estimate the hydrodynamic conditions in packed and expanded bed mode as well as the influence of the feed viscosity on the process. Therefore, the residence time distributions RTD of a tracer substance were recorded for the Streamline™ 25 column and 75 ml of Streamline™ SP XL. The experiments were carried out with washing buffer of different glucose content and a buffer based 2% (v/v) acetone solution as a marker. For the packed bed experiments the column adapter was adjusted to the surface of the sedimented resin. In fluidized bed it was lowered to 1 cm above the expanded bed height. All runs were carried out at a linear flow velocity of 300 cm h^{-1} and directed by an integrated FPLC™ software (FPLC™ director). After washing the matrix with 10 sedimented bed volumes of equilibration buffer in upward flow direction, a pulse signal (Dirac-function) of 10 ml of the marker substance was injected through a 50 ml loop. The UV absorption (λ = 280 nm) of the flow-through was continuously measured and recorded as a function of the time by the FPLC™ director. The resulting C-function $c_E = f(t)$ was used as a basis for the characterisation of the back mixing phenomena in the column. The evaluation implied several modulations and calculations, which are described in detail by the literature (Levenspiel, 1972; Fitzer and Fritz, 1982; Weiss, 1985). The following section summarises only the major relations that were required for our application.

Normalising the response function c_E according to its integral enables comparative examinations of the back mixing effects in different columns. The normalised RTD or E-function describes the statistical probability for a tracer molecule to leave the column at the time *t*. On the assumption that all tracer molecules have left the column after $t \Rightarrow \infty$, the integral equals one. The E-function is given by Equation (1):

$$E = \frac{c_E}{\int_0^\infty c_E dt} \Rightarrow \int_0^\infty E dt = 1 \tag{1}$$

The location and spread of the distribution are important descriptive quantities for the characterisation of an RTD. The location is described by the mean residence time $\bar{t}$, whereas the spread is commonly measured by the variance σ_t^2. The definitions are given by the first and the second moment of the normalised response function *E* (Equations (2) and (3)):

$$\bar{t} \int_0^\infty t \cdot E dt \tag{2}$$

$$\sigma_t^2 \int_0^\infty (t - \bar{t})^2 E dt \tag{3}$$

Characterising the distribution with the dispersion model yields the dimensionless Bodenstein number *Bo*, which relates the convective transport of liquid to the dispersion according to Equation (4).

$$Bo = \frac{U_z \cdot L_C}{D_{ax}} \tag{4}$$

The Bodenstein number *Bo* consists of the column length L_C, the coefficient of axial dispersion D_{ax}, and the effective linear flow velocity U_Z, which depends on the hydrodynamic residence time τ according to Equation (5):

$$U_z = \frac{L_C}{\tau} \tag{5}$$

According to the dispersion model for closed systems, without liquid back mixing effects beyond the vessel borders, the Bodenstein number *Bo* can also be related to the normalised variance σ_Θ^2 through Equation (6):

$$\sigma_\Theta^2 = \frac{2}{Bo} - \frac{2}{Bo^2}[1 - \exp(-Bo)] \tag{6}$$

The normalised variance σ_Θ^2 equals the quotient of the variance σ_t^2 and the square of the hydrodynamic residence time τ. Assuming a closed vessel the latter equals the mean residence time $\bar{t}$, so that σ_Θ^2 can be calculated via Equations (2) and (3) according to Equation (7):

$$\sigma_\Theta^2 = \frac{\sigma_t^2}{\tau^2} = \frac{\sigma_t^2}{\bar{t}^2} \tag{7}$$

For the evaluation of the back mixing effects in the Streamline™ column the Bodenstein number *Bo* was calculated by a computer programme, which was written in the programming language C by Huneke (Huneke, 1995). It solves Equations (2) and (3) by numerical integration according to Simpson and Equation (6) iteratively by the Newton method. In addition, the resulting Bodenstein number Bo was used to determine the axial coefficient of dispersion D_{ax} by Equation (4). Although D_{ax} may be reduced with increasing bed height (Bascoul et al., 1988), this quantity can be used to characterise and compare the overall back mixing effects of different columns.

Besides the dispersion model, the tanks-in-series model is a second one-parameter model used to describe back mixing phenomena. In this model the vessel is separated in a series of equal-size ideal stirred tanks. The descriptive quantity for a column is the number of theoretical plates *N*, which is given by the amount of those tanks in the chain. An ideal plug flow is characterised by $N \rightarrow \infty$. The number of theoretical plates *N* can be calculated by the normalised variance σ_Θ^2 of the E-function, as well (Equation (8)).

$$\sigma_\Theta^2 = \frac{1}{N} \tag{8}$$

N depends on the length of the column L_C, which was however varied in the experiments. Consequently, a precise comparative valuation of the numbers of theoretical plates *N* required the knowledge of the height of a single tank *HETP (Height Equivalent of Theoretical Plate)*. Therefore the *HETP* was determined, too (Equation (9)).

$$HETP = \frac{L_C}{N} \tag{9}$$

2.6. Analytical methods

2.6.1. Electrophoretic analyses

Titration curve The pH dependency of the antibody's charge was examined by two-dimensional electrophoretic titration to restrict the working pH range for the cation exchange chromatography. The analysis was carried out with a purified IgG_1 concentrate that contained a small amount of human serum albumin. This sample composition enabled a direct comparison of the acid/base titration curves of the antibody and the main contaminating protein HSA, and thereby an estimation of their competitive behaviours in the chromatography.

The sample was applied parallel to the linear pH gradient of a 'ready for use' polyacrylamide gel (PhastGel$^{\Gamma(R)}$IEF 3-9, Amersham Pharmacia Biotech), the ampholytes of which covered a pH range from 3 to 9. After the separation the gel was treated with a silver stain to mark the protein curves (Heukeshoven and Dernick, 1988). The development as well as the preliminary electrophoresis steps were automated with a programmable PhastSystem™ (Amersham Pharmacia Biotech).

SDS-PAGE SDS polyacrylamide gelelectrophoresis (SDS-PAGE) under non-reducing conditions was used to determine the antibody and albumin content in the flow-through and in the eluate fractions. The samples were diluted 1:1 with Tris/HCl-EDTA-SDS-buffer and applied to a PhastGel$^{\Gamma(R)}$Gradient 8-25 (Amersham Pharmacia Biotech). The separation and following silver staining again run in a PhastSystem™ . An additionally applied marker kit with a molecular weight range from 14 to 96 kDa (LMW-Calibration Kit 14-96, Amersham Pharmacia Biotech) served for the allotment of the bands to the proteins.

2.6.2. Antibody quantification: ELISA

The antibody content in the chromatographic fractions was quantified by a standard sandwich ELISA (*Enzyme Linked Immuno Sorbent Assay*), which included a biotin/avidin based amplification. The used enzyme/substrate system consisted of horseradish peroxidase (Sigma) and 3,3′,5,5′-tetramethylbenzidin (TMB; Kirkegaard and Perry Labs.), which was converted into a di-imine derivative after approx. 10 min of incubation and the addition of acid (3 M H_2SO_4). The yellow derivative was detected in an ELISA reader (EL 311, BioTek, Tecnomara) at a wavelength λ of 450 nm. The samples were applied in a dilution range from 1:100 to 1:50,000 depending on the expected IgG_1 concentration, which was determined by an integrated software (KinetiCalc, BioTek, Tecnomara) via a standard calibration curve.

The ELISA results were used to calculate the recovery and binding rates as well as the yield that was obtained in the ion exchange chromatography.

2.6.3. Determination of the particle size distribution

A CASY$^{(R)I'}$1 cell counter and analyser system (Schärfe System GmbH) was used to determine the particle size distribution in the chromatographic fractions, which gives information about the condition of cells.

The CASY$^{I'(R)}$1 technique is based on the combination of the 'resistance measuring principle' (Glauner, 1991) with a signal processing method called 'pulse area analysis' (Schärfe, 1988). The measuring unit consists of an electrolyte filled capillary (CASY$^{I'(R)}$ton; Schärfe System GmbH) with a ruby precision measuring pore, through which a suspended aliquot of a cell culture sample is sucked in. The displacement of electrolyte solution by passing cells and particles causes a change of the electrical resistance level, which is registered via two electricity supplied platinum electrodes. The measuring signal is scanned at a frequency of 1 MHz and processed with an integrated computer, which determines the temporal integral of the resistance (pulse area analysis). The volume of a passing particle is proportional to it and can be calculated by the software. The resulting diameter is determined on the assumption that the particles are spherical. Since the membrane of viable cells is – in contrast to dead cells – an effective insulator, both kinds of cells can be clearly distinguished from each other in the cumulative particle size distribution, which is calculated by the system from repeated analyses of a diluted cell culture sample.

The CASY$^{I'(R)}$ system was used to examine the extent of cell damage during the downstream process and to quantify the biomass retention on the matrix as well as the clarification grade achieved in the eluate. In this case the device was equipped with a 150 μm capillary covering a particle size ranging from 3.4 to 90 μm. The cell culture samples were diluted 1:100 with CASY$^{I'(R)}$ ton and analysed three times in 400 μl portions. The resulting particle size distribution was interpreted according to the definition of Glauner, who differentiates between three classes: cell debris: < 6 μm, dead cells and cell nuclei: 6 - 10 μm, and viable cells: > 10 μm.

3. Results and discussion

3.1. Optimisation of the working pH

Working with cation exchangers the first point of interest for the determination of the chromatographic parameters is to locate the target protein's′ isoelectric point and to examine its pH dependent charge below the pI. Figure 2 shows the electrophoretic titration curves of the antibody and human serum albumin, the main protein component in the medium. The IgG's pI was found to be 6.3. Due to the only slight increase of the positive charge in the range from pH 6.3 to approx. 5.0, the best binding results could be expected under even more acidic conditions, at which a co-chromatography of HSA (pI 4.8) however was predictable, too. As a result the IgG binding would be simultaneously influenced by the increasing competitive effects of HSA below pH 4.8 and the decreasing charge of the antibody above the albumin's pI. To define the optimal working pH, small chromatographic binding studies, which considered both were necessary. Therefore, 100 ml samples of particle-free cell culture supernatant were applied at different pH to the C10/10 column and analysed concerning the IgG binding. Figure 3 represents the pH optimum curves, which have their maxima at pH 4.6 with 94.8% for the binding rate and 92.7% for the yield.

SDS-PAGE of the flow-through and eluate fractions (Figure 4) explained the circumstances: The

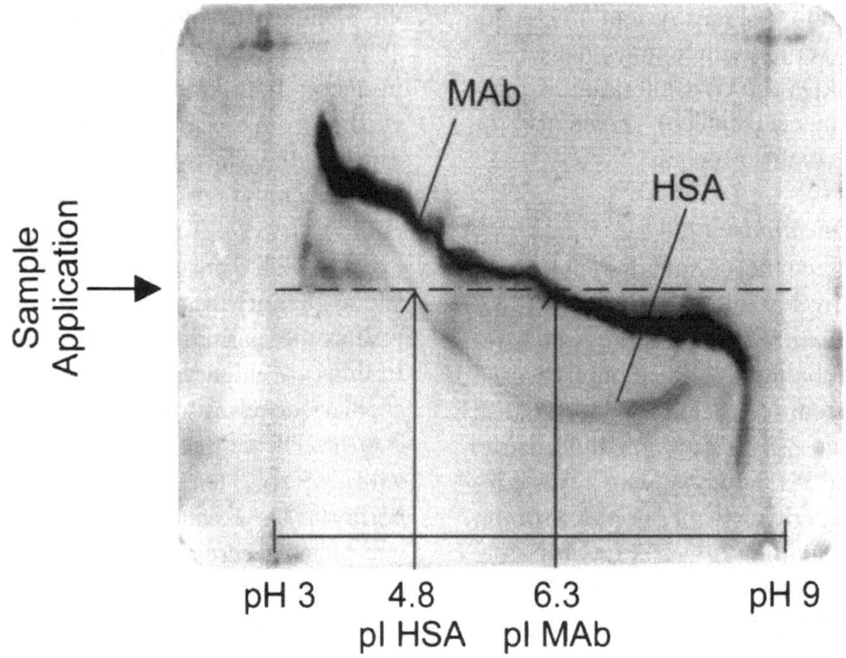

Figure 2. Titration curves of the IgG and albumin. The pI of the antibody was found to be 6.3.

albumin content continuously increased in the flow-through by rising the pH and vice versa in the eluate, indicating that the binding sites were more and more occupied with HSA by lowering the pH. This competitive effect, together with the insufficient positive charge of the antibody at higher pH, resulted in less binding of the IgG at both edges of the examined pH range. The biggest amount of antibody was bound at pH 4.6, when nearly no IgG band could be electrophoretically detected in the flow-through. Although the SDS-PAGE showed that an effective depletion of contaminating proteins (HSA, HTF) would only be achieved at pH 5.0, the working pH for the following chromatographic runs was set to pH 4.6 in order to avoid a loss of product at higher pH.

3.2. Examination of the dynamic binding capacity

In order to estimate the needed amount of Streamline™ SP XL for the 100 L pilot scale runs, it was necessary to characterise the binding capacity of the matrix. Figure 5 illustrates the IgG_1 break-through curves obtained with particle-free supernatant and cell-containing broth. Using the C10/10 column in packed bed mode more than 6.5 mg IgG/ml resin were applied without exceeding 10% of the initial antibody concentration in the flow-through. In comparison with the preceding matrix Streamline™ SP, the capacity was approx. 4.5 fold improved (data not shown), but the results could not be reproduced in expanded bed. Working with the Streamline™ 25 column and the same particle-free supernatant, the dynamic capacity was found to be only 1.5 mg IgG/ml. The processing of unclarified cell culture broth finally resulted in an additional decrease down to 0.5 mg IgG/ml. Starting out from this value and expecting an IgG_1 content of about 70 mg L^{-1} in the harvest, the required matrix volume for the antibody purification in the 100 L pilot scale runs amounted to approx. 14 L. In order to improve the efficiency of the process the possible reasons for the decreased capacity in expanded bed were investigated.

The loss of capacity was assumed to be based on the changed hydrodynamic conditions, which were therefore examined in buffer experiments. Using the Streamline™ 25 column the residence time distribution RTD of an acetone pulse was recorded for both operating modes. Figure 6 shows the resulting C- (diagram A) and the dimensionless E-functions $E(\Theta)$ (diagram B). The latter were calculated by normalising the residence time and the integral of the C-curves. The wider tailing of the E-curve and the displacement of the maximum from the calculated mean residence time $\bar{t}$ indicate that the hydrodynamic conditions in expanded bed were affected by liquid back mixing and channel formations much more than in packed bed. Correspondingly, the calculated values for the Bodenstein number Bo and the number of theoretical plates N were with $Bo = 74.3$ and $N = 37.6$ more than two-

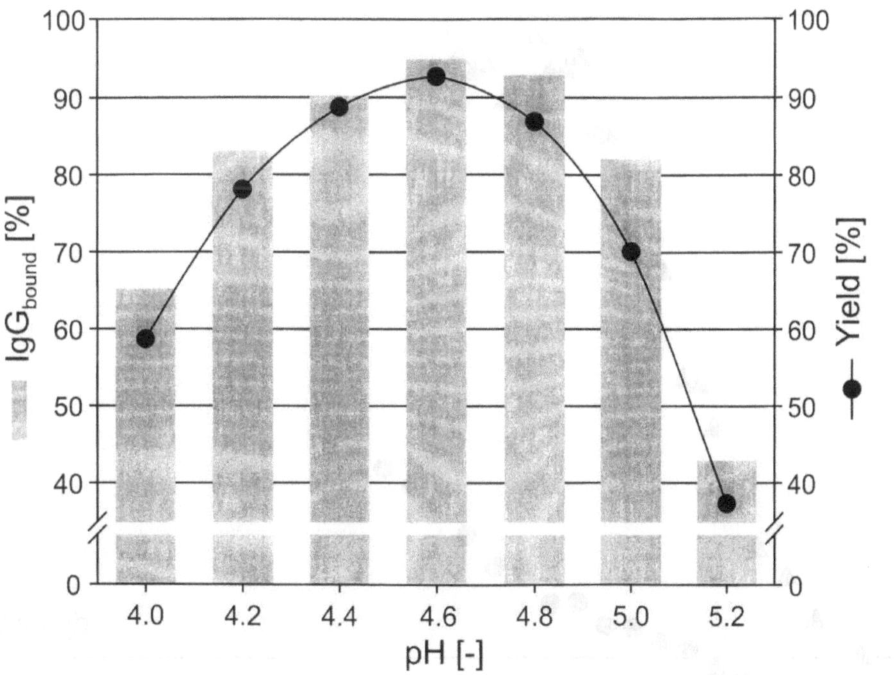

Figure 3. pH dependency of the IgG binding. Sample volumes of 100 ml of hybridoma supernatant were applied onto the C10/10 column at a flow rate of 300 cm h^{-1}. Varying the pH between 4.0 and 5.2 resulted in a maximal IgG binding of 94.8% at pH 4.6.

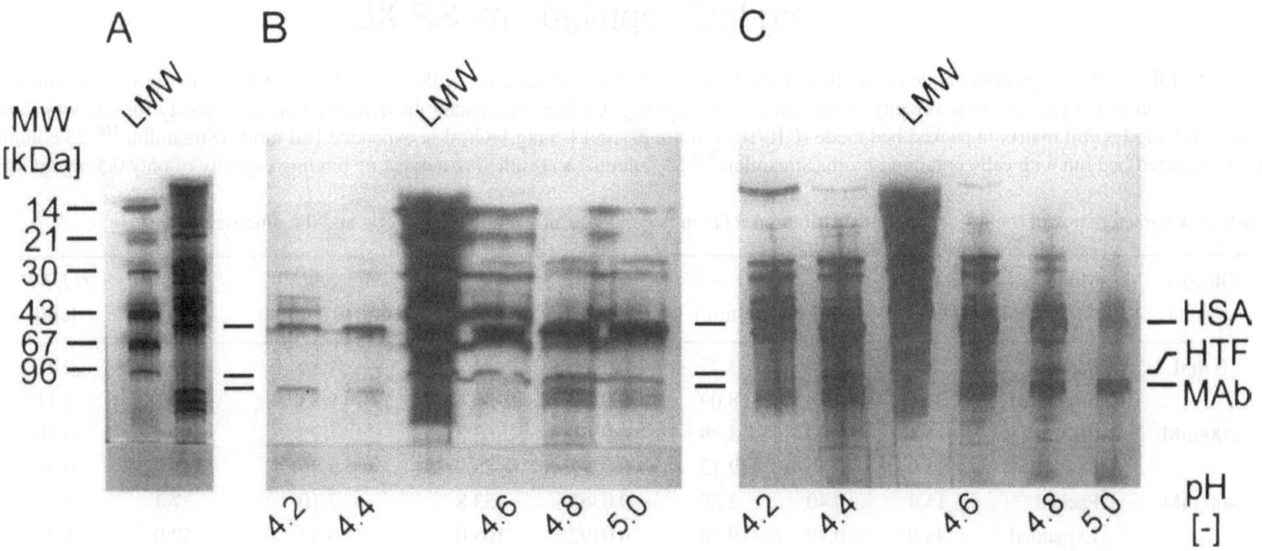

Figure 4. SDS-PAGE analyses: Comparison of the protein content in the initial feed (A), the flow-through fractions (B), and the eluates (C). The samples were taken from the experiments concerning the pH dependency of the IgG binding (see Figure 3). The largest amount of antibody was bound at pH 4.6, whereas an efficient depletion of albumin could only be achieved at pH 5.0.

fold higher for the packed bed than for the expanded bed mode (see Table 2). According to the manufacturer of the matrix the flow in a column is for $N < 25$ mainly characterized by effects of dispersion (Pharmacia, 1997). However, the number of theoretical plates N for the expanded bed amounted to only 18.7 and was thereby already in the critical area. The worse hydrodynamic conditions were additionally emphasized by the determination of the coefficients of axial dispersion D_{AX} and of the plate heights *HETP*. The comparison of the values indicate an approx. 6-fold increase of the axial dispersion in the fluidized bed (see Table 2).

The results of the tracer experiments suggest a relation between the decreased dynamic binding capacity of the matrix and the worse hydrodynamic conditions

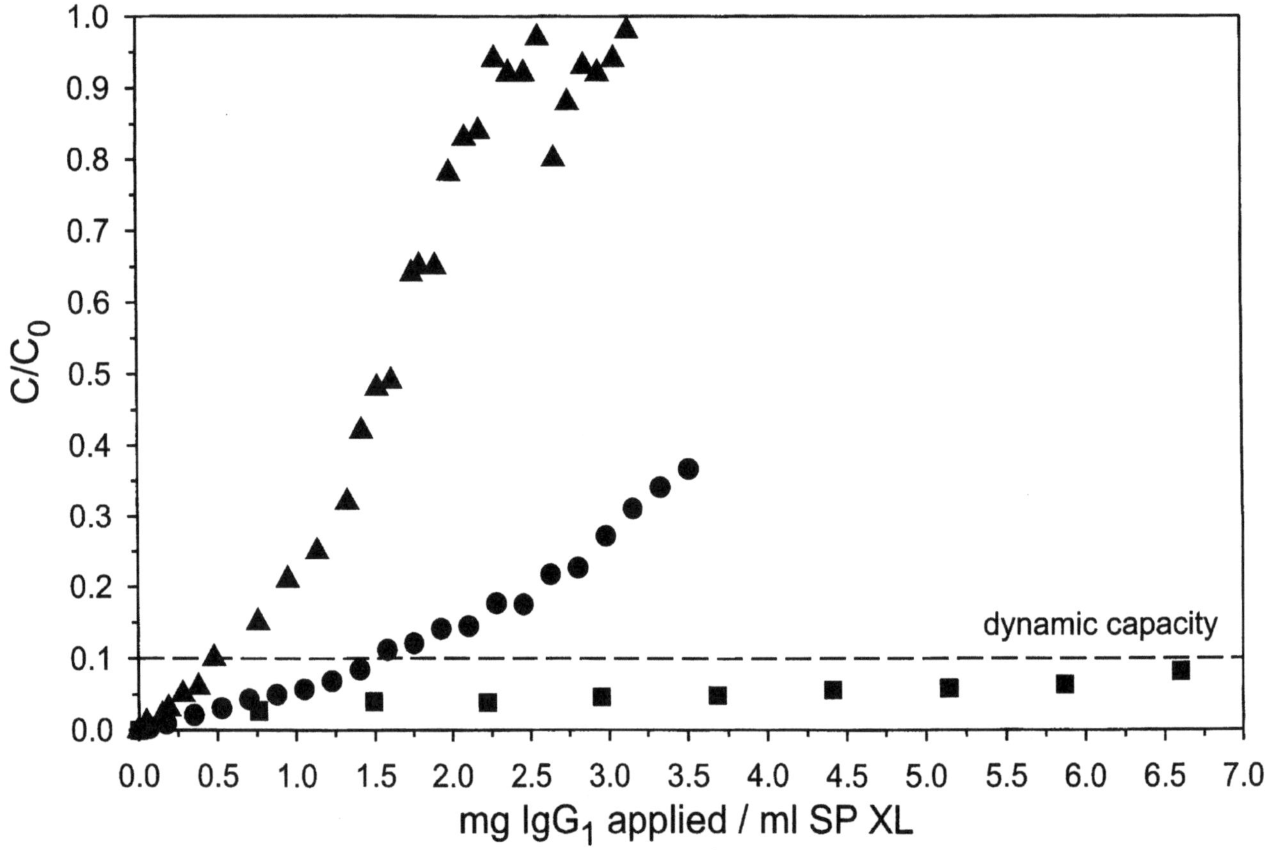

Figure 5. Effects of the operating mode on the IgG break-through. The binding capacity of the Streamline™ SP XL matrix was examined in frontal adsorption analyses at a flow velocity of 300 cm h^{-1}. Processing cell-free supernatant the dynamic binding capacity for IgG was found to be > 6.5 mg IgG/ml matrix in packed bed mode (C10/10 column, ■) and 1.5 mg IgG/ml in expanded bed mode (Streamline™ 25 column, ●). A fluidized bed run with cell-containing broth (Streamline™ 25 column, ▲) resulted in a dynamic binding capacity of only 0.5 mg IgG/ml.

Table 2. Characteristics of the residence time distribution *RTD* in dependency of the operating mode and the glucose content

Glucose content	Mode	L_c [cm]	ε [–]	$\tau = \bar{t}$ [min]	$\sigma(\Theta)^2$ [–]	Bo [–]	D_{ax} [$\times 10^{-5} m^2 \cdot s$]	N [–]	$HETP$ [cm]
0 mM	Packed	15.4	0.40	3.45	0.0266	74.3	1.54	37.6	0.41
	Expanded	39.0	0.76	8.09	0.0534	36.4	8.83	18.7	2.11
200 mM	Packed	15.4	0.40	3.49	0.0312	63.3	1.79	32.1	0.48
	Expanded	43.0	0.78	9.12	0.0222	88.9	3.80	45.0	0.96
400 mM	Packed	15.4	0.40	3.50	0.0365	53.8	2.10	27.4	5.6
	Expanded	45.0	0.79	9.58	0.0192	103.0	3.42	52.0	8.7

in expanded bed mode. The additional loss of capacity by the processing of cell-containing solutions was presumably caused by the massive adherence of biomass to the ion exchanger, which disturbed the stability of the fluidized bed.

3.3. Influence of glucose on the EBA process

In order to compensate the occuring instabilities of the expanded bed we decided to change the feed properties. However, most of the parameters were fixed by the required chromatographic conditions for the optimal product recovery. As one promising modification we decided to increase the feed viscosity. Since each change of the feed conditions in an EBA process had to consider the integrity of cells the viscosity was only moderately enhanced by the addition of 200 and 400 mM glucose. The influences of the glucose on the physical parameters of the buffer system and the cell

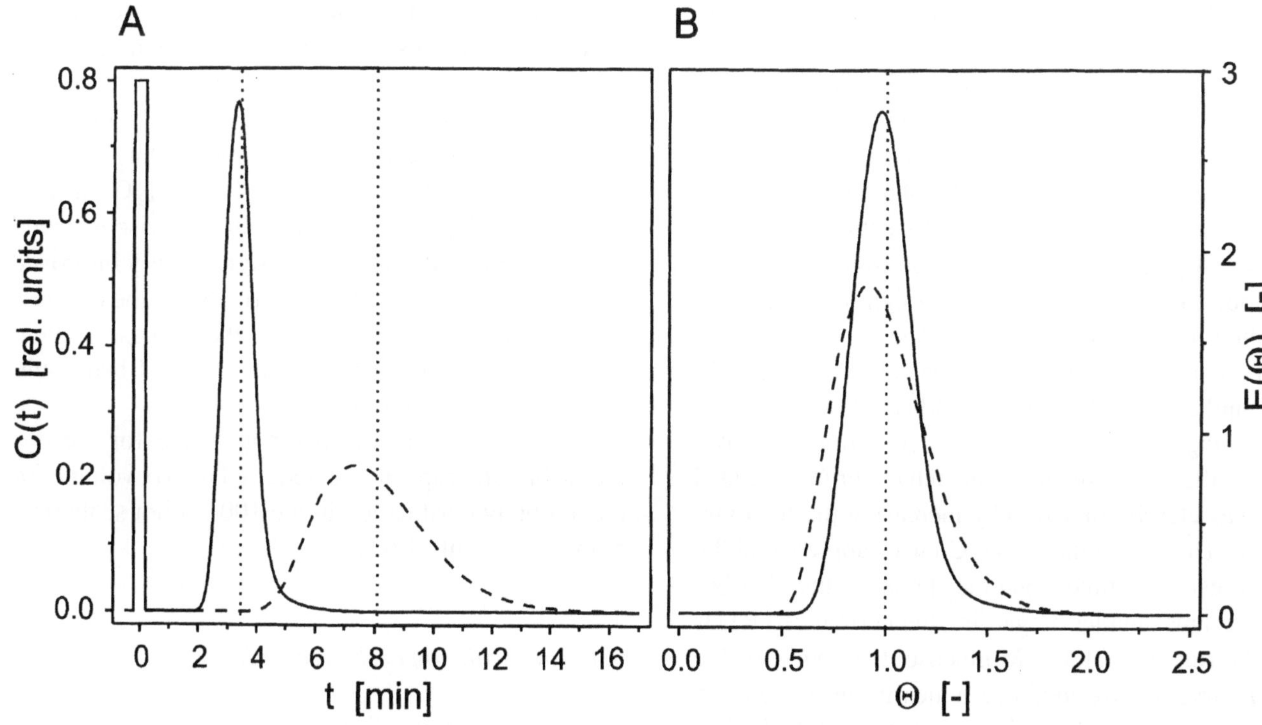

Figure 6. Examination of the hydrodynamic conditions in packed and expanded bed mode. The residence time distributions *RTD* of an acetone pulse were recorded with the Streamline™ 25 column at a flow velocity of 300 cm h^{-1}. The resulting C-curves are represented in diagram A, whereas diagram B compares the normalized *RTD*s. Solid lines: packed bed, broken lines: expanded bed, dotted lines: mean residence time $\bar{t}$.

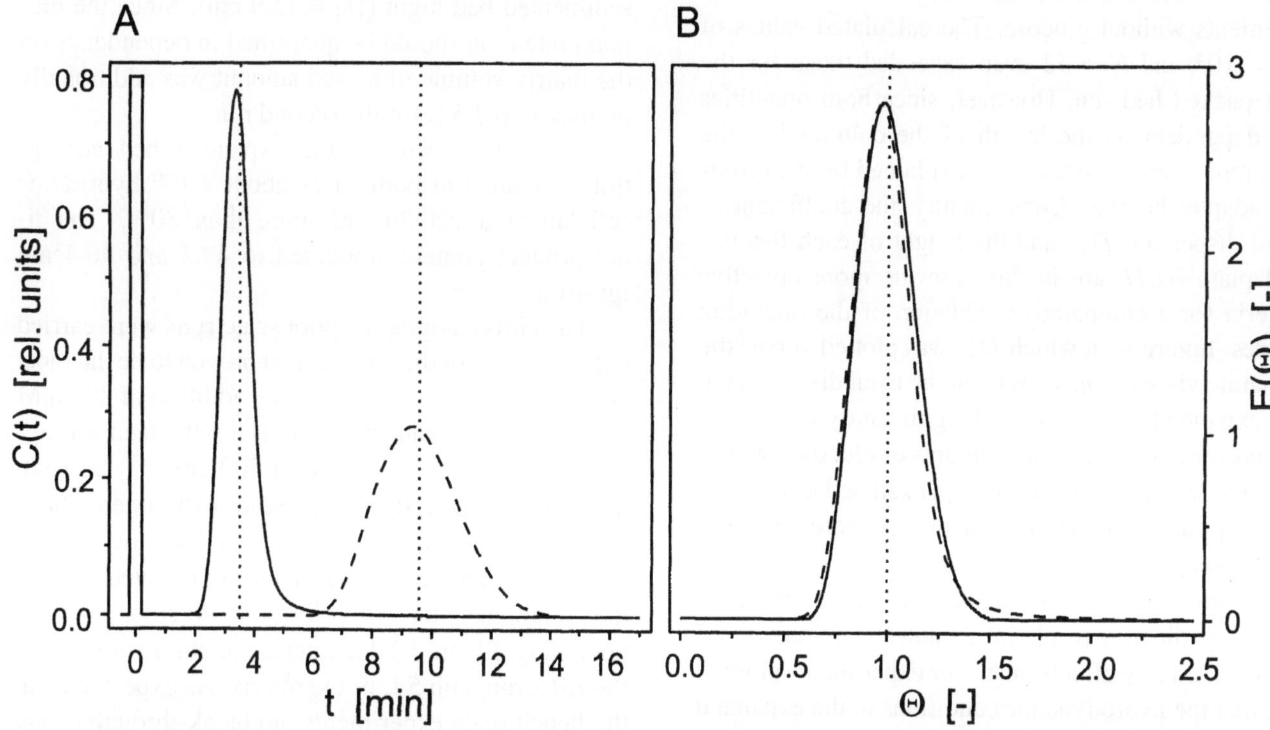

Figure 7. Influence of glucose on the hydrodynamic conditions in packed and expanded bed mode. The represented C- (A) and E-curves (B) resulted from tracer experiments, which were carried out with the Streamline™ 25 column and a 400 mM glucose solution at a flow velocity of 300 cm h^{-1}. Solid lines: packed bed; broken lines: expanded bed; dotted lines: mean residence time $\bar{t}$.

culture solutions are summarized in Table 3. The addition of 400 mM glucose increased the viscosity by approx. 20%.

For a preliminary estimation of the changed fluidization behaviour the hydrodynamic conditions were again first examined in tracer experiments. Figure 7 represents the residence time distributions of an acetone pulse, which was injected into a stream of buffer A with 400 mM glucose. In comparison to the RTDs without glucose (see Figure 6) no significant changes resulted for the packed bed run. The C- and E-curves were nearly congruent and the calculations of *Bo*, *N*, D_{ax} and *HETP* (see Table 2) indicated even a slight worsening of the hydrodynamic conditions. In contrast to that, the residence time behaviour in expanded bed was clearly improved by increasing the feed viscosity. The maximum of the C-curve approached the mean residence time $\bar{t}$ and the spread of the distribution became smaller. Regarding the normalized RTDs for the runs with 400 mM glucose, the course of the $E(\theta)$-curve in expanded bed is nearly the same as in packed bed. The achieved independence of the hydrodynamics from the operating mode indicates that the increased viscosity has actually stabilized the fluidized bed. In comparison with the previous runs the Bodenstein number *Bo* as well as the number of theoretical plates *N* were about 2.8 times higher than in the experiments without glucose. The calculated values of $Bo = 103$ and $N = 52$ even exceeded those for the first packed bed run. However, since both quantities are dependent on the length of the column L_C, the apparent improvement can be explained by the different adapter heights. Consequently, the coefficient of axial dispersion D_{ax} and the height of each theoretical plate *HETP* are in this case the more objective criteria for a comparative valuation of the operating modes. Figure 8, in which D_{ax} was plotted versus the dynamic viscosity η, shows that the axial dispersion in the expanded bed was in fact significantly reduced in the presence of increasing amounts of glucose, but not down to the values achieved in packed bed mode. The same applies to the HETP, which was decreased from 2.11 down to 0.87 cm.

In spite of the fact that the increase of the feed viscosity did not result in a complete reduction of the back mixing effects the tracer experiments pointed out that the hydrodynamic conditions in the expanded bed could be improved enormously. Since the characterization of the residence time behaviour does not directly consider any adsorption phenomena, the influence of the improved hydrodynamics on the stability of the expanded bed and the binding capacity of the matrix had to be examined, too. Therefore, frontal adsorption analyses were carried out with glucose-supplemented cell culture broth. Figure 9 represents the resulting IgG break-through curves. By the addition of 400 mM glucose the dynamic binding capacity increased from the initial 0.5 to 4.5 mg IgG/ml matrix. This result proved that the increased viscosity and thereby reduced dispersion has significantly improved the binding behaviour of the matrix. Since the biomass retention was independent from the glucose content the occurring instabilities of the fluidized bed could be obviously compensated.

On the basis of the increased value for the dynamic binding capacity the required matrix volume for the antibody purification in the 100 L pilot scale runs amounted to only 1.6 L.

3.4. Pilot scale ion exchange EBA

Two 100 L pilot scale runs were carried out with a Streamline™ 200 column. Although the optimization of the binding capacity led to a required matrix volume of only 1.6 L the column was filled with 5 L of Streamline™ SP XL in order to achieve an adequate sedimented bed hight ($H_0 = 15.9$ cm). Since the biomass retention should be quantified in dependency on the matrix volume, the used amount was additionally increased to 7.5 L for the second run.

The initial feed for the expanded bed adsorption contained in both cases about $2 \cdot 10^6$ hybridoma cells/ml at a viability of more than 80%. The final product content amounted to 49.5 and 36.4 mg IgG_1/ml.

The chromatographic pilot scale runs were carried out under the conditions that had proven to be the most efficient in bench scale. After the addition of 400 mM glucose and the adjustment to pH 4.6 the feed was applied with a linear flow rate of 300 cm h^{-1} ($\equiv$ 1570 ml min^{-1}). Since the bench scale experiments have shown that the expanded bed height increases during the load to approx. 5 fold of the sedimented matrix, the column adapter was adjusted to 90 cm at the very beginning. Figure 10 documents the chromatogram of the EBA run with 5 L of the matrix. As expected from the bench scale experiments, no break-through of the target protein occurred (data not shown).

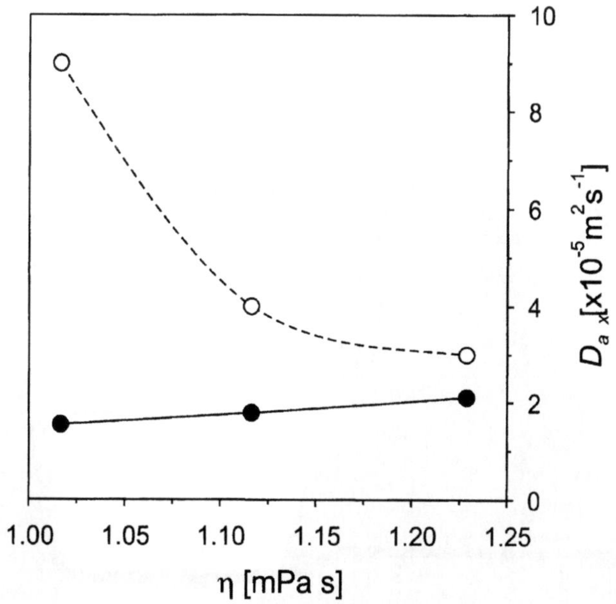

Figure 8. Dependency of the axial dispersion on the glucose content. By increasing the glucose concentration in the buffer from 0 to 200 and 400 mM (≡ 1.02; 1.12; 1.23 mPa·), the coefficient of axial dispersion D_{ax} for the expanded bed (○) approached to the one in packed bed (●).

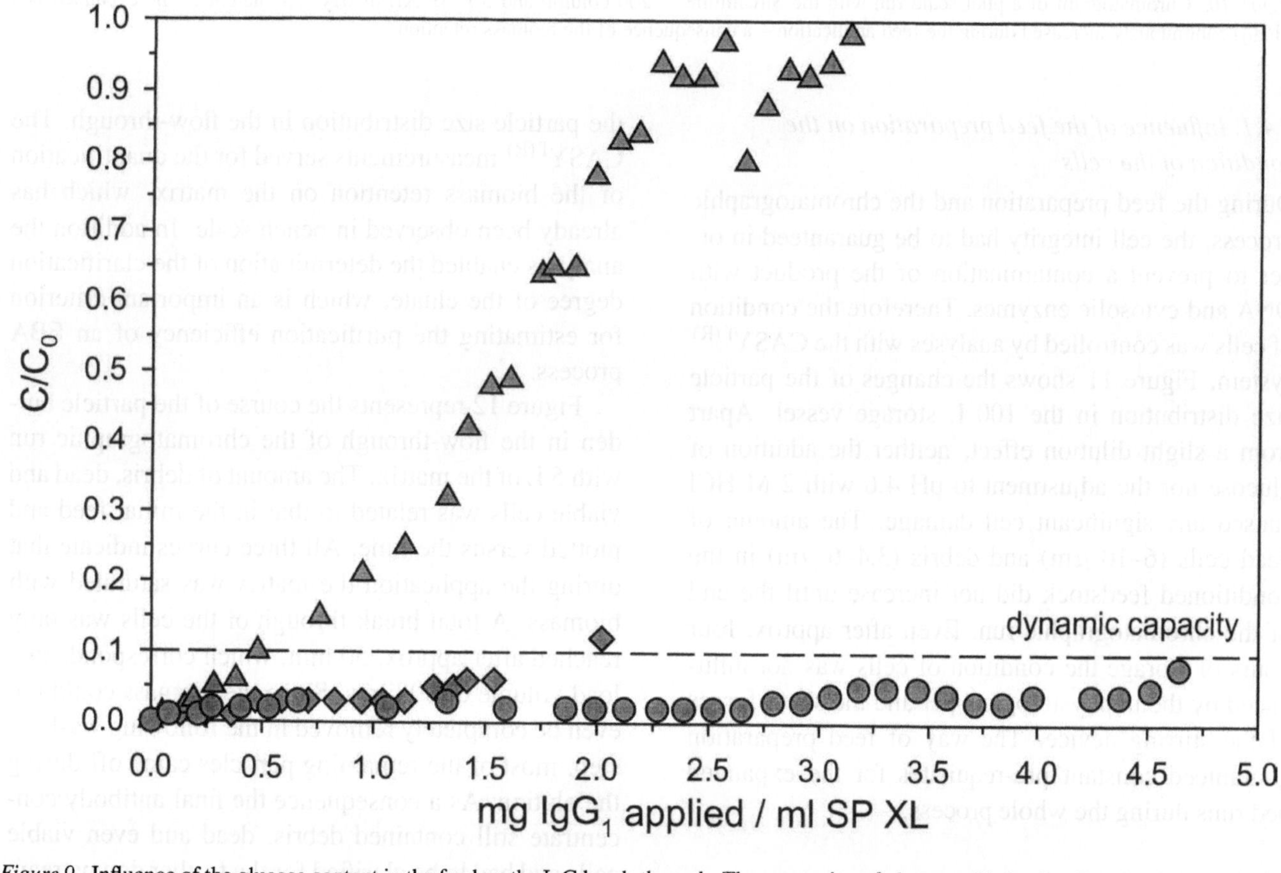

Figure 9. Influence of the glucose content in the feed on the IgG break-through. The processing of glucose-added cell culture broth (▲ 0, ♦ 200, ● 400 mM glucose) resulted in an enhanced dynamic binding capacity of the matrix from 0.5 to more than 4.5 mg IgG/ml.

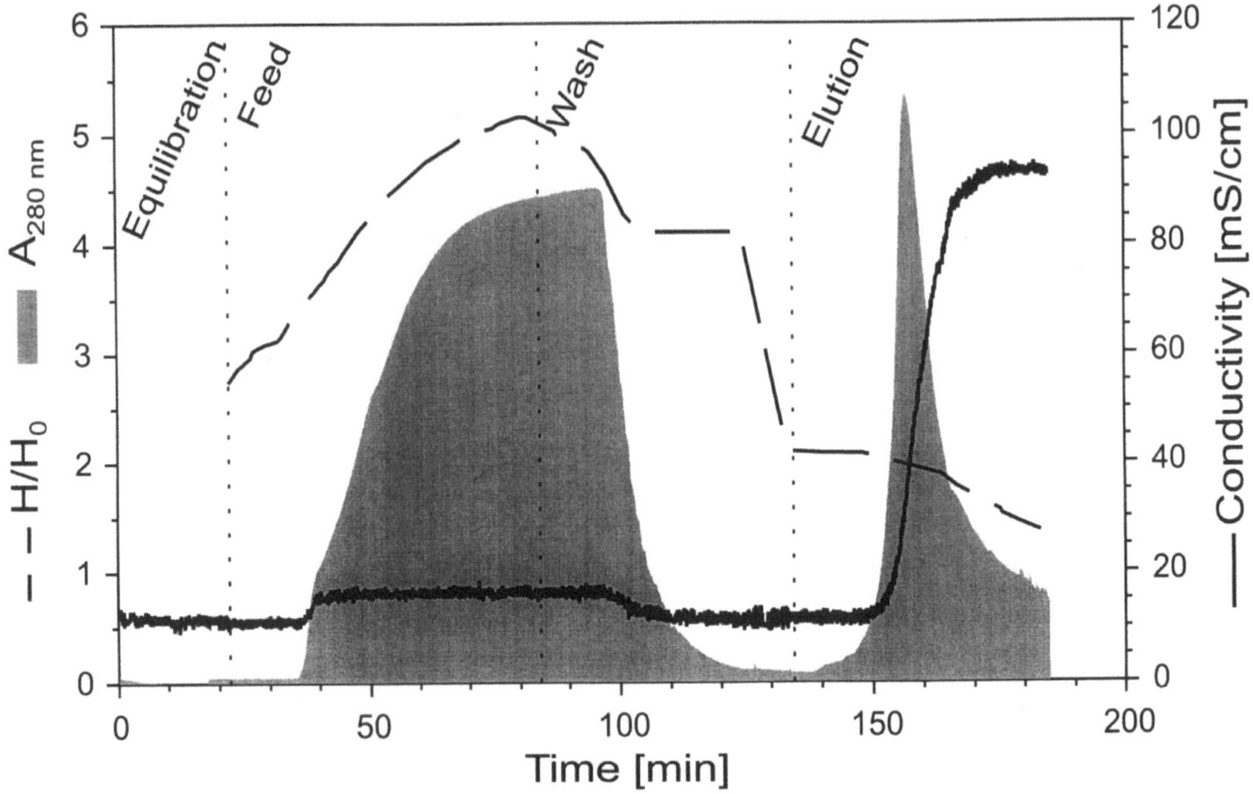

Figure 10. Chromatogram of a pilot scale run with the Streamline^{TM} 200 column and 5 L SP XL matrix. The height of the expanded bed (H/H_0) continuously increased during the feed application – a consequence of the biomass retention.

3.4.1. Influence of the feed preparation on the condition of the cells

During the feed preparation and the chromatographic process, the cell integrity had to be guaranteed in order to prevent a contamination of the product with DNA and cytosolic enzymes. Therefore the condition of cells was controlled by analyses with the CASY$^{I\,(R)}$ system. Figure 11 shows the changes of the particle size distribution in the 100 L storage vessel. Apart from a slight dilution effect, neither the addition of glucose nor the adjustment to pH 4.6 with 2 M HCl caused any significant cell damage. The amount of dead cells (6–10 μm) and debris (3.4–6 μm) in the conditioned feedstock did not increase until the end of the chromatographic run. Even after approx. four hours of storage the condition of cells was not influenced by the unphysiological pH and the sheer forces of the stirring device. The way of feed preparation guaranteed constant pre-requisites for the expanded bed runs during the whole process.

3.4.2. Biomass retention on the matrix

During the EBA chromatography the integrity of the hybridoma cells was controlled by regular analyses of the particle size distribution in the flow-through. The CASY$^{I\,(R)}$ measurements served for the quantification of the biomass retention on the matrix, which has already been observed in bench scale. In addition the analyses enabled the determination of the clarification degree of the eluate, which is an important criterion for estimating the purification efficiency of an EBA process.

Figure 12 represents the course of the particle burden in the flow-through of the chromatographic run with 5 L of the matrix. The amount of debris, dead and viable cells was related to that in the initial feed and plotted versus the time. All three curves indicate that during the application the matrix was saturated with biomass. A total break-through of the cells was only reached after approx. 50 min, which corresponds to a load volume of 80 litres. Since the biomass could not even be completely removed in the following washing step, most of the remaining particles came off during the elution. As a consequence the final antibody concentrate still contained debris, dead and even viable cells and had to be clarified for the further downstream process.

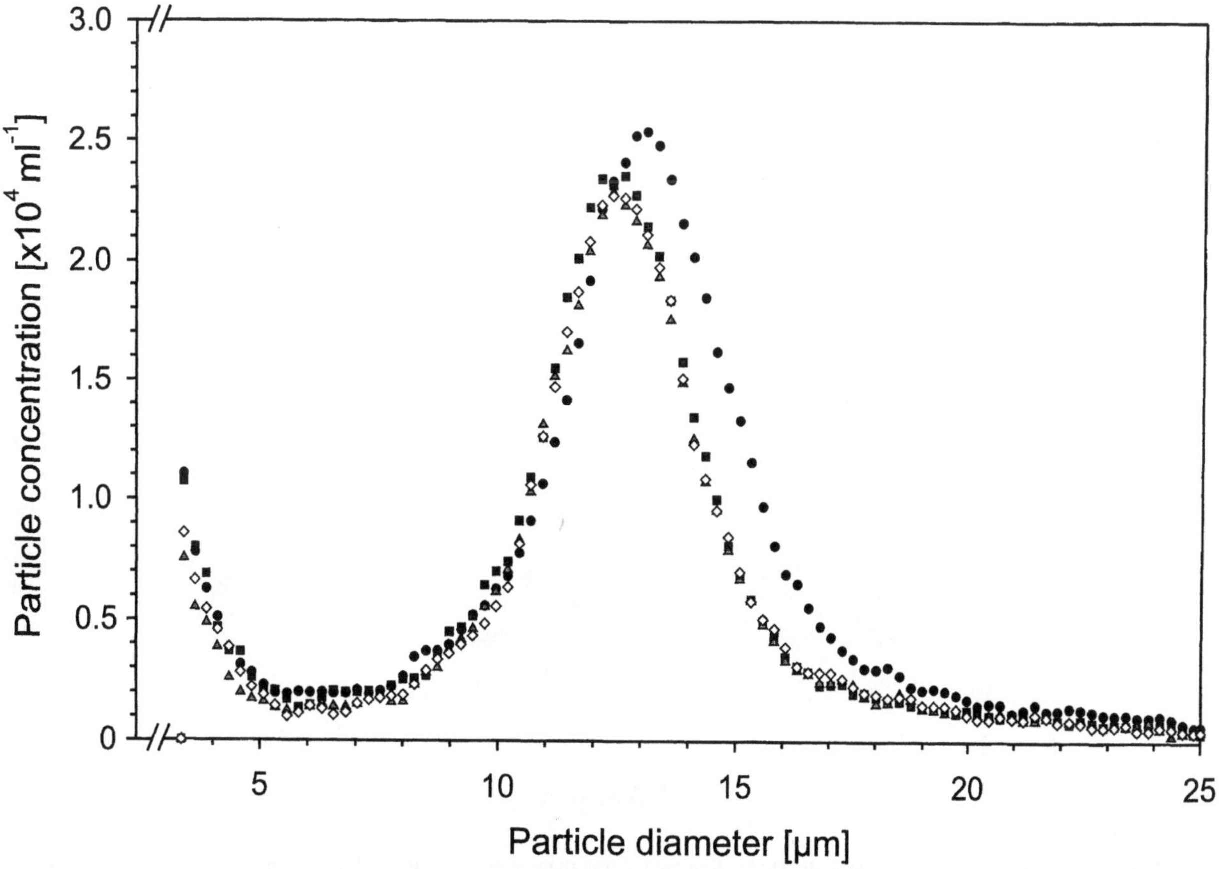

Figure 11. Influence of the feed preparation on the particle size distribution in pilot scale. Initial harvest pH 7.1 (●), glucose-added harvest pH 7.1 (■), pH 4.6 adjusted harvest at the beginning (▲) and at the end of the run (◇).

Although the debris content in the flow-through increased to more than 160% compared with the initial feed, neither the passage of the chromatographic system nor the harsh elution conditions caused any remarkable loss of the cells' viability. Since the amount of dead cells did never exceed 100%, the generation of debris could not be causally related to a damage of the cells.

In order to estimate the influence of the biomass/matrix interactions on the EBA process, the break-through behaviour of the cells was additionally examined with an increased matrix volume. Keeping all the other conditions constant the second EBA run was carried out with 7.5 L of Streamline™ SP XL. Figure 13 compares the CASY$^{I(R)}$ analyses of both runs concerning the burden of dead and viable cells in the flow-through during the load. Running the column with 7.5 L of the matrix (diagram B) instead of 5 L (diagram A) resulted in a significant increase of the cell retention, although the void volume was equal. In both cases the cell content in the flow-through increased only slowly. However, using 5 L of Streamline™ SP XL the cells already occurred in the flow-through after 10 L, whereas in the experiment with more matrix 40 L had to be applied to detect the first cell peak. At the end of the first run the CASY$^{I(R)}$ diagram of the flow-through became nearly congruent with that of the initial feed, indicating that a complete saturation of the matrix with biomass had been achieved. On the other hand, in the second run no comparable total break-through occurred even after 90 L were applied.

Experiments in bench scale pointed out that the continuous rise of the expanded bed height during the load depends on the cell retention and that the fluidization remains nearly constant after the complete saturation with biomass (data not shown). That means for the use of large matrix amounts that the column length might limit the load volume due to the biomass attachment and the thereby changed expansion behaviour. In case of biomass/matrix interactions the amount of resin used should consequently not exceed a defined value, which is dependent on the new fluidization characteristic. In addition, an up scale can

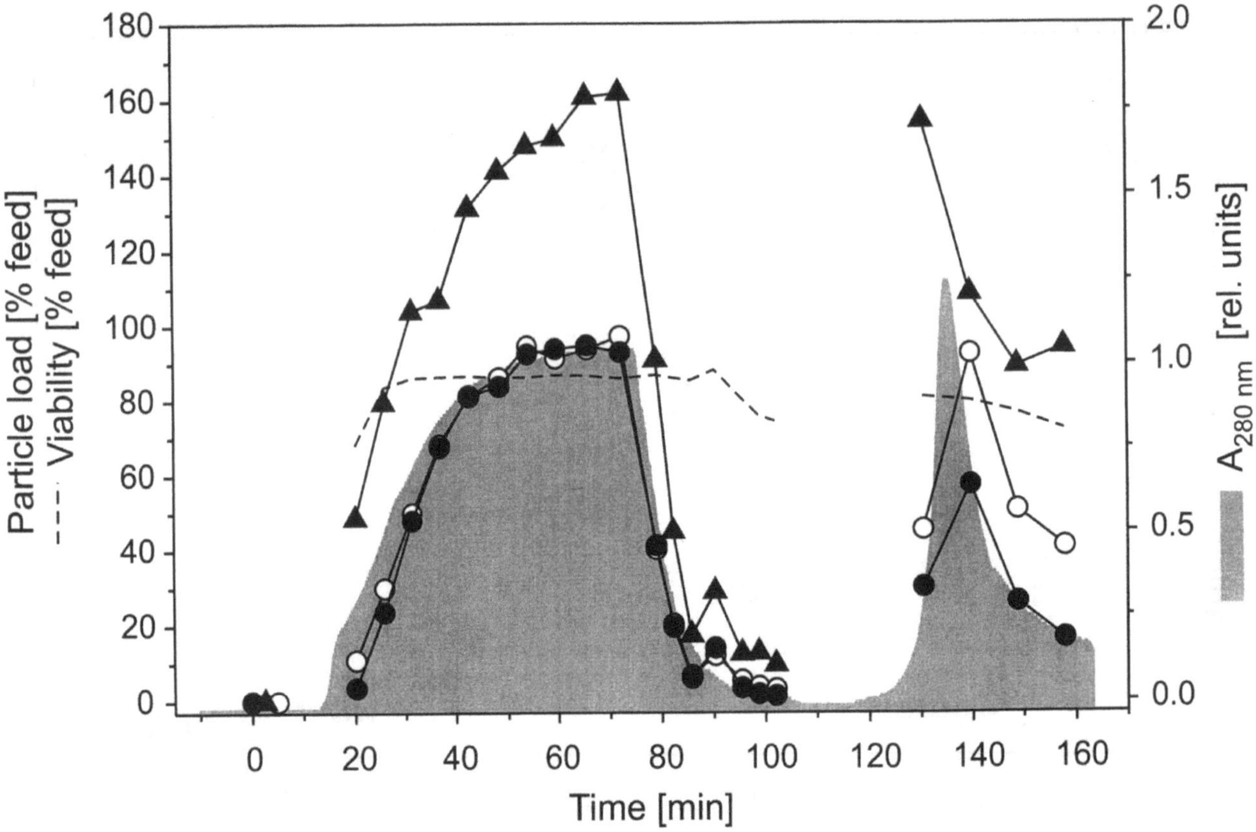

Figure 12. Normalized particle burden in pilot scale. By loading the column the matrix was saturated with debris (▲), dead (○) and viable cells (●), which were recovered in the eluate.

Table 3. Influence of the presence of glucose on the physical parameters of the solutions

	Glucose content [mM]	Dyn. viscosity η [mPa·s]	Density ρ [kg·m^{-3}]	Osmolarity [mOs·kg^{-1}]
Buffer A	0	1.017	1002	281
	400	1.229	1030	602
Buffer B	0	1.094	1039	1726
	400	1.338	1066	2198
Cell culture broth	0	1.059	1007	328
	200	1.163	1021	521
	400	1.258	1035	724

only succeed at a given column length by a simultaneous enlargement of the matrix volume and the column diameter. However, since the complete saturation of the resin with biomass was already achieved in the run with 5 L of Streamline™ SP XL the load volume could have been enlarged without expecting worse conditions.

3.4.3. *IgG$_1$ purification balance*

The success of each downstream process is dependent on the product purification balance. Table 4 summarises the results of the two pilot scale runs, in which a total volume of approx. 220 L hybridoma cell culture broth were processed. During the application the IgG content in the flow-through did not exceed the dynamic binding capacity of the matrix. The IgG binding rate amounted to approx. 95% independent from the column load, which achieved 23% of the dynamic binding capacity in the run with 5 L and 12% in the experiment with 7.5 L of the matrix. The elution was aborted after 15.3 and 15.4 L respectively, which led in both cases to an IgG concentration factor of 4.2 and a volume reduction of about 85%. In order to minimise the peak tailing, the flow velocity was reduced at the end of the washing step down to 100 cm h^{-1} (≡ 0.52 L min^{-1}), followed by the adjustment of the adapter to the surface of the decreased expanded bed height. However, the compression of the fluidized matrix led to visible channel formations and aggregations of the beads, which were presumably caused by the sticking biomass. As a consequence, parts of the gel were no longer accessible for the buffer stream, which resulted in an incomplete desorption of the bound substances. The unfavourable elution conditions explain the low

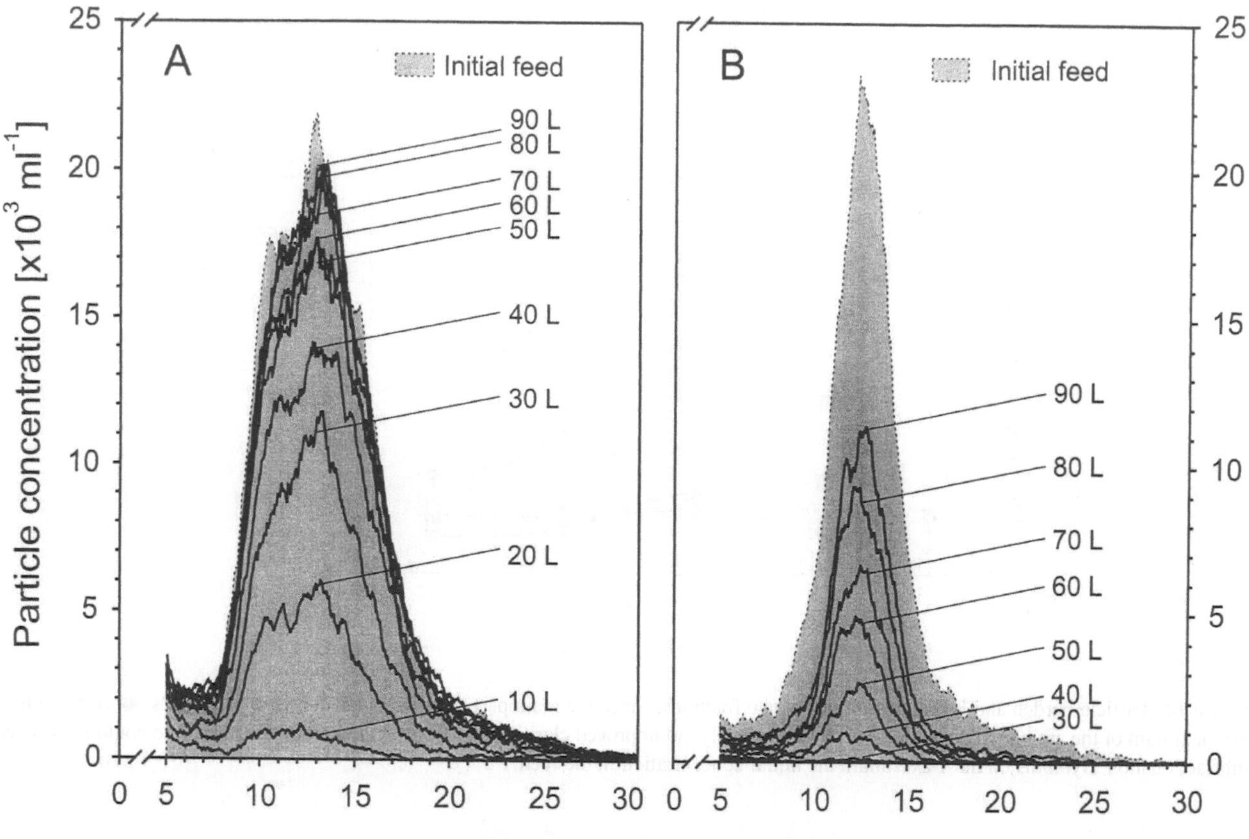

Figure 13. Dependency of the cell retention on the amount of matrix. The break-through of cells was recorded by regular CASY[Γ(R)] analyses of the flow-through. Diagram A: run with 5 L matrix; diagram B: run with 7.5 L matrix.

Table 4. Balance sheet of the pilot scale runs. The processing of approx. 100 L of hybridoma cell culture broth took less than 3.5 hours and led to a 4.2 fold concentrated IgG eluate.

		Run 1 (5 L SP XL)		Run 2 (7.5 L SP XL)	
		Load	Eluate	Load	Eluate
Volume [L]		105.4	15.3	112.1	15.4
IgG concentration [mg·L^{-1}]		49.5	208.3	36.4	154.2
IgG total [mg]		5215	3187	4086	2374
IgG bound [%]		95.6		94.0	
Recovery [%]		69.0		61.8	
Yield [%]		65.9		58.1	
IgG concentration factor		6.9		7.3	
Volume reduction [%]		85.5		86.3	
Particle depletion total [%]	Debris	93.9		77.7	
	Dead cells	91.0		86.0	
	Viable cells	94.5		89.4	
Column load [% dyn. bind. capacity]		23		12	

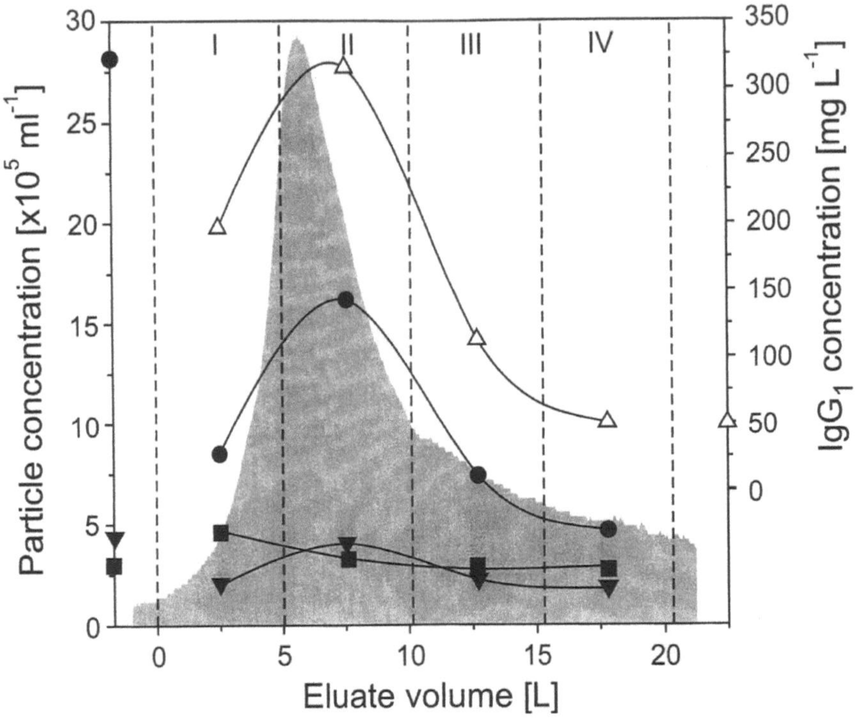

Figure 14. Particle burden and IgG content in the eluate fractions. Since the main part of biomass (■ debris, ▼ dead cells, ● viable cells) and the maximum of the antibody (△) were eluted simultaneously, no improved clarification degree of the product concentrate could be reached by subfractionation. Symbols on the x-axes mark the initial concentration in the feed.

recovery rates and IgG yields of 69 and 66% for the run with 5 L of Streamline™ SP XL and only 62 and 58% for the other run. The additional loss of product in the second experiment was caused by the enlarged matrix volume, which led to a wider tailing of the elution peak, so that parts of it exceeded the range of quantification.

Comparing the initial feed with the eluate, the total amounts of debris, dead and viable cells were depleted between 77 and 95%. In spite of that, the concentration of biomass was more or less the same in the eluate as in the feed. Figure 14 illustrates the detailed courses of the particle burden and of the IgG content during the elution. Since the main part of biomass and the maximum of the antibody were eluted simultaneously, no improved clarification degree of the product concentrate could be reached by subfractionation.

At the end of the elution the matrix was still contaminated with biomass. Consequently, the regeneration required a time- and material-consuming CIP procedure. The performance, described in Table 1, was carried out after each run and led to comparable preconditions for the following experiment.

Since the biomass attachment is a saturation process, the total amount of retained biomass becomes independent from the amount of applied cell culture broth, so that the results would not get worse by increasing the load volume. In view of the fact that only up to 23% of the dynamic binding capacity were used, at least an improvement of the concentration factor and the IgG yield can be expected, if the column load would actually be maximised.

In spite of all difficulties, the EBA process succeeded in concentrating the IgG content of approx. 100 L unclarified and undiluted hybridoma broth in less than 3.5 hours.

4. Conclusion

The aim of the investigations was to establish an integrative expanded bed process for the pilot scale recovery of monoclonal antibodies from undiluted hybridoma cell culture broth. Using the cation exchange matrix Streamline™ SP XL, the main point of interest was to achieve a high binding capacity at gentle conditions, which should guarantee the integrity of the cells.

The product specific optimization of the method was limited by the high amount of contaminating proteins and the unfavourable characteristic of the antibody's charge. In addition, the cell culture broth should be processed without a preliminary dilution step. Consequently, the salt concentration amounted to approx. 150 mM, which in general is a disadvantageous prerequisite for the use of ion exchangers. However, optimizing the chromatographic parameters resulted in a dynamic binding capacity of more than 6.5 mg IgG/ml matrix in packed bed mode. Comparative examinations with the precursor matrix Streamline™ SP showed that the capacity was improved by approx. 4.5 fold, but the results could not be reproduced in expanded bed mode. In fluidized bed the dynamic binding capacity only reached a satisfying value of 4.5 mg IgG/ml matrix after modifying the feed by the addition of glucose. Residence time analyses proved a relation between the resulting enhancement of the feed viscosity and improved hydrodynamic conditions in the column. Presumably, stabilizing effects of the glucose on the expanded bed had favoured the binding behaviour.

The major problem of using ion exchange EBA for the processing of hybridoma cell culture broth is the biomass retention. The adherence of debris and cells to the matrix drastically changed the fluidization behaviour, which caused a deterioration of the binding capacity and an enlargement of the elution volume. Since the eluates still contained biomass and even viable cells, a following clarification step was required for the further downstream process.

As a consequence of the mentioned difficulties, the pilot scale runs entailed a relatively high loss of product and a time- and material-consuming regeneration process for the matrix. However, with regard to the timesaving potential of the method and the low overall expenses, ion exchange EBA might be a profitable alternative for a conventional downstream process in cell culture technology, if the problem of biomass retention could be solved.

Acknowledgement

We gratefully acknowledge Amersham Pharmacia Biotech for the financial support and for kindly providing the pilot scale equipment. We would especially like to thank Dr L.-J. Larsson and Dr P. Girard.

References

Arshady R (1991) Beaded polymer supports and gels: 1. Manufacturing techniques, 2. physico-chemical criteria and functionalization. J. Chromatogr. 586: 181–219.

Barnfield Frej AK, Hjorth R and Hammarström A (1994) Pilot scale recovery of recombinant annexin V from unclarified *Escherichia coli* homogenate using expanded bed adsorption. Biotechnol. Bioeng. 44: 922–929.

Bascoul A, Delmas H and Couderc JP (1988) Caracteristiques Hydrodynamiques de la Fluidisation Liquid-Solid: Influence du Distributeur. Chem. Eng. J. 37: 11.

Chang YK and Chase HA (1996) Development of operating conditions for protein purification methods using expanded bed techniques: The effect of the degree of bed expansion on adsorption performance. Biotechnol. Bioeng. 49: 512–526.

Chase HA (1994) Purification of proteins by adsorption chromatography in expanded beds. Trends Biotechnol. 12: 296–303.

De Luca L, Hellenbroich D, Titchener-Hooker NJ and Chase HA (1994) A study of the expansion characteristics and transient behaviour of expanded bed adsorbent particles suitable for bioseparations. Bioseparation 4: 311–318.

Draeger NM and Chase HA (1991) Liquid fluidized bed adsorption of proteins in the presence of cells. Bioseparation 2: 67–80.

Eketorp R (1992) Affinity chromatography in industrial plasma protein fractionation. Engineering Foundation Conference, Sep. Interlaken (CH); Recovery of Biological Products 6: 20–25.

Fitzer E and Fritz W (1982) Technische Chemie: Einführung in die chemische Verfahrenstechnik. Springer-Verlag, Berlin, Heidelberg, New York.

Glauner B (1991) Vitalitätskontrolle in der Zellkulturtechnik. Charakterisierungvon Zellen durch hochauflösende Bestimmung der Größenverteilung. BioTec 3(5): 58–61.

Goochee CF (1992) Bioprocess factors affecting glycoprotein oligosaccaride structure. Dev. Biol. Stand. 76: 95–104.

Heidemann R, Riese U, Lütkemeyer D, Büntemeyer H and Lehmann J (1994) The SuperSpinner: A low cost animal cell culture bioreactor for the CO_2 incubator. Cytotechnology 14: 1–9.

Heukeshoven J and Dernick R (1988) Improved silver staining procedure for fast staining in PhastSystem™ development unit. Electrophoresis 9: 28–32.

Hjorth R (1997) Expanded bed adsorption in industrial bioprocessing: Recent developments. Trends Biotechnol. 15(6): 230–235.

Huneke FU (1995) Biokatalyse im Flüssig-Flüssig-Fest-Dreiphasen-Bioreaktor: Reaktionstechnische Untersuchungen des Reaktorverhaltens. Dissertation, Universität Bielefeld.

Jockusch BM et al. (1986) Probing the cytosceleton by microinjection. Cell and Molecular Biology of the Cytoskeleton, J.W. Shay (ed.), Plenum Publishing Corp., New York: 1–40.

Johansson HJ, Jägersten C and Shiloach J (1996) Large scale recovery and purification of periplasmic recombinant protein from *E. coli* using expanded bed adsorption chromatography followed by new ion exchange media. J. Biotechnol. 48: 9–14.

Karau A, Benken C, Thömmes J and Kula MR (1997) The influence of particle size distribution and operating conditions on the adsorption performance in fluidized beds. Biotechnol. Bioeng. 55(1): 54–64.

Köhler G and Milstein C (1975) Continuous cultures of fused cells secreting antibody of predefined specificity. Nature 256: 495–497.

Le MS and Gollan KL (1989) Fouling of microporous membranes in biological applications. J. Membr. Sci. 40: 231–242

Lehmann J, Piehl GW and Schulz R (1987) Bubble-free cell culture aeration with porous moving membranes. Dev. Biol. Stand. 66: 227–240.

Levenspiel O (1972) Chemical Reaction Engineering. John Wiley and Sons, New York.

Lind W, Lietz M, Jäger V and Wagner R (1991) Proteolytic activities in serum-free supernatants of mammalian cell lines. Animal Cell Culture and Production of Biologicals. Kluwer Academic Publishers, Dordrecht.

Lütkemeyer D, Bretschneider M, Büntemeyer H and Lehmann J (1993) Membrane chromatography for rapid purification of recombinant Antithrombin III and monoclonal antibodies from cell culture supernatant. J. Chromatogr. 639: 57–66.

Noda M, Sumi A, Ohmura T and Yokoyama K (1996) European Patent Application EP 0699687A2.

Nowinski RC et al. (1983) Monoclonal antibodies for diagnosis of infectious diseases in humans. Science 219: 637–644.

Pharmacia (1997) Expanded bed adsorption: Principles and Methods.

Schärfe J (1988) Impulsflächenanalyse. LaborPraxis 10: 1096–1103.

Tebbe H, Lütkemeyer D, Gudermann F, Heidemann R and Lehmann J (1996) Lysis-free separation of hybridoma cells by continuous disc stack centrifugation. Cytotechnology 22: 119–127.

Thömmes J (1997) Fluidized bed adsorption as a primary recovery step in protein purification. Advances in Biochemical Engineering/Biotechnology 58: 185–230.

Weiss S (1985) Verfahrenstechnische Berechnungsmethoden 8: Experimente in der Verfahrenstechnik. Verlag Chemie, Weinheim.

Werner RG (1990) Cultivation of mammalian cells from the view point of the industry. Proceedings of the German-Japanese workshop on animal cell culture technology in Hamburg.

Address for correspondence: N. Ameskamp, Institute of Cell Culture Technology, Faculty of Technology, University of Bielefeld, P.O. Box 100131, 33501 Bielefeld, Germany Tel. 521-106-6234; Fax: 521-106-6328; E-mail: nam@zellkult.techfak.uni-bielefeld.de

Bioseparation **8:** 189–193, 1999.
© 1999 *Kluwer Academic Publishers.*

Capture of human Fab fragments by expanded bed adsorption with a mixed mode adsorbent

Marie Bendix Hansen*, Allan Lihme[1], Mari Spitali[2] & David King[2]
[1] *UpFront Chromatography A/S, Lersø Parkallé 42, DK-2100 Copenhagen, Denmark. Homepage: www.upfront-dk.com; *E-mail: mbh@upfront-dk.com;* [2] *Celltech Therapeutics, Slough, U.K.*

Received 8 September 1998; accepted in revised form 8 May 1999

Abstract

A novel group of mixed mode adsorbents has been developed for purification of monoclonal and polyclonal antibodies from a broad range of raw materials such as hybridoma cell culture, ascites fluid, animal sera, milk, whey and egg yolk. The aim of this study was to determine whether such mixed mode adsorbents were also useful for the recovery of recombinant proteins from microbial feedstocks. This paper describes the performance of one of these adsorbents for expanded bed capture of a human Fab fragment from recombinant *E. Coli* cell extracts.

It is concluded that the mixed mode adsorbent binds the Fab fragment efficiently from crude extracts without any requirement for preconditioning the extract by for example de-salting or dilution. The capacity of the mixed mode adsorbent is approx. 12 mg Fab/ml matrix.

The novel mixed mode adsorbent can be useful during production of highly purified Fab fragments as the first step in a purification scheme. In this respect the mixed mode adsorbent is advantageous over alternative commercially available ion-exchange materials which require pre-conditioning of cell extract for Fab' capture. Together with the concentration and clarification effect a significant enrichment of the Fab fragment is obtained in one single high yield operation.

Introduction

Large scale isolation of Fab fragments may be a troublesome task since traditional tools for antibody purification such as protein A and ion exchangers have draw backs that make them difficult to apply on crude raw materials. Protein A and protein G binds immunoglobulins primarily through the Fc-region, although the low affinity binding sites in the Fab region can also be used for purification of a subset of Fab fragments (Carter et al., 1992; Proudfoot et al., 1992).

The binding of proteins to ion exchangers is very dependent on ionic strength. Most non-treated feedstocks have a relatively high ionic strength and often result in a generally low binding efficiency. In this case it is necessary to pre-condition the feedstock before isolation of the Fab fragment. For large scale production dilution increases the volumes, which need to be handled and consequently the process time significantly and also increases the problem of waste disposal. In addition the binding selectivity of ion-exchangers is often not very high.

Mixed mode adsorbents comprise low molecular weight ligands that have hydrophobic and hydrophilic functionalities within the same molecule (Yon, 1972; Hofstee, 1973; Yon and Simmonds, 1975; Sasaki et al., 1979; Bischoff and McLaughlin, 1984; Ruann et al. 1988; Kasche et al., 1990; Teichberg 1990; Olander, Hansen and Lihme, 1996; Burton et al., 1997; Olander and Lihme, 1998; Hansen and Lihme, 1996). The ligands comprise a hydrophobic core e.g. an aromatic or hetero-aromatic ring structure, onto which hydrophilic or ionic substituents e.g. amino groups or carboxylic acid groups are attached. The protein binding mechanisms seem to be rather complex, but mixed mode ligands generally show optimum binding under conditions where the substituents are on a non-charged form. The binding strength is therefore strongly pH dependent but in contrast to ion-exchangers, largely independent of ionic strength in the raw material. Binding will occur at pH values characteristic the particular mixed mode ligand and the specific protein, and elution is most often performed by a simple change in pH. The number and specific nature of the substituents influences the overall binding specificity towards different proteins.

The mixed mode ligand used in this study comprises an aromatic ring-system with a carboxylic acid substituent (Patent application Lihme and Hansen, 1997). The ligand is bound covalently to 6% cross-linked agarose comprising a solid glass core giving the beads a high density (Lihme et al., 1997). The resulting mixed mode adsorbent shows a selective binding of immunoglobulins in complex protein mixtures. IgG's bind to the adsorbent in the pH range of 4.0 to 6.0 and are typically eluted at pH 7–9. A high binding constant makes the adsorbent well suited for purification of monoclonal antibodies from dilute ($<$ 100μg Mab/ml) hybridoma cell culture supernatants (Lihme & Hansen, 1997). The mixed mode ligand is furthemore stable to regeneration with 1 M sodium hydroxide.

In this study we have shown that this mixed mode adsorbent also has affinity towards Fab fragments. The isolation of a human Fab fragment from a recombinant *E. coli* cell extract, was achieved using expanded bed technology. For this EBA application the adsorbent employed was of a high density (approx. 1.4 g/ml).

Materials and methods

Preparation of feedstock

Recombinant human Fab' was expressed in *E. coli* such that Fab' was accumulated in the cell periplasm. Fermentations were carried out and periplasmic cell extracts were prepared as previously described (Humphreys et al., 1998).

Adsorbents

The adsorbent used in this study is an experimental product produced by UpFront Chromatography A/S, Denmark. The adsorbent is a variant of the standard product FastMabs AD. Cat. No.: 150-0200, wherein the ligand concentration has been optimised. The density of the agarose/glass composite matrix was 1.4 g/ml. *"Red matrix":* The same adsorbent additionally coupled with a red Remazol dye. About 10% of the bed volume was "red matrix" to make it possible to visually monitor the expansion of the adsorbent during capture of the Fab fragments from the non-clarified cell extract.

Equipment

Initial determination of the optimal binding pH of the Fab fragment to the mixed mode matrix was carried out in a 0.5 cm diameter Poly-Prep Chromatography Columns (BioRad USA).

An FPLC column with a diameter of 5 mm (HR 5/20, Amersham Pharmacia Biotech) was modified for the expanded bed adsorption experiments. The top and bottom filters were removed so that the cell containing raw materials could pass freely through the column.

A HiLoad^{TM} pump P-50 (Amersham Pharmacia Biotech) and a dual path monitor UV-2 (Amersham Pharmacia Biotech) were used to control the experiment.

Determination of optimal binding pH vs. Purity

Small-scale experiments were performed at five different pH values to determine the optimal binding pH. The experiments were performed as batch incubations, where the matrix was eluted as a packed bed column. The tests were performed at pH 4.0, 4.5, 5.0 and 5.2.

0.25 ml mixed mode matrix was equilibrated with a 10 mM sodium citrate buffer with respective pH values. The *E. coli* extract was adjusted to the respective pH values with 1 M HCl. The matrix was incubated for 2 h with the pH adjusted cell extract at 8mg Fab (6.45 ml) per ml matrix. After incubation the matrix was transferred to a small column and washed with 5 ml 10 mM sodium citrate with the respective pH values. Elution was carried out with 2.5 ml 30 mM lysine pH 11.5 and the purity of the eluate was estimated by SDS-page gel electrophoresis.

Expanded bed capture of Fab fragments with mixed mode matrix

After determination of the optimal pH at pH 5.2, expanded bed chromatography was carried out using 2 ml mixed mode matrix + 0.2 ml "red matrix" equilibrated with 10mM sodium citrate pH 5.2. 87 ml Fab extract was adjusted to pH 5.2 with 0.1 m HCl and loaded onto the column in expanded bed mode (linear flow rate: 2 cm/min).

The column was washed with 30 ml 10 mm sodium citrate pH 5.2 (Linear flow rate: 4 cm/min). The matrix was eluted in expanded bed mode with 10 ml 30 mm lysine pH 11.5. (Linear flow rate: 2 cm/min)

The purity of the Fab fragments was estimated by SDS-Page with precast 4–20% tris/glycine gradient gels. Precast gels (Novex, USA). Samples were applied under non-reduced conditions and the gels were stained with coomassie brilliant blue. The purity of the Fab fragments was estimated by scanning the gels using an Agfa flatbed scanner and CREAM software (Kem-En-Tec A/S, Denmark).

Biological activity

Kinetic analysis to determine the on and off rate for the Fab binding to its antigen was performed using a BIACORE 2000 (BIACORE AB). The assay involves capture of the antigen by an immobilised capture molecule on a CM5 sensor chip followed by an injection of the Fab fraction. Kinetic parameters were calculated using BIAevaluation 2.1 software.

Fab was quantitated using a Fab assembly ELISA, based on the capture of the Fab' onto microtitre plates coated with anti-Fd antibody (ATCC) and then revealing the captures Fab using an anti-kappa horseradish peroxidase conjugated antibody (The Binding Site).

Results and discussion

Optimal binding pH vs. Purity

Initial studies have shown that this particular mixed mode adsorbent shows optimal binding of immunoglobulins in the range of pH 4.0 to 6.0. The binding selectivity is strongly pH dependent, where the most selective binding occurs at the upper end of the binding-pH range (Patent application, Lihme and Hansen, 1997).

The intention in this study was to work with maximum binding capacity rather than maximum purity. The influence of binding pH on binding efficiency and resulting purity is shown in Table 1.

Table 1. Optimal binding pH vs. Purity: Performed as described in materials and methods. Binding efficiency was determined by quantitative ELISA of the Fab'content af the feedstock and eluent. Selectivity was estimated from SDS-PAGE.

Binding pH	Binding efficiency,%	Selectivity
4.0	100	no selectivity
4.5	100	no selectivity
5.0	>95	low selectivity
5.2	approx. 95	high selectivity

The purity of the eluted Fab fragments depends on the binding pH. At low pH other proteins also bind strongly to the column (i.e. matrix has no selectivity). The pH for this specific Fab fragment with the highest selectivity was pH 5.2 and therefore this was used for further experiments. At this pH approx. 95% of the Fab' bound on the matrix.

Elution conditions/biological activity

Using the standard elution conditions described for the corresponding FastMabs AD adsorbent (i.e. neutral elution buffer) (FastMabs A^{TM}, User Guide) it was found that only 60–70% of the bound Fab-fragment were eluted from the column. Investigation of alternative pH conditions demonstrated that in this particular case it was necessary to use an elution buffer with pH 11.5 (i.e. 30mM lysine/NaOH pH 11.5). This high pH may lead to the risk of denaturation of the antibody fragment and the destruction of the biological activity. However, due to the buffer capacity of the ligand on the adsorbent and the buffer capacity of the bound protein itself, the eluate only reaches approx. pH 9.5. The ability of the eluted Fab fragments to bind antigen determined to be fully intact when eluted at this relatively high pH, with no apparent loss of binding affinity.

Expanded bed capture

Based on the results of the initial determination of optimal binding pH and elution conditions as described above, the adsorbent was used for expanded bed adsorption. Due to limited availability of the crude raw material it was chosen to scale down the experiment using a column with a diameter of 5 mm.

EBA columns in that size are not commercially available which led to a modification of a standard FPLC column by removing the bottom and top filters from the adapters. With this modification it was possible to pass the crude raw material through the column without clogging. Although this was not an optimal solution from a technical point of view it worked to such a degree that we were able to demonstrate the feasibility of expanded bed adsorption. However, due to the small diameter of the column combined with the very high cell density it was necessary to apply a mild centrifugation step to the raw material in order to lower the viscosity before loading (1500 g for 5 min). The centrifugation did not

Table 2. Purification of Fab fragments Main details frc ı EBA purification of human Fab fragments from E. coli cell extract.

Column diameter:	0.5 cm
Packed bed height: (H_0)	11.2 cm (2.2 ml adsorbent)
Flow rates:	Adsorption: 2 cm/min Washing: 4 cm/min Elution: 2 cm/min
Expansion, H/H_0: (Adsorption)	1.8
Raw material:	87 ml cell extract containing 0.31 mg Fab/ml pH 5.2
Volume of eluate:	10 ml (Expanded bed elution)
Concentration factor:	8.7
Yield:	90%
Capacity:	Approx. 12 mg Fab/ml matrix
Total process time:	Approx. 5 hours.

clarify the raw material, but merely eliminated the bulk of cells and lowered viscosity significantly. The consequence of not performing the centrifugation step was a break down of the stable expanded bed and the formation of channels. This resulted in a very poor adsorption efficiency and early break through of the Fab fragments. Parallel tests have shown that the centrifugation step is *not* required if the adsorption is performed in a 2-cm dia. UpFront EBA column with proper (stirred) distribution of the raw material in the bottom of the column (data not shown) (Lihme et al., 1990). The 5mm-dia. column worked well with centrifuged raw material and a stable expansion of the adsorbent was easily achieved.

Table 2 describes the main details and results obtained from this experiment. 87 ml of cell extract was applied to 2.2 ml mixed mode adsorbent and the bound Fab fragment was eluted with 90% yield in a volume of 10 ml. Since the last fraction of the raw material coming through the column showed a break through of the Fab fragment this corresponds to a binding capacity of approx. 12mg Fab/ml adsorbent.

Elution was performed in expanded bed mode as opposed to reversal of the flow and elution in the packed bed mode. This resulted in an efficient elution in a volume of 10ml corresponding to a concentration factor of approx. 9. This factor is expected to increase significantly when scaling up the process to larger columns with more optimal design for expanded bed elution (Lihme et al., 1998). The eluate was a clear liquid devoid of cells and cell debris.

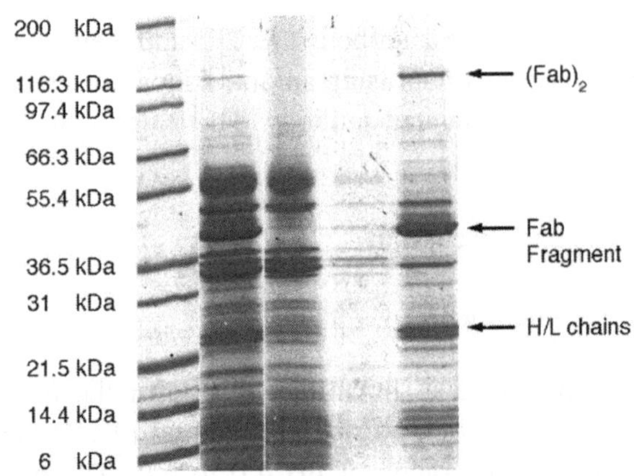

Figure 1. SDS-page gel electrophoresis: Non-reduced coomassie stained SDS-PAGE of EBA purification of human Fab fragments from E. coli cell extract. Lane 1. Mw marker; lane 2. Crude E. coli cell extract; lane 3. Run through from EBA column, lane 4. Wash, lane; 5. Eluate from Mixed mode matrix.

The SDS-page (Figure 1) demonstrated that the EBA procedure leads to a significant increase in the purity of the Fab fragment. Calculations show that the purity has been increased by a factor of 3–4. As well as Fab' other Fab' related bands known to be present including Fab'$_2$ and some unassociated heavy and light chain were also observed in the eluent.

Figure 1 shows that the major non-Fab proteins remain in the run-through (non-bound fraction), while all the intact Fab fragments have been bound.

Conclusion

A novel mixed mode matrix comprising an aromatic ring-system and a carboxylic acid group has been tested for use in expanded bed adsorption. It is concluded that it can be utilized for capture of Human Fab fragments. The fragments are bound in the slightly acidic range (pH 4.0-6.0) and eluted by a change of pH. For production of highly purified Fab fragments the adsorbent seems suitable for a "first capture/concentration step" due to the high yield and enrichment obtained together with a significant volume reduction and clarification in this single step operation.

Due to the relative insensitivity of the binding of proteins to the mixed mode ligand towards high ionic strength, the binding of the Fab fragments can take place without desalting or diluting the raw material. This is a significant advantage when compared to ion exchangers and the associated demand for low ionic strength during the binding step.

Acknowledgements

The authors wish to thank the technicians Anja Dideriksen and Anne Skovgaard for valuable technical assistance during this project.

References

Carter P, Kelley RF, Rodrigues ML, Snedcor B, Covarrubias M, Velligan MD, Wong WLT, Rowland AM, Kotts CE, Carver ME, Yang M, Bourell, JH, Shepard HM and Henner D, (1992b), High level Esherichia coli expression and production of a bivalent humanized antibody fragment, Bio/Technol 10, 163–167.

Proudfoot KA, Torrance C, Lawson ADG and King DJ, (1992), Purification of recombinant chimeric B72.3 Fab' and F(ab')2 using Streptococcal protein G, Protein Expression & Purification 3: 368–373.

Yon, RJ (1972) Chromatography of lipophilic proteins on adsorbents containing mixed hydrophobic and ionic Groups. Biochem J. 126: 765–767.

Hofstee BHJ (1973) Protein binding by agarose carrying hydrophobic groups in conjunction with charges, Biochem. Biophys. Res. Commun., 50: 751–757.

Yon RJ and Simmonds RJ, (1975) Protein chromatography on adsorbents with hydrophobic and Ionic Groups. Biochem. J., 151: 281–290.

Sasaki I et al (1979) Hydrophobic-ionic chromatography. J. Biochem., 86: 1537–1548.

Bischoff R and McLaughlin L, (1984) Nucleic acid resolution by mixed-mode chromatography, J. Chromatogr., 296: 329–337.

Ruann RC et al. (1988) Dual-functional affinity protein purification. Biotechnol. Prog. 4: 107–112.

Kasche V et al. (1990) Rapid protein purification using phenylbutylamine-eupergit: A novel method for large-scale procedures. J. Chromatogr. 510: 149–154.

Teichberg VI, (1990) Affinity-repulsion chromatography. J Chromatogr. 510: 49–57.

Olander M, Hansen MB and Lihme A (1996) Mixed mode ligands for ionic strength independent high capacity expanded bed adsorption of proteins. Poster at the First International Conference on Expanded Bed Adsorption, Queens' College Cambridge, UK 8–10 December 1996.

Burton SC, Haggarty NW and Harding RK, (1997) One step purification of chymosin by mixed mode chromatography. Biotech and Bioeng., 56 (1): 45–55.

Olander M Aa and Lihme A (1998) Mixed mode ligands – sophisticated 'ion exchangers' for efficient capture of proteins from crude raw materials. Poster at the Second International Conference on Expanded Bed Adsorption, Napa Valley, CA, USA, 21–23 June 1998. Paper submitted for publication in Bioseparation.

Hansen MB and Lihme A (1996) A new protein A mimicking affinity matrix for expanded bed purification of monoclonal antibodies. Poster at the First International Conference on Expanded Bed Adsorption, Queens' College Cambridge, UK 8–10 December 1996.

Patent application: A. Lihme and M.B. Hansen, "Isolation of Immunoglobulins", PCT/DK97/00359

Lihme A, Nielsen CS and Bøg-Hansen TC (1990) Substance carrying conglomerate. Patents: EP 0538.350, EP 0607998.

Lihme A and MB Hansen (1997) American Biotechnology Laboratory, July, "Protein A mimetic for large-scale monoclonal antibody purification".

Humphreys DP, Vetterlain OH, Chapman AP, King DJ, Antoniw P, Suitters AS, Reeks DG, Parton TAH, King LM, Smith BJ, Lang V and Stephens PE, (1998), Fab'2 molecules made from Escherichia coli produced Fab' with hinge sequences conferring increased serum survival in an animal model. J. Immunol. Methods 217: 1–10.

FastMabs A^{TM}, User Guide. UpFront Chromatography A/S, p. 3.

Lihme A, Nielsen CS and Bøg-Hansen TC (1990) Method of distributing a liquid in a fluid bed reactor. Patent application: EP 0722771 A1.

Lihme A, Zafirakos E, Hansen MB and Olander M (1998) Simplified and more robust EBA processes by elution in expanded bed mode. Poster at the Second International Conference on Expanded Bed Adsorption, Napa Valley, CA, USA, 21-23 June 1998. Paper accepted for publication in Bioseparation

Bioseparation **8:** 195–200, 1999.
© 1999 *Kluwer Academic Publishers.*

Purification of Recombinant Human Serum Albumin
Efficient purification using STREAMLINE

Akinori Sumi[1], Kouzou Okuyama[2], Kaoru Kobayashi[2], Wataru Ohtani[3], Takao Ohmura[1] & Kazumasa Yokoyama[2]
[1]*BIPHA Corporation, 1007-124, Izumisawa, Chitose, Hokkaido 066-0051, Japan;* [2]*Yoshitomi Pharmaceutical Industries Ltd., Pharmaceutical Research Division, Osaka Laboratories, 2-25-1 Shodai-Ohtani, Hirakata, Osaka 573-0053, Japan*

Received 7 July 1998; accepted in revised form 13 February 1999

Key words: albumin, HSA, rHSA, STREAMLINE

Abstract

Two inter-related challenges must be overcome to develop a recombinant human serum albumin process. One is purity; the other is cost. Regarding cost, our goal is to produce recombinant human serum albumin at least as economically as plasma derived human serum albumin. To control production costs, maximum quantities of albumin must be produced from minimum volumes of cell culture, followed by high efficiency, high-yield purification methods. By introducing STREAMLINE technology, we have improved productivity by roughly 50% in terms of processing time and 45% in terms of yield. Furthermore, the life time of the gel and column are very long. After more than 1000 process cycles, performance remained unchanged.

Abbreviations: pHSA – Plasma derived Human Serum Albumin; rHSA – Recombinant Human Serum Albumin

Introduction

Two major problems arise with respect to the supply of plasma proteins for medical use. The first is an anticipated shortage of source materials. The second problem is the risk of transmission of pathogenic viruses such as hepatitis, HIV, and others not yet identified. Genetic engineering is the best approach toward solving these problems. Plasma derived human serum albumin (pHSA) is one of the most useful plasma proteins (Peters, 1985). However, producing recombinant human serum albumin (rHSA) commercially via genetic engineering is not entirely straightforward. Two inter related challenges must be overcome: cost and quality. Albumin is typically administered in tens of gram quantities. At a purity level of 99.999% (a level considered sufficient for other recombinant protein preparations such as vaccines or cytokines), rHSA impurities of the order of one mg will still be injected into patients. So impurities from the host organism, in this case yeast, must be reduced to a minimum. Furthermore, purified rHSA must be identical to pHSA. Despite these stringent requirements, costs must be kept in check. One gram of pHSA costs a few dollars. Our goal is to produce rHSA at least as economically as pHSA. Thus to control rHSA costs, maximum quantities of albumin must be produced from minimum volumes of cell culture, and this crude albumin must be refined via high efficiency, high-yield purification methods.

In the initial phase of the project, we used a filter-press to separate yeast cells from the culture broth. Although the filter-press is more efficient than centrifugation, the filtering process is slow, and automated operation and scale up are also highly problematic. Furthermore, containment of the recombinant yeast is uncertain with the filter-press. In contrast, the STREAMLINE™ system (Chase, 1994; Hjorth, 1997, 1998; Thömmes, 1997), which operates on the same principle as general chromatography, is easy to scale up and can be operated automatically. Since

STREAMLINE is a closed system, the recombinant yeast cannot readily escape.

In this paper we describe a purification method for rHSA using STREAMLINE system. The purification efficiency is compared with a method utilizing a filter-press. Furthermore, we discuss the life time of the STREAMLINE SP gel and a STREAMLINE column.

Materials and methods

Preparation of rHSA producing strain

A cDNA library was constructed using mRNA extracted from human liver cells and a human serum albumin (HSA) cDNA was isolated. An expression vector was prepared putting the HSA-gene under the control of a modified Alcohol Oxidase 2 (mAOX2) promoter derived from *Pichia pastoris* (Ohi et al., 1994). An rHSA producing strain was prepared by inserting the expression vector into a chromosome of *P. pastoris* [Strain GTS115].

Fermentation

Methanol served as inducer and carbon source, causing rHSA to be secreted into the culture broth. Fermentation was performed at high density using a fed-batch technique.

Purification of rHSA

In Figure 1, purification of rHSA using STREAMLINE system is described and compared to an alternative method using a filter-press (Sumi et al., 1993). The buffer conditions of the chromatographic steps are shown in Table 1.

The longevity of STREAMLINE SP medium was investigated using a handmade acrylic pipe column of 4 mm id. and 900 mm height. The column connections for buffer and sample application were made using silicon tubing of 2 mm id. Switching among buffers and sample were performed using solenoid pinch valves. These valves were controlled via a small programming logic controller. The experimental set-up is shown in Figure 2.

In order to investigate the useful life length of the O-rings mounted on the adapter, we performed an endurance test. The experimental set-up used is shown in Figure 3. Any damage to the O-ring mounted between the column upper cover and the adapter shaft would be visually apparent. Any destruction of the O-ring

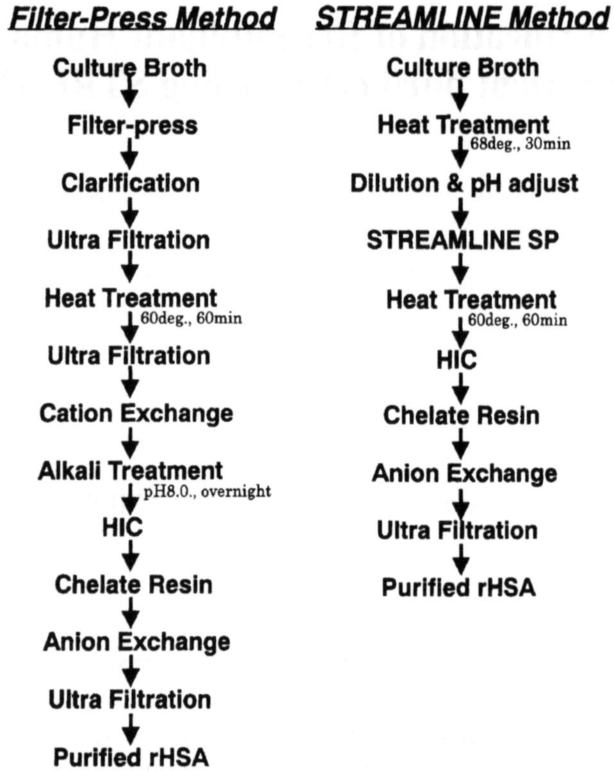

Figure 1. Purification methods based on either Filter-press or STREAMLINE.

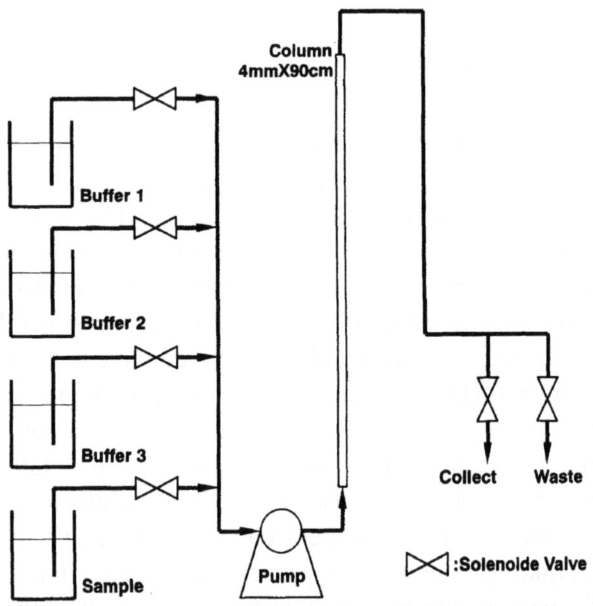

Figure 2. Experimental set-up for testing STREAMLINE SP gel longevity/durability. The chromatographic conditions used were: id, 4 mm id; bed height, 90 cm; volume of media used for packing, 2.5 mL (before expansion); linear flow rate, 200 cm/h.

Table 1. Conditions of Chromatographs

Chromatography step	Apply condition	Elute condition	Remarks
Cation Exchange or STREAMLINE (SP)	pH 4.0–4.8	pH 6.8–9.5	
	Cond. <10mS/cm	Cond.6–40 mS/cm	
HIC (Phenyl)	pH 6.3–7.2 Cond. 15-20 mS/cm		rHSA is passed through
Chelate (Poly-amine)	pH 4.1–4.9 Cond. 1–3 mS/cm		rHSA is passed through
Anion Exchange (DEAE)	pH 6.3–7.3 Cond. <1 mS/cm	pH 5.8–7.8 Cond. 4.4–6.6 mS/cm	

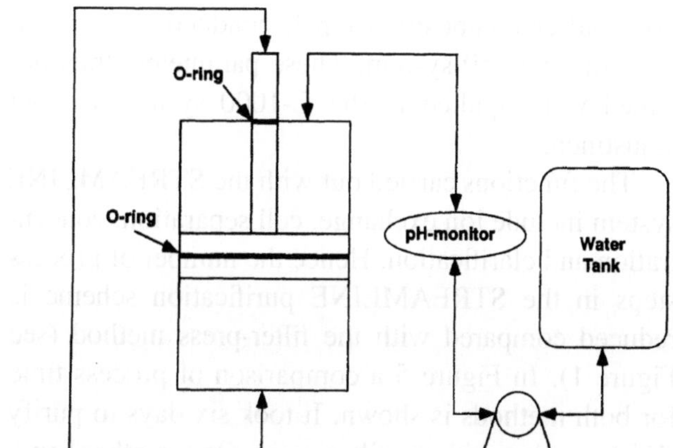

Figure 3. Equipment layout for endurance testing of STREAMLINE C-1000 column O-ring.

mounted between the adapter and the column would be detectable as follows. The part which normally contains the gel was filled with alkali solution and the section with the drive adapter was filled with pure water. If the O-ring became damaged, pure water and alkali solution would mix, thus increasing the pH of the initially pure water.

Results and discussion

In production scale fermentation of the yeast *P. pastoris* containing the rHSA expression vector, the amount of rHSA obtained in the culture broth exceeded 10 g/L. This was achieved by optimizing the feed of methanol.

The number of steps in the purification method were reduced using STREAMLINE. We confirmed that the rHSA purified using this method was identical to pHSA (Ohtani et al., 1997; Ohtani et al., 1998a; Ohtani et al., 1998b).

Keeping the process requirements described in the introduction in mind, the rHSA method was developed with two major objectives in mind.

- Ease of automatic control
- Ease of scale up

Meeting the first objective was necessary for holding down costs and also for obtaining uniform quality. The second objective was dictated by the way we chose to work. In the first phase of rHSA development, we examined purification methods based on a 3 to 10 L fermentation scale. In the next phase, we confirmed and optimized the purification methods established earlier using a 1000 L fermentation scale. In the final phase we moved to commercial scale production. The plan was to use 50 m^3 fermentors.

The purification methods of rHSA should be identical throughout all three phases of development. Hence, easy scale up is a very important factor. We believe that column chromatography is the best approach to solve these problems (see Figure 1).

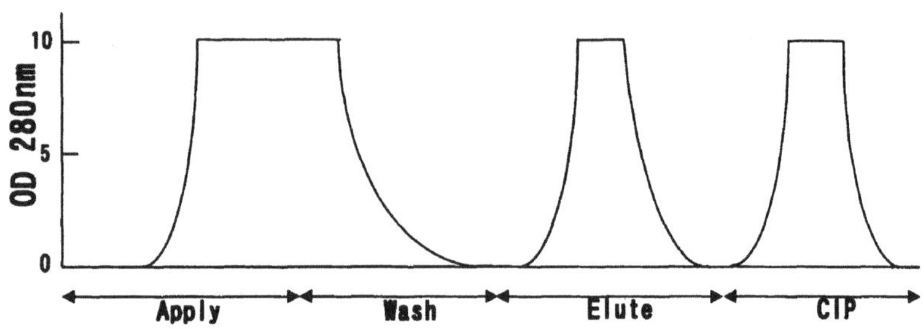

Figure 4. Chromatogram pattern of STREAMLINE.

Furthermore, the method developed should also adhere to the two following limitations:

- No use of special materials
- No use of complicated methods

These two restrictions are necessary to keep the process simple and the process economy healthy. Affinity chromatography is usually considered a very efficient method for capture, but in the case of rHSA, the affinity media available at the time of development were not efficient in terms of cost or capacity. In most cases of large scale chromatography, absorption profile is not useful. As shown in Figure 4., almost the whole chromatogram is out of range. So we could not control substantially using UV absorption profile and also could not collect the target protein from complex partially resolved peaks.

Based upon the reasoning described above we selected simple and commonly used chromatography steps, such as ion exchange-, hydrophobic interaction- and gel permeation chromatography. In all these methods rHSA is adsorbed on the media and impurities are washed out. Then rHSA is eluted, usually by a gradient step. In this case scaling up is easy. The same bed height and linear flow can be used to yield identical results in spite of gel volume changes.

In isolating recombinant proteins, cell debris and other insoluble materials must be separated from the secreted product or inclusion body. Therefore, we used a filter-press in early experiments to separate yeast cells from the culture broth (see Figure 1). Although the filter-press is more efficient than centrifugation, the filter-press process is slow, and automated operation and scale up are also highly problematic. Furthermore, exposure of the operator to the recombinant yeast is an ever present danger. Considered in this light the filter-press process fails to meet our objectives for purification design mentioned earlier. In contrast, expanded bed adsorption with STREAMLINE, which operates on the same general principles as chromatography, is easy to scale up and can be automated. The STREAMLINE system is therefore entirely consistent with our purification goals. Furthermore, since the STREAMLINE system is closed, the recombinant yeast cannot escape.

Initially we optimized various STREAMLINE parameters, such as linear flow rate, adapter position, buffer composition (pH, conductivity) and so on, using a C-50 system. These parameters thus obtained were applied to the C-1000 system without adjustment.

The functions carried out with the STREAMLINE system include ion exchange, cell separation, concentration and clarification. Hence the number of process steps in the STREAMLINE purification scheme is reduced compared with the filter-press method (see Figure 1). In Figure 5 a comparison of process time for both methods is shown. It took six days to purify rHSA starting with the filter-press. On the other hand, it took only four days to purify rHSA using the STREAMLINE-based method (see Figure 5). The productivity with respect to process time improves about 50% by introducing a STREAMLINE system.

Figure 6 shows the comparison of rHSA recovery between the methods with STREAMLINE and with a filter-press. The total recovery of rHSA using STREAMLINE is 14.5% better than from the filter-press method. The total recovery from the STREAMLINE method was roughly 60%. In contrast, the filter-press method reached only 45%. The yield is greatly improved by introducing the STREAMLINE system. As for total productivity, judging from process time and rHSA recovery, the overall advantage of STREAMLINE is clear.

In order to estimate the production cost we needed to determine the life expectancy of the gel. We tested the life length of STREAMLINE SP medium using a column with a 4 mm i.d. made in-house.

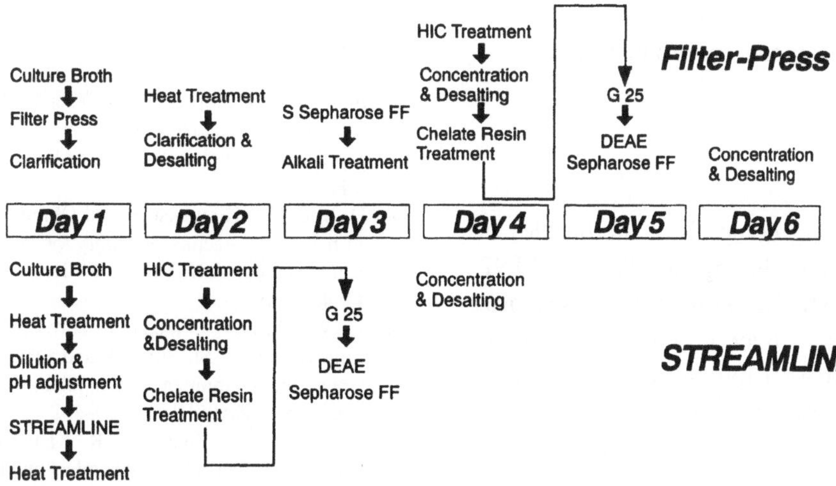

Figure 5. Process time comparison between the STREAMLINE and Filter-press based methods.

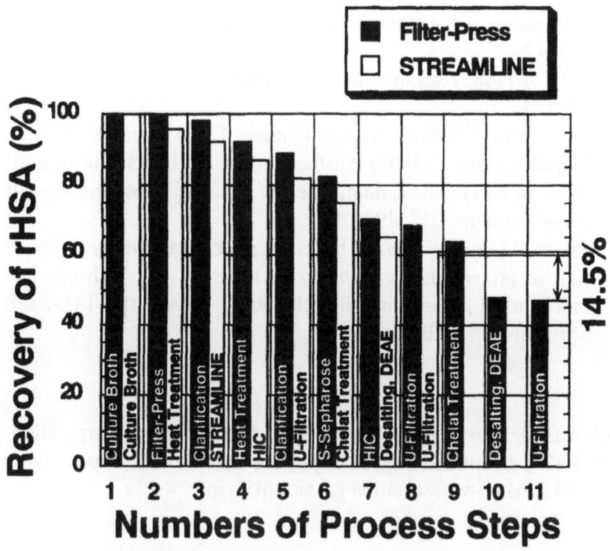

Figure 6. Comparison of step by step as well as total rHSA recoveries between the two methods.

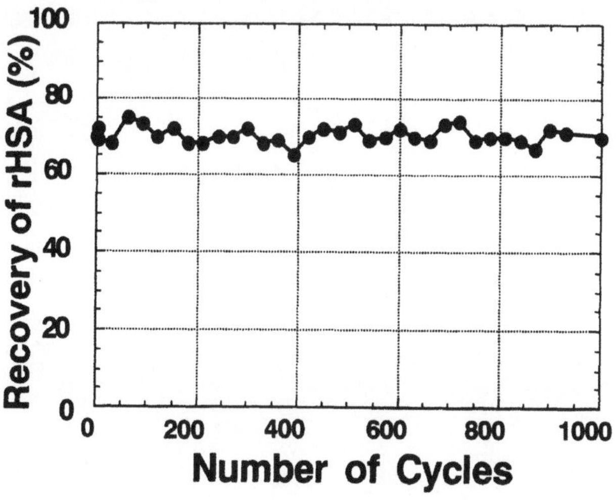

Figure 7. Results of the STREAMLINE SP gel fatigue testing.

Figure 7 shows the results of life length testing we carried out on the STREAMLINE SP medium. We found no impairment of performance over the course of 1000 usage cycles. The recovery of rHSA was relatively low in the early runs, though the amount of applied rHSA was below gel capacity. We suspect that the reason for this was the column orientation. It was very hard to set the narrow acrylic column vertically. Furthermore, the bottom of the hand-made column was covered only with a nylon mesh. Because of these reasons, the medium in the hand-made column probably could not expand optimally.

During operation of chromatography columns other than STREAMLINE, there are no moving parts. In contrast, in the STREAMLINE system the adapter moves in every chromatography cycle. Moreover, the moving parts are sealed only by a thin O-rings. The life expectancy of these O-rings is an important factor for rHSA production. The wear-resistance of two O-rings at different positions was investigated (Figure 3). One is located between the column upper cover and the adapter shaft. The other is located between the adapter and the column wall. The first one is less important. If this O-ring fails, the process solution will not be contaminated. Therefore we made only visual inspections of this O-ring. The other O-ring though is very important. If this O-ring fails, process solution will mix with the water which drives the adapter. This means that we must guarantee the function of this O-ring in order to guarantee the quality of the entire process. The analysis method relied upon detection of pH increase as described in the methods section and was rather sensitive. Nevertheless, even after 1200 cycles we could not detect any pH change. Our conclusion is that this O-ring meets our requirements for product safety.

Conclusions

We have successfully overcome all of the hurdles of rHSA development by introducing the STREAMLINE system. The STREAMLINE system increases productivity and gives better overall economy, or to be more precise, the STREAMLINE system reduces processing time and increases yield. The STREAMLINE system longevity (gel and column) is satisfactory from a commercial perspective.

Acknowledgements

The authors gratefully acknowledge Shouji Ishikawa, Mitsuo Sakaguchi and Masahide Kondo (research division of Yoshitomi Pharmaceutical Industries Ltd.) for useful discussions. The authors also gratefully acknowledge Dr. Björn Hammarberg (Amersham Pharmacia Biotech KK) for useful comments on the manuscript.

References

Chase HA (1994) Purification of proteins by adsorption chromatography in expanded beds. Trends in Biotechnology 12: 296–303.

Hjorth R (1997) Expanded-bed adsorption in industrial bioprocessing: recent developments. Trend in biotechnology 15: 230–235.

Hjorth R, Leijon P, Frej A-KB and Jagersten C (1998) Expanded bed adsorption chromatography In: (pp 199–226). Bioseparation and Bioprocessing, Vol. 1 Weinheim. Subramanian G (ed). Wiley-VCH.

Ohi H, Miura M, Hiramatsu R and Ohmura T (1994) The positive and negative cis-acting elements for methanol regulation in the *Pichia pastoris* AOX2 gene. Molecular and General Genetics 243: 489–499.

Ohtani W, Masaki A, Ikeda Y, Hirose M, Chuganji M, Takeshima K, Kondo M, Sumi A and Ohmura T (1997) Structure of recombinant human serum albumin from *Pichia pastoris.* YAKUGAKUZASSHI 17: 220–232.

Ohtani W, Nawa Y, Takeshima K, Kamuro H, Kobayashi K and Ohmura T (1998a) Physicochemical and Immunochemical properties of recombinant human serum albumin from *Pichia pastoris.* Analytical Biochemistry 256: 56–62.

Ohtani W, Ohda T, Sumi A, Kobayashi K and Ohmura T (1998b) Analysis of *Pichia pastoris* components in recombinant human serum albumin by immunological assays and by HPLC with pulsed amperometric detection. Analytical Chemistry 70: 425–429.

Peters T Jr (1985) Serum Albumin, Advances in protein chemistry 37: 173.

Sumi A, Ohtani W, Kobayashi K, Ohmura T, Yokoyama K, Nishida M and Suyama T (1993) Purification and physicochemical properties of recombinant human serum albumin. Biotechnology of Blood Proteins. 227: 293–298.

Thömmes J (1997) Fluidized bed adsorption as a primary recovery step in protein purification, In: Scheper T (ed), Advances in biochemical engineering/biotechnology, Vol. 58 (pp 185–230). Springer-Verlag, Berlin.

Address for correspondence: Akinori Sumi, BIPHA Corporation, 1007-124, Izumisawa, Chitose, Hokkaido 066-0051, Japan. (Fax: 12 349 2161; e-mail: asumi@yoshitomi.co.jp).

Bioseparation **8:** 201–207, 1999.
© 1999 *Kluwer Academic Publishers.*

Direct coupling of expanded bed adsorption with a downstream purification step

Joanne T. Beck, Brian Williamson & Barbara Tipton
Purification Process Development, AMGEN, 1840 DeHavilland Drive, Thousand Oaks, CA 91320-1789

Received 2 September; accepted in revised form 15 February 1999

Abstract

In the course of developing a cost-effective, scaleable process for the purification of a recombinant protein from Chinese hamster ovary (CHO) suspension cell culture, we investigated direct capture of this molecule using expanded bed adsorption (EBA). EBA combines clarification, purification, and concentration of the product into a single step. The unclarified bioreactor material was directly applied to a STREAMLINE 25 column containing an affinity STREAMLINE adsorbent. This work focused on simplifying the EBA operations and minimizing the overall processing time by running the EBA column unidirectionally, eluting in the expanded bed mode, and coupling the EBA column directly with ion exchange or hydrophobic interaction chromatography. Unidirectional EBA was clearly a simpler unit operation and did not require the use of specialized equipment. The increase in the elution pool volume was insignificant, especially when the EBA column was eluted directly onto the downstream column. Scale-down was simple and could be automated. Coupling of unidirectional EBA with a downstream purification step reduced processing time, equipment requirements and cost.

Introduction

Since feedstocks containing cells or other particulates cannot be applied directly to packed-bed chromatographic columns, the recovery of biological products begins with a clarification step. The clarified material can then be applied to a chromatography column for further purification. Well-established methods of producing clarified feedstocks include centrifugation, microfiltration and dead-end filtration (Lee, 1989). Recently, expanded bed adsorption chromatography (EBA) has emerged as a method that combines particulate removal with chromatographic separation (Barnfield et al., 1994; Batt et al., 1995; Chang et al., 1995; Chase et al., 1994; Chang et al., 1996).

As purity requirements for the manufacturing of protein pharmaceuticals become more stringent, it is advantageous to use at least one highly selective chromatography step, such as affinity chromatography, in the purification process (Clonis, 1987). EBA allows the use of affinity chromatography at the primary capture step, eliminates the need for other time-consuming and expensive clarification steps, and delivers a highly purified feedstream which can be applied directly onto the next chromatography column (Chase et al., 1992; Thömmes et al., 1996).

Even though harvest by expanded bed adsorption offers obvious savings in time and cost over traditional harvest technologies, its use in the biotechnology industry has been limited. The main reason is that the traditional EBA unit operation is complicated, and both industrial and small-scale experiments are very difficult to automate. Furthermore, commercially available expanded bed columns are dedicated pieces of equipment and require large capital investment.

The traditional EBA column is equilibrated, loaded, washed and cleaned-in-place in the expanded bed mode. Elution, however, is performed in the packed-bed mode in order to minimize the pool volume. Just prior to elution, the moveable flow adapter is lowered to the surface of the packed bed and the direction of flow through the column is reversed. Unidirectional EBA, where elution is performed in the expanded bed mode, is simple and easy to automate, which is especially important in performing small-scale characterization and validation studies.

Since affinity resins and large-scale EBA systems are expensive, there are obvious advantages to recovering the contents of one bioreactor in more than one EBA cycle. Even though cycling packed-bed columns is common practice in industrial settings, cycling of EBA chromatography has not been previously reported. In this paper we investigated the possibility of cycling an EBA affinity column.

The conventional bioseparation process consists of several discrete unit operations, each performing a single function. The integrated process, whereby two or more unit operations are combined, aims to maintain product integrity and purity to high levels, to increase process compactness and yield, and to reduce processing time, equipment numbers, and labor costs (Kaufmann, 1997). In the scenario involving cycling of unidirectional EBA, coupling the EBA column with the downstream purification step would have an additional advantage. Since expanded bed elution would yield a large pool volume, which would then have to be stored until the next cycle was processed, the direct coupling of the EBA column with a downstream chromatography step would eliminate the need for a large holding tank.

Materials and methods

Chromatographic adsorbents and column

The expanded bed adsorbent used in these experiments was a STREAMLINE affinity resin obtained from Amersham Pharmacia Biotech AB (Uppsala, Sweden). The column used in these experiments was the STREAMLINE 25 column (internal diameter of 25 mm and height of 1 m) equipped with a hydraulic adapter (Amersham Pharmacia Biotech Inc., Piscataway, USA). Residence time distribution (RTD) tests were performed as described by the manufacturer using 2% acetone in water.

Cell culture

The protein of interest was produced by a Chinese hamster ovary cell line (CHO). The contents of 20-liter bioreactors were used in the experiments with the STREAMLINE 25 column. Cell viability in the culture was determined by trypan blue and was in the range of 70–90%. The temperature of the feedstock was maintained at 37 °C.

Assays

The concentration of the protein of interest in the cell culture fluid (CCF) and the various chromatography pools was determined using a reversed phase HPLC (RP-HPLC) assay. CCF samples were centrifuged to remove cells and debris, and filtered through a 0.22 μm filter. The concentration of the protein of interest was calculated by fitting the data to a standard curve generated using pure protein.

Elution in the expanded bed mode

To compare elution in the expanded bed mode with elution in the packed-bed mode three STREAMLINE 25 columns were run simultaneously. In all three cases, the packed-bed height of the resin was 15 cm, the flow rate was 400 cm/hr and the bed expanded 3-fold to 45 cm. During the control packed-bed elution experiments, the equilibration, the load, and the wash were performed in expanded bed mode, then the flow adapter was lowered to the top of the packed bed and the resin was eluted. The adapter was raised again for the regeneration of the resin. In unidirectional EBA, all the steps were performed in expanded bed mode. The flow adapter was positioned either at 10 cm above the surface of the expanded bed or at the top of the column, 48 cm above the surface of the expanded bed.

Measurement of breakthrough curves and determination of product purity during cycling of the EBA column

Breakthrough curves for the protein of interest were measured in order to determine the feasibility of cycling the EBA column without regeneration. The EBA was equilibrated, loaded, washed and eluted three consecutive times and regenerated only after the third cycle. Flow-through samples were collected at regular intervals during the loading of the column and analyzed for product content by RP-HPLC. The elution pool from each of the cycles was collected separately and analyzed for purity by RP-HPLC and SDS-PAGE.

Integrated process design

Figure 2 illustrates the experimental setup for experiments in which the STREAMLINE 25 EBA column was directly coupled with the downstream packed-bed column. Table 2 depicts the relevant process parameters. The downstream packed-bed column could be operated at a maximum linear flow rate of 120 cm/hr,

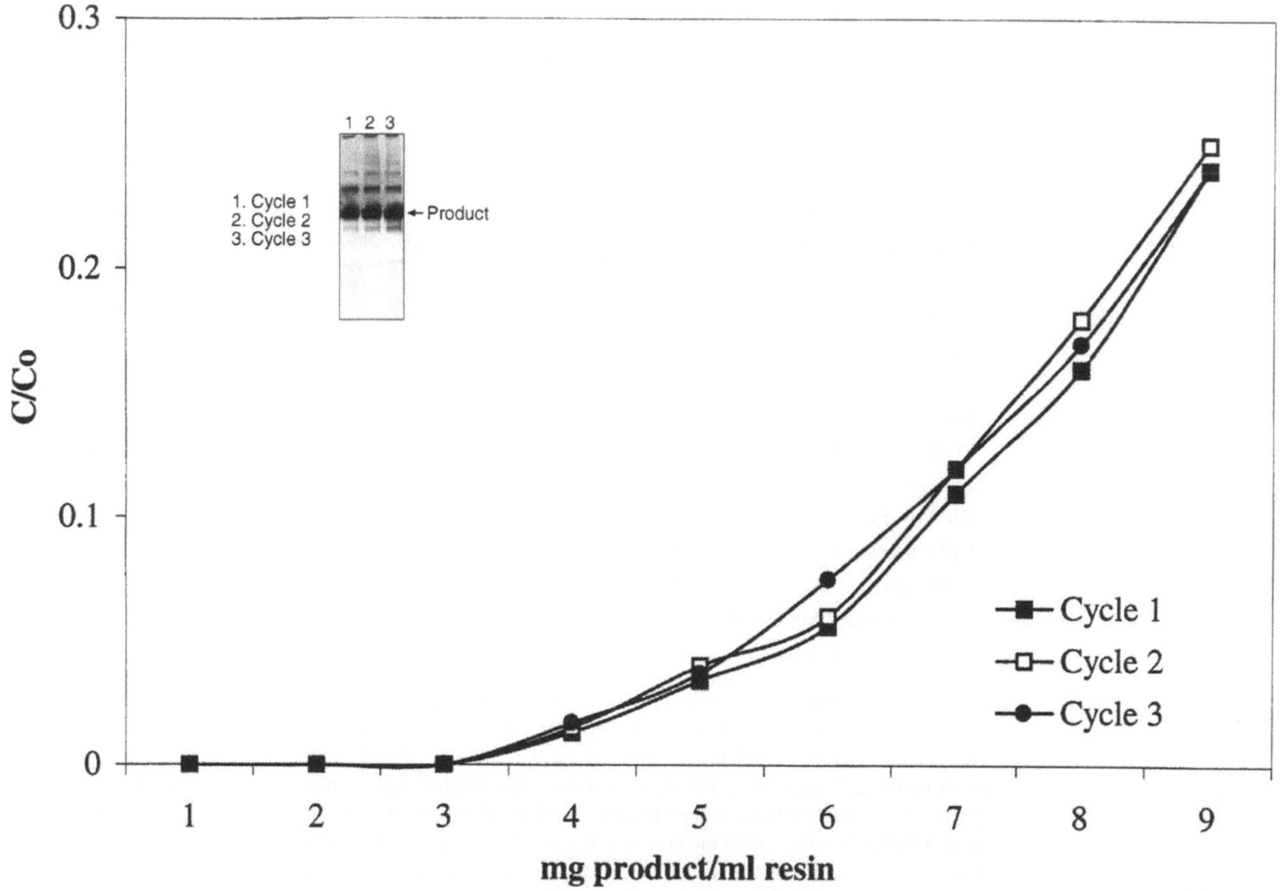

Figure 1. Product breakthrough capacity curves for three consecutive cycles of EBA without transitional regeneration and SDS-PAGE of the product recovered from each cycle. The experiments were conducted using a STREAMLINE 25 column (inner diameter 2.5 cm). The packed bed height of the resin was 15 cm and the expanded bed height at the operational flow rate of 400 cm/h was 45 cm. C is the concentration of product in the flow-through and Co is the concentration of product in the feed. Five micrograms from the eluate pool generated by each EBA cycle was analyzed by SDS-PAGE. The gel was silver-stained using the (Wake Silver Stain II kit).

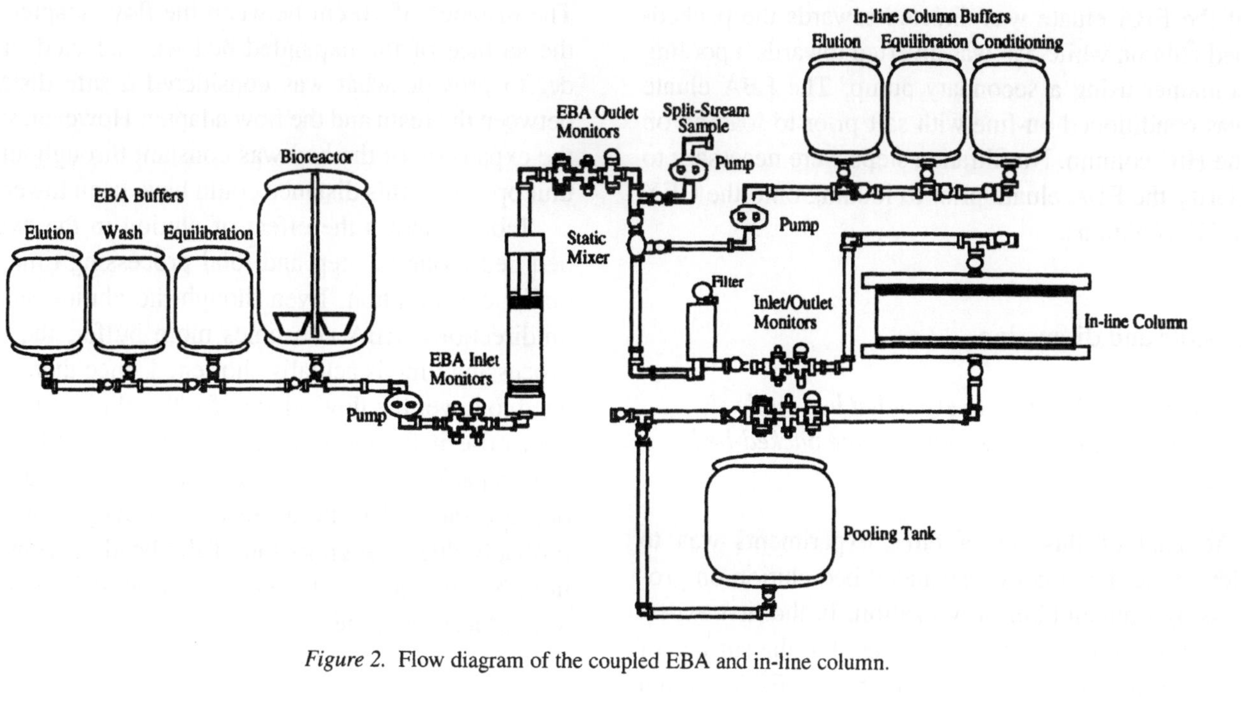

Figure 2. Flow diagram of the coupled EBA and in-line column.

Table 1. Process step duration and buffer consumption.

	PB elution	EB elution	EB elution
Expanded bed height	46	46	46
Flow adaptor position (cm)	55	55	93
Flow rate (L/h)	2.0	2.0	2.0
Equilibration volume (L)	0.7	0.7	1.0
Equilibration time (h)	0.3	0.3	0.5
Feedstock volume (L)	8.0	8.0	8.0
Loading time (h)	3.9	3.9	3.9
Wash volume (L)	0.9	0.9	1.8
Wash time (h)	0.4	0.4	0.9
Elution flow rate (L/h)	1.0	2.0	2.0
Elution volume (L)	0.3	0.6	1.6
Elution time (h)	0.3	0.3	0.8
CIP flow rate (L/h)	1.0	1.0	1.0
CIP volume (L)	1.4	1.2	2.1
CIP time (h)	1.4	1.2	2.1
Total Processing Volume (L)	3.3	3.4	6.5
Total Processing Time (h)	6.3	6.2	8.1

Three different experiments were conducted concurrently to determine the effects of unidirectional EBA on processing time and buffer consumption. The UV absorbance, conductivity and pH of the column effluent were monitored to determine when the resin had been either washed free of non-specifically bound proteins and particulate material or equilibrated in the applied buffer, at which point the next step in the process was initiated. The flow rate during CIP was lowered to allow for at least 1 h of exposure of the resin to the cleaning buffer.

therefore, the EBA column was also eluted at this flow rate. In order to monitor the yield and purity of the EBA column, a representative EBA pool was made by splitting the eluate stream in two. In this set-up, 95% of the EBA eluate was directed towards the packed-bed column while 5% was diverted towards a pooling container using a secondary pump. The EBA eluate was conditioned on-line with salt prior to loading on the HIC column. No filtration steps were necessary to clarify the EBA eluate prior to loading onto the CEX or HIC columns.

Results and discussion

Comparison of elution in expanded bed mode (unidirectional EBA) to elution in the packed-bed mode

The goal of this set of three experiments was to determine the effect of expanded bed elution on processing time and buffer utilization. In the first experiment, the flow adapter was lowered to the surface of the bed, the flow was reversed, and the EBA column was eluted in the packed bed mode. In the other two experiments the EBA column was eluted in the expanded bed mode with the flow adapter positioned either 10 or 48 cm above the surface of the expanded bed. The distance of 10 cm between the flow adapter and the surface of the expanded bed was selected in order to provide what was considered a safe distance between the resin and the flow adapter. However, since the expansion of the bed was constant throughout the unit operation this distance could have been lower.

Table 1 shows the effect of elution in the expanded bed mode on step and total processing time and buffer consumption. Even though the elution step in unidirectional EBA consumes more buffer, the total processing time is actually shortened since there is no need to lower the flow adapter for the elution step and then raise it for the CIP step. Furthermore, the CIP step in packed-bed elution EBA had a longer duration because the bed re-fluidized more slowly. This was probably due to aggregation of the beads during elution. Nevertheless, CIP consumed at least 75% of the total processing time.

Table 2. Relevant process parameters for EBA and in-line column

	EBA	In-line Column
Feedstock	Cell culture fluid	Affinity pool
Column type	STREAMLINE 25	XK 26
Resin type	STREAMLINE	HIC, CEX
Resin volume (ml)	69	80
Resin packed-bed height (cm)	14	15
Column I.D. (cm)	2.5	2.6
Linear flow rate (cm/hr)	400	240
Expansion factor (H/Ho)	3	n.a.
Flow adaptor during elution (cm)	55	n.a.
Elution flow rate (cm/hr)	120	120
Loading capacity (mg/ml)	12	10

Used to assess the effect of directly coupling the EBA column with the downstream purification step.

The total buffer consumption for unidirectional EBA was higher, especially in the scenario where the flow adapter was positioned at 45 cm above the surface of the expanded bed. Elution in the expanded bed mode resulted in an elution pool that was twice the pool size of packed bed elution when the flow adapter was positioned 10 cm above the surface of the bed. When the flow adapter was positioned 45 cm above the surface of the bed the pool size was 23 times the size of the packed bed elution pool. This excessive increase in the volume of the elution pool indicated that it was advantageous to set the flow adapter as close to the surface of the fluidized bed as possible.

Cycling the EBA column without CIP

In order to minimize EBA affinity resin costs, perform product capture can be performed in several cycles. However, CIP of the EBA resin consumed a significant amount of time since the length of exposure to the CIP solution was important for complete removal of contaminants. Consequently, the effect of performing 3 column cycles on product recovery and purity, without regenerating the resin in between. Figure 1 shows breakthrough capacity curves generated with three consecutive cycles of the EBA column where the resin was not regenerated after the first and second cycles was investigated. The fact that the curves are overlapping indicates that the dynamic binding capacity of the resin remains unchanged following three consecutive cycles without regeneration. The inset in Figure 1 shows the SDS-PAGE gel of the product purified from each of the consecutive cycles. No additional bands were detected, which indicates that there was no detectable carryover of protein contaminants from one cycle to the next. Finally, the amount of product recovered from all three cycles was identical.

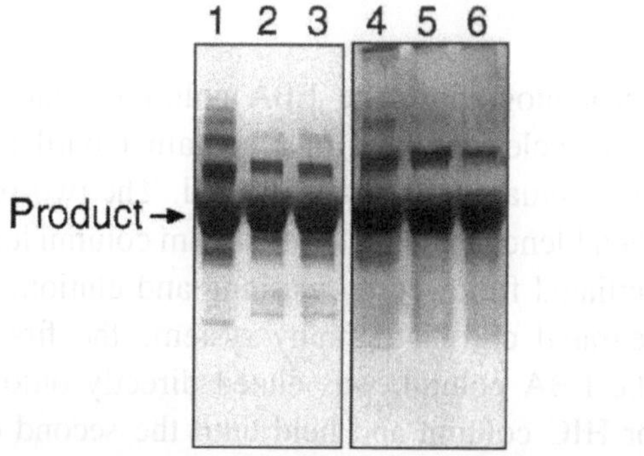

Figure 3. SDS-PAGE analysis of the pools generated by batch and integrated processing. Lane 1: affinity pool for CEX experiments, lane 2: CEX pool, continuous processing, lane 3: CEX pool, batch processing, lane 4: affinity pool for HIC experiments, lane 5: HIC pool, continuous processing, lane 6: HIC pool, batch processing. The gel was silver-stained using the Wako Silver Stain II kit.

Coupling of EBA to the downstream chromatography step

The goal of these experiments was to compare integrated with batch chromatography. Table 2 shows the relevant process parameters in these experiments and Figure 2 shows the experimental set up for the integrated chromatography. The set up for the batch chromatography experiments was similar, however it required a tank for the hydraulic fluid used to lower the flow adapter, a pump, to pump this fluid into the hydraulic compartment of the column, and a holding tank for the expanded bed elution pool. In batch

Table 3. Processing time, recovery and relative purity in batch vs. integrated processing

	EBA/CEX Batch	EBA/CEX Integrated	EBA/HIC Batch	EBA/HIC Integrated
2 cycle EBA processing time (hrs)	7.3	7.3	7.3	7.3
In-line column processing time (hrs)	3.7	3.3	5.5	5
Total processing time (hrs)	11	10.6	12.8	12.3
% recovery of EBA step	100	100	100	100
% recovery of in-line column	74	76	57	39
Relative purity after column 2	93	93	97	96

Four concurrent experiments were performed to assess the effect of directly coupling the EBA column with the downstream purification step. Process parameters are described in Table 2. Processing time included equilibration, feed application, wash, and elution for both chromatography steps. In batch processing the EBA column was cycled twice without intermediate regeneration of the resin. The eluate from both cycles was pooled, conditioned, applied to the CEX or HIC columns and processed. In integrated processing the eluate from the EBA column was conditioned on-line and directly applied to the CEX or HIC column. The eluate was held *in situ* until the second EBA cycle was similarly eluted at which point processing of the in-line column continued.

chromatography, the EBA column eluate from the first cycle was held in a container until the second cycle eluate pool was collected. The two pools were then blended and the downstream column loading was initiated followed by washing and elution. In the integrated chromatography system, the first cycle of the EBA column was eluted directly onto the CEX or HIC column and held until the second cycle was similarly eluted. The CEX or HIC column was then washed and eluted. Table 3 shows the difference in processing time, percent recovery and relative purity between batch vs. integrated processing. Performing the chromatography in an integrated format shortened the total processing time. The product recovery of the CEX column was slightly higher in the integrated processing format. However, when the HIC column was operated in the integrated mode recovery was decreased significantly. This was probably due to the prolonged exposure of the product to the resin under high salt conditions. Therefore, in future experiments the HIC column was operated in cycles corresponding to the cycles of the EBA column.

Conclusions

In this work, we describe the use of unidirectional EBA linked to a downstream purification step, which results in the simplification of operations, a decrease in equipment use, and an increase in product yield.

The protein of interest was extracted from CHO cell culture fluid by direct application onto an affinity EBA column. In order to simplify operations, the EBA column was eluted in the expanded bed mode. Two cycles of the EBA column were used to harvest the contents of one bioreactor, and, since adsorbent fouling was not a problem, the inclusion of a CIP stage between cycles was not necessary. Thus, cycling of the EBA column was both cost and time-efficient. The EBA elution pool was completely devoid of any particulate matter and product purity was greater than 95%. The volume of the unidirectional EBA elution pool was increased by at least 50% compared to a packed bed mode elution pool. However, the concentration of the product in the pool was increased 13-fold as compared with the feed.

The EBA column was directly linked to a second column containing either a CEX or HIC resin. This integrated processing was effective in decreasing equipment use, since there was no need for a storage tank large enough to accommodate the EBA pool. Furthermore, processing time was decreased. Overall yield was slightly improved when the downstream CEX step was coupled to the cycled EBA column instead of being operated as a discrete step. However, since the protein of interest in this case was labile in high concentrations of salt, the overall HIC yield decreased when HIC was coupled to the cycled EBA elution, therefore, the HIC was also cycled.

The work described here was performed at laboratory scale. Industrial scale-up of this work will require more detailed study of the link between the EBA and downstream purification step. For example, no filtration step was applied between the two steps at small scale but it will be required at industrial scale. Furthermore, the regulatory aspects of linking fermentation and purification are currently under investigation.

References

Barnfield Frej A-K, Hjorth R and Hammarström A (1994) Pilot scale recovery of recombinant annexin V from unclarified *Escherichia coli* homogenate using expanded bed adsorption. Biotechn. Bioeng. 44: 922–929

Batt CB, Yabannavar VM and Singh V (1995) Expanded bed adsorption process for protein recovery from whole mammalian cell culture broth. Bioseparation 5: 41–52.

Chang YK and Chase HA (1995) Development of an expanded bed technique for an affinity purification of G6PDH from unclarified yeast cell homogenates. Biotechn. Bioeng. 48: 355–366.

Chase HA, Draeger NM (1992) Affinity purification of proteins using expanded beds. J. Chromatogr., 597: 129–145.

Chase HA (1994) Purification of proteins by adsorption chromatography in expanded beds. Trend Biotechnol. 12: 296–303.

Chang YK and Chase HA (1996) Development of operating conditions for protein purification using expanded bed techniques: The effect of the degree of bed expansion on adsorption performance. Biotechn. Bioeng. 49: 512–526.

Clonis YD (1987) Large scale affinity chromatography Bio/Technol. 5: 1290–1293.

Hansson M, Ståhl S, Hjorth R, Uhlen M and Moks T (1994) Single-step recovery of a secreted recombinant protein by expanded bed adsorption. Bio/Technol. 12: 285–288.

Kaufmann M (1997) Unstable proteins: How to subject them to chromatographic separations for purification procedures. J Chromatog. B: Biomed. Appl. 699: 347–369.

Lee S-M (1989) The primary stages of protein recovery. J. Biotechnol. 11: 103–118.

Thömmes J, Halfar M, Lenz S and Kula M-R (1995) Purification of monoclonal antibodies from whole hybridoma fermentation broth by fluidized bed adsorption. Biotechn. Bioeng. 45: 205–211.

Thömmes J, Bader A, Halfar M, Karau A, Kula MR (1996) Isolation of monoclonal antibodies form cell containing hybridoma broth using a Protein A coated adsorbent in expanded beds. J. Chromatogr. 752: 111–122.

Bioseparation **8:** 209–217, 1999.
© 1999 *Kluwer Academic Publishers.*

Production of plasmid DNA for human gene therapy using modified alkaline cell lysis and expanded bed anion exchange chromatography

D.L. Varley, A.G. Hitchcock, A.M.E. Weiss, W.A. Horler, R. Cowell, L. Peddie, G.S. Sharpe, D.R. Thatcher & J.A.J. Hanak
Cobra Therapeutics Ltd, The Science Park, Keele, Staffs., ST5 5SP, U.K.

Received 7 July 1998; accepted in revised form 12 November 1998

Key words: alkaline cell lysis, bag filtration, expanded bed chromatography, gene therapy, GMP manufacture, plasmid DNA

Abstract

We describe a process for the commercial manufacture of therapeutic grade plasmid DNA. The industrially scaleable unit operations employed in this process are: (i) optimized alkaline lysis; (ii) bag filtration; (iii) expanded bed anion exchange chromatography; (iv) ultrafiltration, and (v) size exclusion chromatography. These steps are scaleable alternatives to current approaches to plasmid DNA isolation such as high speed centrifugation for feedstock clarification and solvent precipitation for plasmid concentration, and an efficient alternative to conventional low through-put packed bed chromatography.

The process produces plasmid DNA characterized by low level chromosomal DNA, RNA and endotoxin contamination without the use of flammable solvents or toxic reagents and is suitable for therapeutic administration.

Introduction

As regulatory approval is granted for increasing numbers of non viral human gene therapy clinical trials there has been a corresponding rise in demand for pharmaceutical grade plasmid DNA (pDNA).

The source of plasmid DNA for these trials is exclusively recombinant *Escherichia coli. E. coli* has a history of safe use as a production organism in the biotherapeutics industry and has several obvious advantages as a source of DNA for gene therapy: the genetics of the host are completely defined, high copy number plasmids are available and high biomass fermentation approaches increase productivity. On the other hand, the isolation of pDNA on a large scale from *E. coli* and at a quality suitable for parenteral injection, poses a number of novel challenges. The biological properties of the *E. coli* production system dictate that the plasmid product will be heavily contaminated with anionic polymers of very similar structure, charge and physical behaviour (chromosomal DNA, RNA and lipopolysaccharide); each of these contaminants constituting real and significant product safety issues. The number of separation technologies applicable to plasmid isolation is therefore reduced. Plasmid DNA is also sensitive to mechanical damage and is highly sensitive to nuclease degradation. The options for recovery and handling of pDNA are therefore also reduced. Whilst well established laboratory methods have overcome such problems in the isolation of milligram quantities of pDNA (Birnboim and Doly, 1979; Sambrook et al., 1989), these methods do not meet requirements for the manufacture of pharmaceutical grade injectables and, as with all other drugs, gene based medicines must be manufactured by processes which reproducibly deliver material to tight specifications. For example, many of the most popular laboratory methods require the use of toxic chemicals (e.g. ethidium bromide, phenol, chloroform and caesium chloride) or non-scaleable unit operation such as high speed ultracentrifugation.

Previously published protocols for the production of pharmaceutical grade pDNA (Horn et al., 1995; Schorr et al., 1995), although effective, are limited in their industrial scale-up potential. Like most laboratory protocols they employ relatively large volumes of flammable solvents (e.g. ethanol, propan-2-ol or butan-2-ol) for the precipitation and concentration of pDNA. Solvent precipitation in the production of biopharmaceuticals, in particular when high speed centrifugation is required to collect the precipitated material, is undesirable in terms of cost and facility design for scale-up.

We descibe here a simple efficient process for the production of pDNA. This process strategy is fully and directly scaleable to commercial operation and produces pDNA of a quality acceptable for parenteral use.

Materials and methods

Construction of therapeutic expression plasmid vector pTX0161

pTX0161 is a high copy number (ColE1 origin of replication) plasmid vector carring the *Escherichia coli* B nitroreductase gene (*ntr*) driven by the cytomegalovirus immediate early promoter. Delivery of this vector to mammalian tumour cells allows specific cell killing by the prodrug CB 1954 (Drabek et al., 1997), a form of enhanced cancer chemotherapy described as gene directed enzyme-prodrug therapy or GDEPT. pTX0161 was constructed by the removal of the *ntr* cassette from pTX0160 (Williams et al., 1998) as an *Apa*I-blunted, *Xba*I restriction enzyme digested fragment which was then inserted into pTX0003 (a pUC19 derived vector encoding tetracycline resistance).

Generation of the Production Strain; DH1 pTX0161

10 ng of pTX0161 plasmid DNA was used to transform calcium competent *E. coli* DH1 cells (Hanahan, 1983; Bachmann, 1987). A single transformant was selected to generate a Master Cell Bank (MCB) of the Production Strain under conditions of Good Manufacturing Practice. Cells harvested from an exponentially growing culture, were cryopreserved in 20% glycerol and stored at −80 °C to constitute the MCB.

Fermentation of DH1 pTX0161

Inoculum preparation

A 1 ml vial of DH1 pTX0161 was taken from the MCB and thawed rapidly at 37 °C. The contents of this vial was aseptically inoculated into 5 ml of LB medium (Peptone 10 gl^{-1}, Yeast Extract 5 gl^{-1}, NaCl 5 gl^{-1}) supplemented with tetracycline (12 $\mu g\ ml^{-1}$) in a 30 ml universal bottle. This was incubated at 37 °C in an orbital shaker (200 rpm). During mid-exponential growth phase, 4 ml of this culture was used to inoculate 46 ml of Terrific broth (Tartof and Hobbs, 1987) supplemented with tetracycline (12 $\mu g\ ml^{-1}$) in a 500 ml Erlenmeyer flask. This was grown, as above, before being subcultured in two stages (200 ml and 2000 ml cultures in Terrific broth with tetracycline) and used to inoculate the fermenter once grown to a cell density of 2 OD_{600nm}.

Fermentation

A rich complex batch medium consisting of final concentrations of KH_2PO_4 (3 gl^{-1}), Na_2HPO_4 (6 gl^{-1}), NaCl (0.5 gl^{-1}), Peptone (2 gl^{-1}), $(NH_4)_2SO_4$ (10 gl^{-1}), polypropylene glycol, molecular weight 2025 g (0.2%), Trace element solution (0.05% [consisting of $CoCl_2.6H_2O$ (2 gl^{-1}), $CuCl_2.2H_2O$ (1.9 gl^{-1}), H_3BO_3 1.6 gl^{-1}), $MnSO_4.H_2O$ (1.6 gl^{-1}), $Na_2MoO_4.2H_2O$ (2 gl^{-1}), $ZnCl_2{\cdot}7H_2O$ (2 gl^{-1}), $Fe_2(SO_4)_3.xH_2O$ (1 gl^{-1}), $CaCl_2.2H_20$ (1 gl^{-1}), and Citric acid (60 gl^{-1})]), $CaCl_2.2H_2O$ (0.03 gl^{-1}), $FeSO_4.7H_2O$ (0.04 gl^{-1}), citric acid (0.02 gl^{-1}) and yeast extract (20 gl^{-1}) was prepared and made up to a volume of 46 l in the fermentation vessel. After sterilisation *in situ*, aseptic additions of glycerol (2.8%), $MgSO_4$(0.5 gl^{-1}) and vitamin solution (5 ml/l [consisting of biotin 0.6 gl^{-1}, folic acid 0.04 gl^{-1}, pyridoxine-HCl 1.4 gl^{-1}, riboflavin 0.42 gl^{-1}, pantothenoic acid 5.4 gl^{-1}, niacin 6.1 gl^{-1}]) and tetracycline (12 mgl^{-1}) were made.

The vessel was then inoculated as above. Fermentation was carried out at 37 °C, pH 6.8 with the dissolved oxygen setpoint 50% of air saturation.

The vessel was sampled regularly and the OD_{600nm} recorded to allow calculation of the specific growth rate. Plasmid DNA was isolated from samples during culture for qualitative analysis and from fermentation end-points for quantitative comparision (Qiagen Maxi plasmid purification kits, Qiagen Ltd.). Dry weight determination was used to estimate biomass.

Plasmid copy number estimation

The molecular weight of pTX0161, a 7796 base plasmid, is 5.06×10^6 Daltons. The specific recovered plasmid yield can be converted to a recovered plasmid copy number, knowing that 1 OD_{600nm} unit is equivilant to 2×10^8cells ml^{-1} for this fermentation culture (Kech and Brent, 1995).

Fermentation cell harvesting

Cells were harvested when in late exponential growth phase by centrifugation at 5000 g for 25 min at 4 °C in sterile 1000 ml centrifuge bottles. The clarified supernatant was discarded and the cell pellets frozen at −80 °C.

Primary plasmid DNA recovery by modified alkaline cell lysis

Preparation of a cell suspension

A total of 720 g (wet weight) of cells were removed from storage at −80 °C and thawed at 18 °C for approximately 20 min. Cell suspension buffer (50 mM Tris 10 mM EDTA at pH 8.0) was added to each centrifuge bottle sufficient to suspend the cell pellet. Resuspended cells were pooled and the volume adjusted (by weight) with cell suspension buffer to 6 litres, a final cell concentration of approximately 120 g cell wet weight l^{-1} cell suspension buffer.

Cell lysis

The cell suspension was aliquotted into 2 l units in clean stainless vessels. Molecular biology grade bovine RNAaseA (Sigma Aldrich Company) was added at a final concentration of 100 mg l^{-1}. 4 l of lysis buffer containing 0.155 M NaOH and 1% SDS at 20–22 °C were added to each 2 l aliquot before gentle mixing and incubation at 18–22 °C for 10 min. 2 l of 3 M potassium acetate with 10 mm EDTA at pH 5.5 were then poured into each vessel and the contents mixed carefully to form a uniform suspension of precipitated debris. The neutralised lysate was then pooled in stainless steel holding vessels and held at 18–22 °C for 1 hour.

Bag filtration

The contents of the holding vessels was pumped through a 50 μm woven nylon bag filter in custom made stainless steel filter housing (Plastok Holdings Ltd). The filtrate was collected in 20 l stainless steel holding tanks. Bovine RNAaseA was added as before to a final concentration of 80 mgl^{-1} and the solution held at room temperature for 1 h prior to chromatographic purification.

Expanded Bed Adsorption Chromatography

A Streamline 200 chromatography column (Pharmacia) was packed with 7.5 l of DEAE Streamline (Pharmacia). The chromatography bed was expanded using 0.1 M NaOH by upward linear flow at a rate of 105 cm hr^{-1}, stopped to allow overnight sanitisation and re-expanded in equilibration buffer (0.8 M KAc, 10 mm EDTA pH 5.5) to prepare the column for loading. The plasmid containing filtrate was pumped onto the column at the same linear flow rate. Once loaded, the column was washed with equilibration buffer until the on-line absorbance detector optical density at 254 nm or OD_{254nm}, was reduced to 20% of the maximum value or less. The column was then washed with a low salt buffer (25 mM potassium acetate, 10 mM EDTA pH 5.5) and the fluid flow direction was reversed. This allows packing of the chromatography gel in the column and reduces the buffer volumes required for subsequent steps. Washing was continued until 5% of the maximum OD_{254nm} value was reached before residual RNA fragments were eluted from the column using 0.5 M salt wash buffer (25 mM potassium acetate, 10 mM EDTA, 0.5 M NaCl pH 5.5). Plasmid DNA was then eluted with elution buffer (0.75 M NaCl, 25 mM KAc, 10 mM EDTA pH 5.5) and stored overnight at +4 to +8 °C.

Concentration of the Streamline Eluate

An Amicon CH2 ultrafiltration device was sanitised with 0.1 M NaOH and flushed with elution buffer. The plasmid enriched Streamline column eluate was concentrated approximately 25 fold by ultrafiltration using a 30 Kilodalton molecular weight cut off spiral cartridge. The concentrate was decanted and the system flushed with buffer to remove residual plasmid DNA which was pooled with the concentrate. The pool volume of 900 ml was stored at 4–10 °C overnight.

Size exclusion chromatography

To remove residual contaminants such as digested RNA fragments, and in order to exchange the plasmid into a suitable formulation buffer, the concentrated Streamline eluate was loaded onto an Amicon Vantage 130 column packed with approximately 7 litres of Pharmacia S500HR media (bed height 56 cm). The column, associated bubble trap and pipework had been previously sanitised with 0.1 M NaOH, and equilibrated with 0.3 M NaCl. The entire concentrate was loaded onto the column at 50 ml/min. Fraction collection was started once the OD_{254nm} increased. Fractions containing in excess of 20 μg/ml plasmid (by OD_{260} measurement) were pooled. The volume of the pool was approximately 1500 ml. Samples for quality control analysis were removed and the remainder of the bulk plasmid pool aliquoted and frozen at −80 °C.

DNA analysis for quality control

Appearance
The solution was examined by eye for colour, clarity and the presence of extraneous matter (for example, fibres or particles).

pH
The pH was determined using a calibrated glass combination pH electrode.

Product concentration
The total nucleic acid concentration (strength of solution) was determined spectrophotometrically by recording the optical density at 260 nm and converting this value to mg/ml assuming that the extinction coefficient of a 0.0005% solution of plasmid DNA in a 1 cm path length cell is 0.02 units. Highly concentrated samples were diluted by weight such that the recorded optical density was within the linear response range of the chromophore.

Purity
Plasmid purity, or percentage covalently closed circular plasmid, was estimated after separation of all the plasmid forms present in the sample (covalently closed circular, denatured, open circular, and concatamers) by 0.8% agarose gel electrophoresis and visualisation by staining with ethidium bromide and UV transillumination. An initial gel loading of 1 μg plasmid DNA was serially diluted (by a factor of two for each dilution in the series) and each dilution loaded onto the gel. Estimation of plasmid purity was made by comparison of the series dilution number (n_p) at which the band due to covalent closed circular was no longer visible, with that of an observed contaminant (n_c). The reciprocal of 2 to the power of the difference ($n_p - n_c$) multiplied by 100% is a measure of the relative abundance of that contminant.

$$\text{Observed Contaminant (\%)} = 100 \times 1/[2^{(n_p - n_c)}]$$

Covalently closed circular DNA content was quoted as 100% minus the sum of the percentage values for each observed contaminant.

Identity
Plasmid identity was confirmed by 5–10 separate restriction enzyme digests; *Eco*RV, *Bgl*II, *Bam*HI/*Hind*III, *Hae*II and *Hae*III. Appropriate buffer conditions and enzyme activities were used for each enzyme with digestion incubations carried out for 1.5 h.

Endotoxin contamination
Endotoxin contamination was assessed using the KQCL kinetic chromogenic assay (Biowhittaker UK Ltd). Product was diluted to a range known to overcome interference. Positive product controls were included to confirm that no inhibition or enhancement of the reaction cascade by sample or buffer components had occurred.

Host protein
Samples for the assessment of host protein contamination were prepared according to the method of Laemlli (Laemmli, 1970)for SDS-PAGE employing a silver staining technique for detection. Comparison to known standard quantities of RNAase A (100-1000 ng/ml) enabled a maximum value to be quoted.

Host chromosomal DNA
Host chromosomal DNA contamination was assessed using PCR analysis, employing oligonucleotide primers complementary to an ubiquitous element of the *E. coli* genome. Standards of pTX0161 purified by double denstiy gradient ultracentrifugation (Sambrook et al., 1989) were spiked in duplicate with varying percentages (w/w) of DH1 host chromosomal DNA. An identical quantity of the pTX0161 was also subjected to the reaction and analysed alongside the other samples by 1% agarose gel electrophoresis. Maximum contamination levels were derived based on a comparison of the intensity of the host chromosomal DNA spiked standards against the sample being tested.

Host RNA
RNA contamination is generally a heterogeneous, low-level contaminant. This is generally detected on agarose gels using ethidium bromide staining to a detection level of approximately 3.2% w/w. This detection level can be increased to 1.6% w/w by the use of Syber Green II dye (Molecular Probes, Inc.) to stain an agarose gel. Quantitation at lower levels has been achieved by the development of a sensitive HPLC method using material digested using an RNA specific protocol (Cowell et al., manuscript in preparation).

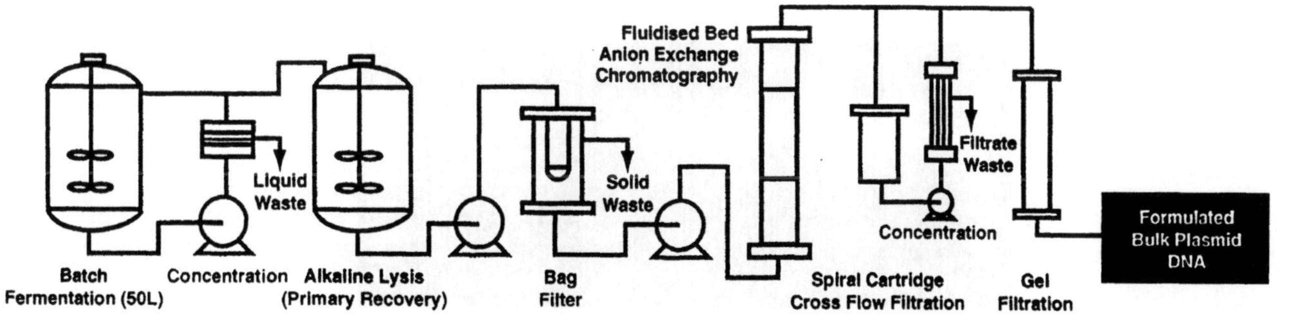

Figure 1. Schematic diagram of the large scale plasmid purification.

Results and Discussion

A process fulfilling the scale-up and quality criteria required for commercial pDNA manufacture has been developed (Figure 1). The process involves optimized alkaline extraction of plasmid DNA, clarification by filtration, capture and purification by anion exchange expanded bed chromatography and finally gel permeation chromatography. Plasmid extractions routinely employ alkaline detergent cell lysis, followed by precipitation, upon neutralisation, for the primary recovery of the plasmid product. This step, although key to the success of the purification strategy, is seldom fully optimised and is generally carried out using conditions developed for the recovery of recombinant plasmids in the research laboratory. We describe optimised cell lysis conditions and a subsequent neutralisation which efficiently precipitates host chromosomal DNA, bacterial proteins and other host derived contaminants in a reproducible manner. Plasmid DNA was then separated from the precipitated debris by bag filtration. The resulting filtrate, free of larger particles, is 'captured' directly, without further clarification, by expanded bed anion exchange chromatography. The eluted plasmid DNA is further purified and buffer exchanged by size exclusion chromatography to yield pharmaceutical grade pDNA.

The total yield from this large scale production batch was 235 mg of pharmaceutical grade plasmid DNA (see Table 1) produced in 3 days from frozen cell paste.

The harvested *E. coli* DH1 cells containing the plasmid pTX0161 gave a specific recoverable yield of 1.4 μg plasmid/OD ml, or a volumetric yield of 15.8 mg/l. Plasmid recoveries obtained using small scale pDNA laboratory purification kits (Qiagen Ltd.) are high and assumed to be approaching 100%. This allows a rapid copy number determination, estimated for this fermentation to be a minimum of 833 copies per cell. Such high specific plasmid yields greatly facilitate purification of pDNA from contaminating host derived impurities, particularly host derived chromosomal DNA. The *E. coli* chromosome is 4.6×10^6 base pairs in length (Blattner et al., 1997) equivalent to 3.1% total dry cell weight (Neidhardt, 1987). A copy number of 833 plasmids per cell (or a total of 6.49×10^6 bp) is a 1.4 fold excess of product over the contaminating chromosomal DNA (assuming one chromosome per cell). Such relative plasmid abundance is critical for the successful plasmid production from an intensified fermentation process.

The subsequent purification of plasmid DNA from the harvested *E. coli* cell paste (summarised in Figure 1) begins with the optimised lysis of evenly resuspended cells by alkaline detergent. The maximum pH at cell lysis was carefully controlled using larger volumes of lysis buffer containing a lower sodium hydroxide concentration than previously reported (Sambrook et al., 1989; Kech and Brent, 1995) to prevent formation of irreversibly denatured plasmid species (Rush and Warner, 1970). Small scale lysis experiments with this host and vector combination demonstrated that this hydroxide concentration allowed an extended lysis time as well as preventing the formation of irreversibly denatured plasmid or causing loss in yield (Thatcher et al., 1997). Irreversibly denatured plasmid is difficult to separate from the covalently closed circular and open circular forms using scaleable technologies such as anion exchange chromatography. Whilst this species does not interfere with many research molecular biology applications, its presence in clinical material is unacceptable. Agarose gel electrophoresis analysis of plasmid purity clearly shows the absence of irreversibly denatured species in the product generated using this method (Figure 2), even at high sample loadings (1000ng pDNA gel loading).

Irreversibly denatured plasmid is resistant to digestion by endonucleases and thus can be identified by

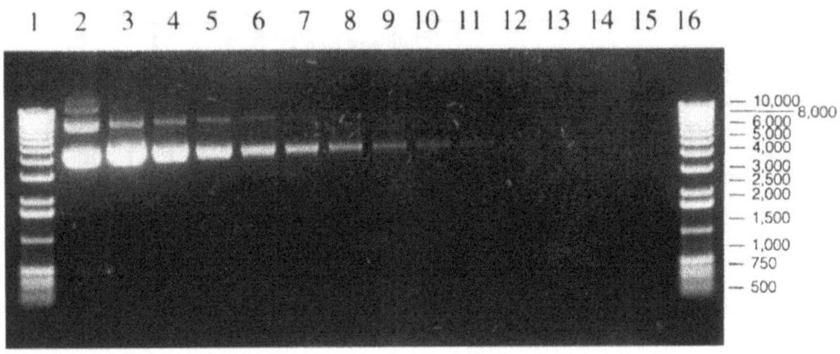

Figure 2. Gel purity test for the purified plasmid 0.8% agarose gel representing the gel purity test. Lanes 1 and 16: 1 Kb ladder molecular size markers. Lane 2: Caesium chloride purified pTX0161. Lanes 3–14: purified pTX0161 at 1000 ng, 500 ng, 250 ng, 125 ng, 62.5 ng, 32 ng, 16 ng, 8 ng, 4 ng, 2 ng, 1 ng, 0.5 ng mass loadings. Lane 15: loading buffer.

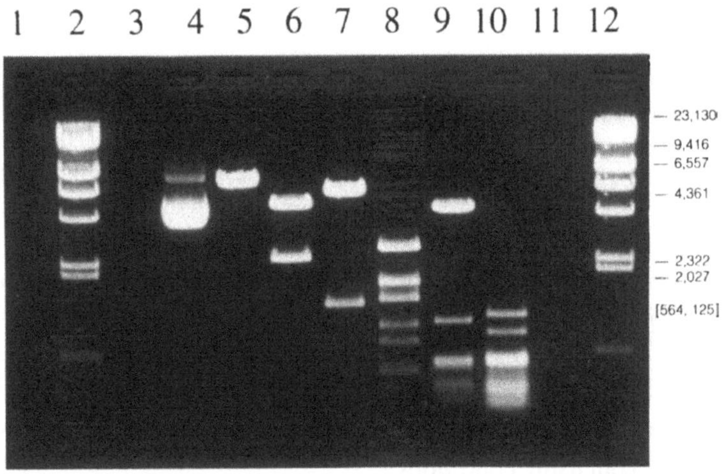

Figure 3. Restriction enzyme assay for plasmid identity. 1% agarose gel. Lanes 2 and 12: λ *Hind*III molecular size markers. Lane 3: loading buffer. Lane 4: Purified pTX0161 (uncut). Lane 5: *Kpn*I digest. Lane 6: *Eco*RV digest. Lane 7: *Bgl* II digest. Lane 8: *Hind*III/*Bam*HI double digest. Lane 9: *Hae*II digest. Lane 10: *Hae*III digest. Lane 11: loading buffer.

failure to be 'cut' by restriction enzymes in a plasmid identity test (Figure 3). Once again this assay demonstrates that the plasmid product is free from denatured forms, and has a restriction 'map' consistant with the plasmid DNA sequence.

The lysis pH used is critical and the optimal value has been found to vary for different host and plasmid vector combinations (Thatcher et al., 1997). The extended lysis time also allows for greater destruction of sodium hydroxide labile contaminants, such as endotoxin and RNA fragments (Table 1). The larger process window for the lysis step is also an important factor for further scale up, when addition and mixing of reagents in large scale vessels within a limited timescale becomes an issue. Larger volumes of lysis buffer allows for a more homogeneous lysis and reproducible precipitation of cellular debris, protein (Table 1), chromosomal DNA (Figure 4) and endotoxins (Table 1).

Bag filtration, similar to that used in the food processing industry, allows for rapid processing of high volumes of the neutralised lysate without the need for high speed centrifugation. Large scale particulate separation methodologies such as disc stack centrifuges can expose the product and precipitated contaminants to high shear forces which can be detrimental to product quality. These types of equipment also have high capital and installation costs and, therefore, may not be appropriate for the generation of the relatively small amounts of highly potent gene medicines required for, for example, DNA vaccines. Other effective clarification technologies which rely on floatation in order to gain high quality filtrate (Theodossoiu et al., 1997) suffer from the disadvantage that part of the flocc sediments. Bag filtration has high solids capacity and does not require precipitate floatation.

Another advantage is that the filter matrix contains no leachables and the housings are biocontained and

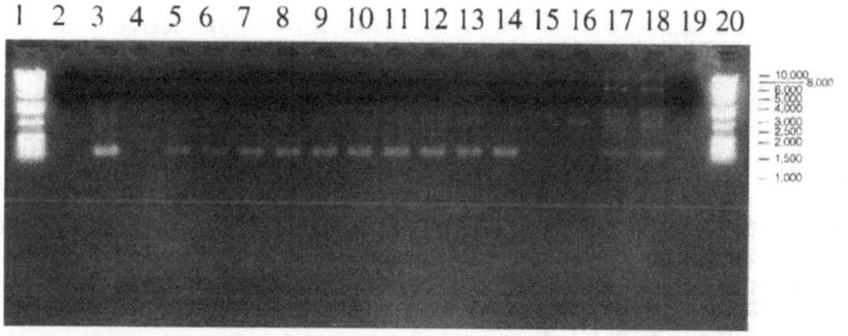

Figure 4. *E.coli* chromosomal DNA assay by polymerase chain reaction (PCR) amplification 1% agarose gel analysis of PCR products. Lanes 1 and 20: 1 Kb ladder molecular size markers. Lane 2: loading buffer. Lane 3: positive control 50 ng host chromosomal DNA. Lane 4: negative control. Lanes 5 and 6: 2% 'spike' of DH1 host chromosomal DNA . Lanes 7 and 8: 4% 'spike' DH1 host chromosomal DNA. Lane 9 and 10: 5% 'spike' DH1 host chromosomal DNA. Lanes 11 and 12: 6% 'spike' DH1 host chromosomal DNA. Lanes 13 and 14: 8% 'spike' DH1 host chromosomal DNA. Lanes 15 and 16: control plasmid sample. Lanes 17 and 18: pTX0161 purified using method described. Lane 19: Loading buffer.

Table 1. Summary of analytical data

Test	Results
Appearance	Clear, colourless solution free of extraneous matter
pH	7.1
Purity	93% covalently closed circular plasmid, 6.25% open circular (Figure 2)
Size	Consistent with plasmid map
Restriction sites	Consistent with plasmid sequence (Figure 3)
E. *coli* chromosomal DNA	<2%w/w, as determined by PCR assay (Figure 4)
RNA levels	<1.6% w/v based on SYBR Green II stain on 1% agarose gel with 1000 ng DNA 0.2% w/w RNA HPLC assay
Concentration (strength)	148.7 μg/ml
Endotoxin	2.5 EU/mg pDNA
Protein content	<100 ng/ml

do not generate aerosols during processing. This offers both increased operator protection, an important consideration during the processing of genetically modified microorganisms, and a higher level of product cleanliness.

The clarity of the bag filtered feedstock is insufficient for reliable performance on conventional anion exchange media in fixed beds without further processing. However, it is ideally suited for fluidized or expanded bed adsorption (EBA). These technologies allow for very rapid capture and purification of the dilute plasmid feedstock and can readily handle far cruder feedstocks such as whole cells and crude cell homogenates (Chang and Chase, 1996a). Using the load generated by bag filtration the resulting EBA column performance is ideal (Chang and Chase, 1996b). Particulates which would foul packed bed chromatography media (causing increase in back pressure or blockage of the column, even at comparatively low linear flow rates) are successfully washed out of the gel in the plug flow expanded bed operations. The column flow can then be reversed and the top flow distributor lowered onto the settled bed to exploit the advantages of conventional packed bed chromatography, such as reduced wash volumes and elution of relatively concentrated product (Figure 5). EBA chromatography also achieves rapid processing in a closed system which, as described earlier, is important during the processing of recombinant pharmaceuticals from genetically modified microorganisms.

Digested RNA fragments are efficiently desorbed using a 0.5 M NaCl buffer wash and the plasmid elutes only in 0.75 M salt (Figure 5). During purification the plasmid extract is reduced to about 40%

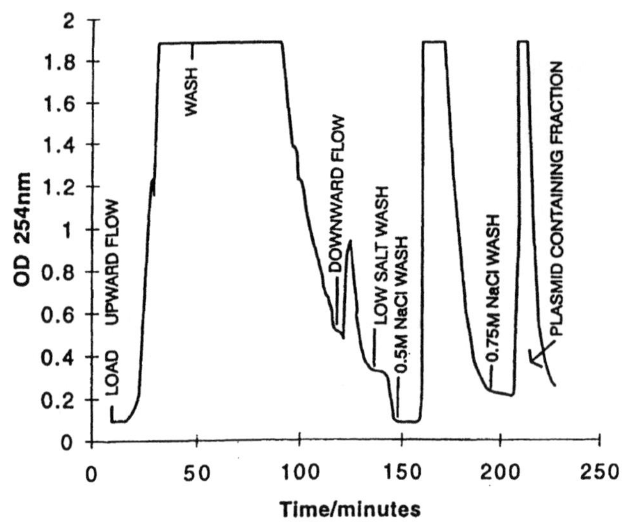

Figure 5. Chromatogram of the fluidised bed anion exchange plasmid capture. Expanded bed anion exchange adsorption column profile. UV data plotted is directly from the UVM1 monitor (Pharmacia) with full scale deflection set at 2 AUF_{254nm}.. Identity of peak content, flow direction and buffer changes are shown.

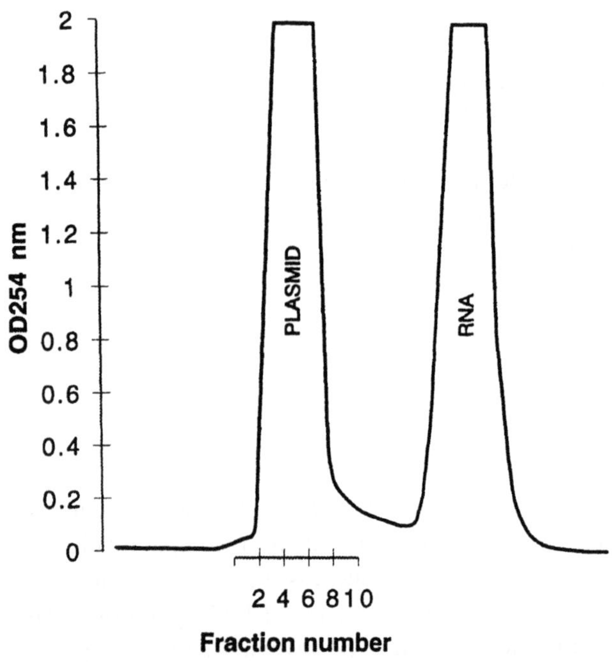

Figure 6. Chromatogram of the size exclusion buffer exchange and product polishing. S500 (Pharmacia) size exclusion column profile. X axis represents time. UV data plotted is directly from the UVM1 monitor (Pharmacia) with full scale deflection set at 2 AUF_{254nm}..

of the original lysate volume. Further concentration, in preparation for size exclusion chromatography, is achieved without large volume ethanol precipitation by the use of spiral cartridge membrane ultrafiltration. There is no visible deterioration in quality of the plasmid despite the high recirculation rate for retentate in this pumped device. The system can be flushed to ensure high recovery at the end of the operation. Tangential flow ultrafiltration also has demonstrated potential for industrial scale up (Bowen, 1993).

The concentrated plasmid loaded onto the size exclusion column is approximately 11% of the S500 column volume. Despite the relatively high loading volume, separation of plasmid and RNA fragments is efficient (Figure 6), indicating that most of the plasmid is excluded from the gel and elutes in the void volume. This step has another important function which is to exchange the plasmid into a more appropriate buffer in preparation for formulation as the gene medicine.

Conclusions

This process has been shown to deliver both the quality and scaleability required for the commercial manufacture of therapeutic grade plasmid DNA. Industrial scale up is now possible using optimized alkaline lysis in conjuction with bag filtration and EBA capture of the plasmid product to replace previously employed technologies. Plasmid concentration by ultrafiltration also overcomes the issues which arise with manufacturing processes involving DNA precipitation in alcohols. These steps are efficient alternatives to current approaches to plasmid DNA isolation and the process produces plasmid DNA suitable for therapeutic administration.

References

Bachmann BJ (1987). In: Neidhardt FC et al. (eds.) *Escherichia coli* and *Salmonella typhimurium* Cellular and Molecular Biology (pp. 1190–1219). ASM.

Blattner RF, Plunkett G the III, Bloch CA, Perna TN, Burland V, Riley M, Wayne Davis N, Kirkpatrick HA, Goeden MA, Rose DJ, Mau B and Shao Y 1997. The complete sequence of *Escherichia coli* K-12. Science 277: 1453–1462.

Birnboim HC and Doly J (1979) A rapid alkaline extraction procedure for the screening of recombinant plasmid DNA. Nucleic Acids Res. 7: 1513–1523.

Bowen R (1993) Understanding the flux patterns in membrane processing of protein solutions and suspensions. Trends Biotechnol. 11: 451–460.

Chang YK and Chase HA (1996a) Ion exchange purification of G6GDP from unclarified yeast cell homogenates using expanded bed adsorption. Biotechnol. Bioeng. 49: 204–216.

Chang YK and Chase HA (1996b) Development of operating conditions for protein purification using expanded bed techniques: the effect of the degree of bed expansion on adsorptive performance. Biotechnol. Bioeng. 49: 512–526.

Drabek D, Guy J and Grosveld F (1997) Expression of bacterial nitroreductase in transgenic mice results in specific cell killing by the prodrug CB1954. Gene Therapy 4: 93–100.

Felgner P (1997) Non-viral Strategies for Gene Therapy. Scientific American June: 86–90.

Hanahan DJ (1983) Studies on transformation of *Escherichia coli* with plasmids. Mol. Biol. 166: 557–580.

Horn NA, Meek JA, Budahazi G and Marquet M (1995) Cancer gene therapy: Purification of DNA for human clinical trials. Human Gene Therapy 6: 565–573.

Kech K and Brent R (1995) Current Protocols in Molecular Biology (pp. 1.2.2). Ausubel FM et al. (eds) John Wiley & Sons Inc., New York.

Laemlli UK (1970) Cleavage of structural proteins during the assembly of the head of bacteriophage T4. Nature 270: 680–684.

Neidhardt FC (1987) Chemical composition of *E.coli*. In: Neidhardt et al. (Eds.) *Escherichia coli* and *Salmonella typimurium* Cellular and Molecular Biology. American Society for Microbiology.

Rush MG and Warner RC (1970) Alkali denaturation of covalantly closed circular duplex deoxyribosenucliec acid. J. Biol. Chem. 245: 2704–2708.

Schorr J, Moritz P, Seddon T and Schleef S (1995) Plasmid DNA for human gene therapy and DNA vaccines. Annals of the New York Acadamy of Sciences 772: 271–273

Sambrook J, Fritsch EF and Maniatis T (1989) Molecular cloning: a laboratory manual (2nd edition). Nolan et al. (eds.) Cold Spring Harbor Laboratory Press, New York.

Tartof KD and Hobbs CA (1987) Improved media for growing plasmid and cosmid clones. Focus 9: 2–12.

Thatcher DR, Hitchcock AG, Hanak JH and Varley DL (1997) Method of plasmid Production and Purification. International Patent Application WO/29190.

Theodossiou I, Collins IJ, Ward JM, Thomas ORT, and Dunnill P, (1997) The processing of a plasmid-based gene from *E. coli*. Primary recovery by filtration. Bioprocess Eng. 16: 175–183.

Williams SG, Cranenburgh RM, Weiss AME, Wrighton CJ Sheratt DJ and Hanak JAJ (1998) Repressor titration: a novel system for selection and stable maintenance of recombinant plasmids. Nucleic Acids Res. 26(9): 2120–2124.

Address for correspondence: Cobra Therapeutics, The Science Park, Keele, Staffs., ST5 5SP, UK (E-mail: julian.hanak@cobrat.com)

Bioseparation **8:** 219–228, 1999.
© 1999 *Kluwer Academic Publishers.*

Development and scale up of a capture step (expanded bed chromatography) for a fusion protein expressed intracellularly in *Escherichia coli*

Michael Brobjer
Pharmacia & Upjohn Process R&D, Stockholm, Sweden Fax: +46 8 6954047

Received 20 October 1998; accepted 20 March 1999

Key words: expanded bed adsorption (EBA), cation exchange chromatography

Abstract

A capture step was developed using the expanded bed adsorption technology to separate a protein of interest on a cation exchanger from a crude *Escherichia coli* homogenate. This method was developed in bench-top scale using a STREAMLINE 25 column (Amersham Pharmacia Biotech, Sweden) and STREAMLINE SP. The development was based on earlier experiments performed in a packed bed column (SP-Sepharose FF) to investigate the conditions for sample application, wash and elution. The packed bed method was transformed into an expanded bed method by slightly modifying the wash procedure and cleaning in place (CIP). This method was then scaled-up to pilot scale and used for production of the fusion protein according to cGMP.

The yield over the step in pilot scale was 70-85% compared with only 30-50% in small scale. Pressure build-up, attachment of biomass to the adsorbent and collapses of the expanded bed were phenomena seen in small scale but not in pilot scale. The scale-up of the step significantly improved the performance of the step.

Abbreviations: CIP – clean in place; GMP – good manufacturing procedure ECP – *Escherichia coli* protein(s); CV – column volume; WFI – water for injection; QC – quality control; DTT – dithiothreitol; ELISA – enzyme-linked immuno sorbent assay; CFU – colony forming unit

Introduction

Two of the most important key issues in the development of an economical competitive purification process are the reduction of volume and high capture of product in the initial chromatography step. In addition to this, an effective reduction of *E. coli* proteins (ECP's) and other host cell derived impurities (DNA and Endotoxins) are essential to achieve early in the process to minimize the complexity of the more high resolution steps downstream. The scalability, robustness and economy are also issues to be considered.

The development of the capture step in this production process was initiated in late 1996. The aim was to capture a fusion protein in an *Escherishia coli* homogenate and to remove cells, cell debris, and host cell derived impurities with a high product recovery. The capture of the product in the initial purification step could either be performed by a cell separation unit operation (i.e. centrifugation) followed by conventional packed bed chromatography or by using the Expanded Bed Adsorption technology. An evaluation between the packed bed chromatography and expanded bed chromatography based on the yield, reduction of host cell contaminants, process time, economy and scalability was to be performed.

During conventional cell removal (i.e. centrifugation or microfiltration) of a homogenate the risk of product loss due to product degradation exists. This together with the fact that centrifugation and filtration both are relatively time consuming and expensive unit operations lead to the introduction of the expanded bed adsorption technology in early 1990 (Grager and Chase, 1991). The expanded bed technology (Hedman and Barnfield Frej, 1992; Barnfield Frej et al.,

1994) exhibits a dual function, both cell separation and protein purification that can result in a decreased number of process steps, increased yield and reduced capital expenditure (Chase, 1994). This made expanded bed adsorption an interesting process alternative to investigate and evaluate.

The target molecule is a 46 kDa large fusion protein that is positive charged at neutral pH. The theoretical calculated pI of the protein is approximate 9.4.

Material & methods

Equipment

The development work was performed on an Äkta explorer (Amersham Pharmacia Biotech, Sweden). The scale up experiments and GMP production were performed on a 10 mM internal diameter piping chromatography station controlled by a PC with Unicorn software (Amersham Pharmacia Biotech, Sweden).

The column used in the development and production were XK 16/20, STREAMLINE 25 and STREAMLINE 200, all from Amersham Pharmacia Biotech, Sweden. All columns were packed with STREAMLINE SP (Amersham Pharmacia Biotech, Sweden)

Chemicals

All chemicals used were of analytical grade or approved by the QC department, Pharmacia & Upjohn. All buffers used were prepared using WFI quality water and the pH was adjusted to within ±0.1 of the stated value and the conductivity was measured and controlled to be within 5% of stated value.

Protein material

The protein material used in the development of the capture step was all from homogenized (three cycles at 600 bar) *E. coli* fermentations. The scale-up experiments and the GMP production were performed by using a homogenate that had been recycled through a homogenizer at 800 bar for 2 hours.

Expanded bed chromatography method

The method was based on experiments performed on packed bed columns in small scale. The buffer system was a 50 mM phosphate buffer at pH 7,8. The bed was expanded under at least 30 minutes prior to sample application. The expansion factor (h_{exp}) was 2,7. The sample application was performed during constant h_{exp} , thus the flow rate varied in the range 250 to 300 cm/h. To wash off unbound material a three CV wash was performed with a 50 mM phosphate buffer. The elution was performed in a packed bed mode by a step gradient addition of NaCl in the phosphate buffer. The target peak was manually collected by monitoring the UV signal (A_{280}) and was 1.5 to 2.5 CV large. The column was regenerated with 2 M NaCl prior to Cleaning In Place (CIP) with 1 M NaOH for approximate 2 h. The linear flow rate used to expand the bed was 300 cm/h. All steps in expanded mode was performed with a flow rate of 300 cm/h. For elution a flow rate of 100-150 cm/h was used and during the CIP the flow rate was 150 cm/h.

Loading capacity measurements

The capacity of the resin was investigated using both homogenized fermentation broth and purified product in batch and chromatography mode. The capacity was measured and evaluated over a pH range from 6.5 to 8.5, Ionic strength from 3 to 15 mS/cm and the linear flow rate from 100 to 300 cm/h.

Wash and elution profile

A linear salt gradient was applied to find the conductivity suitable for elution at a given pH. After selecting a suitable salt concentration the linear salt gradient was converted to a stepwise performance of the elution and applied as a standard elution procedure with a wash step and an elution step. Regeneration and cleaning in place (CIP) followed the elution.

Scale up and GMP production method

A linear scale up was performed by expanding the diameter in proportion to the increase of flow rate, thus keeping the linear flow constant. To increase the amount product processed at each batch, the bed height was doubled. The load and expansion factor (h_{exp}) were kept constant. The buffer system was a 50 mM phosphate buffer at pH 7,8. The bed was expanded for 60 min prior to sample application. The elution was performed in a packed bed mode by a step gradient with 500 mM NaCl in the phosphate buffer. The column was regenerated with 2 M NaCl prior to Cleaning In Place (CIP) with 1 M NaOH for approximately 2 h. The linear flow rate used was 300 cm/h.

For elution a flow rate of 100 cm/hour was used and during the CIP the flow rate was 150 cm/h.

Analytical methods

Normally a silver stained SDS-PAGE was performed to estimate the reduction of ECP's and to identify the fraction containing the product. The SDS-PAGE was performed using a Multiphor II electrophoresis unit (Novex, USA) connected to a power supply. A twelve well pre-cast Excel-gels, 12.5% Tris-Glycine, 0.5 mm thick gels on plastic support was used. The protein separation was performed in reduced mode, using DTT as reducing agent. The proteins were visualized by silver staining using a automated gel stainer (Hoefer)

For determine the product concentration in the starting material and in the eluent a standard ELISA method was used initially, then a more reliable HPLC method was developed and used.

The reduction of host cell derived impurities were monitored by measuring the level of ECP's with a standard ELISA technique, Endotoxins according to the LAL gel cloth method and DNA with the threshold technique method in the starting material, wash fractions and the eluent.

Results

Wash and Elution profile

A series of experiments were performed to investigate the wash and elution profile. Initially the aim was to use a relatively high salt concentration in the wash step to reduce the amount of ECP's in the eluent of the product. A step wise elution profile was identified where the removal of ECP's was high according to SDS-PAGE (Figure 1). A wash step at 220 mM NaCl was used to remove bound ECP's and to receive a fraction in the eluent with a relatively pure product.

The protein fraction occurring during the wash step was applied on the following chromatography step to investigate the importance of reducing those ECP's. None of the ECP's in the wash fraction bound to the following high resolution column at normal conditions and therefore did not effect the capacity or resolution of that step. This led to the exclusion of the wash step from the capture method and wash after sample application was only performed with equilibration buffer. This gave the step a broad conductivity range in the elution profile, eluting almost all bound material in the eluent fraction. It also increased the yield slightly and the step got a more robust set-up.

Loading capacity of resin

The capacity of the resin at relevant pH and conductivity was investigated both with normal starting material (diluted homogenate) and with purified product.

In measurements of the capacity using diluted homogenate at pH 6.5 to 8.5 a low dynamic capacity was noted. A 30% breakthrough was reached already after 3.7 mg/ml resin and in a second experiment no breakthrough was noted at a load of 1.6 mg /ml resin but a 50% breakthrough was noted after 2.12 mg/ml resin. No significant difference was noted at different pH. When capacity measurements were performed with pure product at pH 8, no breakthrough was noted after 7.5 mg/ml resin.

In an eluent from an applied homogenate the fusion protein represents approximately 50% of the total amount of proteins (according to total protein determination using the Bradford technique). The breakthrough of the product when applying diluted homogenate was therefore not only a result of no free functional groups on the resin but also a result of interaction with components in the homogenate. It could also reflect the fact that the kinetics in the diluted homogenate is slower due to higher viscosity and the presence of particles.

The dynamic capacity when applying diluted homogenate containing fusion protein was estimated to 1.8 g/L resin.

Concentration factor

The concentration factor achieved, as measured as mg product per unit volume was 5 to 10 fold in the small scale experiments. In pilot scale and production the concentration factor was approx. 5 fold.

Purity factor

The purity factor over the step was determined as the reduction of host cell proteins, measured as total protein according to Bradford. The concentration of the product in the homogenized fermentation broth corresponded to approx. 1-2% of the total protein according to SDS-PAGE (Figure 1) and to a total protein analysis (Bradford). The eluent held a concentration of the fusion protein that corresponded to about 50% of total protein (Figure 2), thus a reduction of total proteins was in the range of 100 fold.

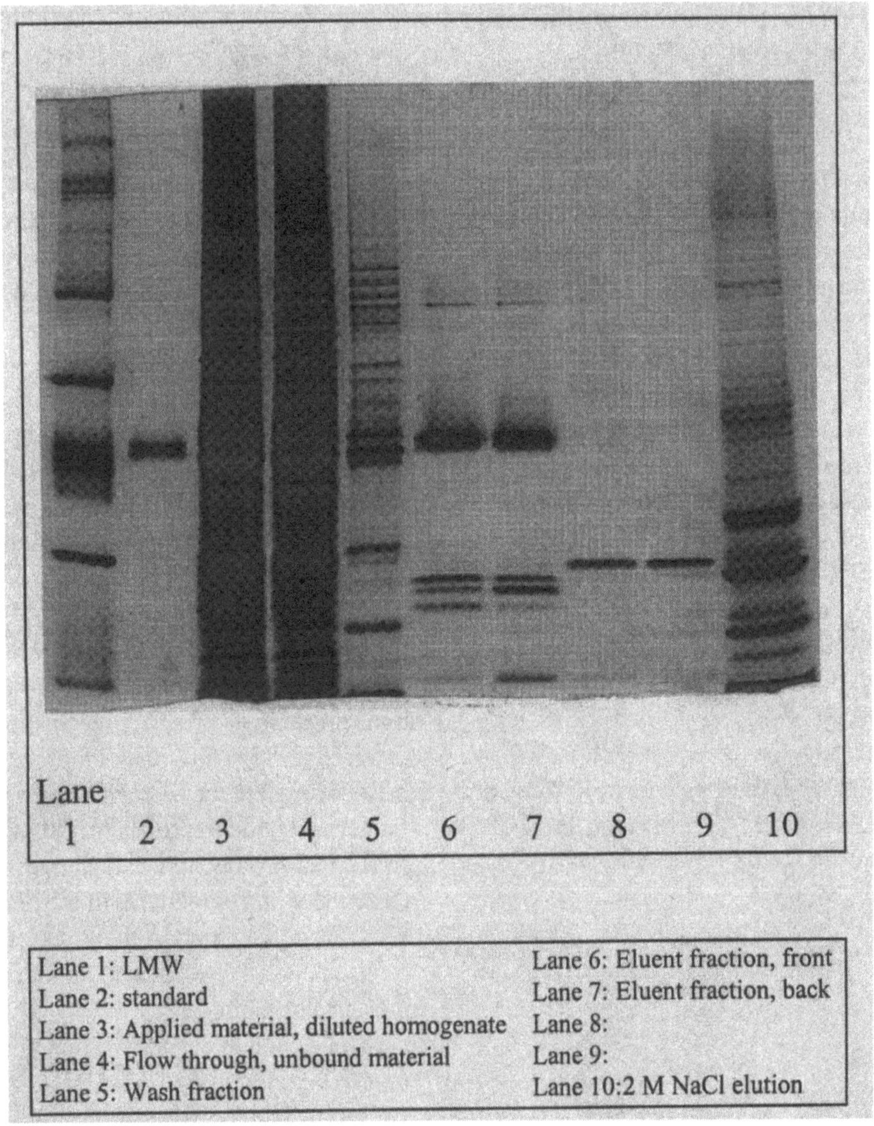

Figure 1. SDS-PAGE from the capture step. This SDS-PAGE is from an experiment with a step wise gradient resulting in four peaks. Peak 1 (lane 5), wash with 220 mM NaCl, peak 2 (lanes 6 and 7) elution of product with 330 mM NaCl, peak 3 (lnes 8 and 9) wash with 470 mM NaCl, peak 4 (lane 10) total elution with 2 M NaCl. The method was modified (see text) and peak 1–3 were included in the eluted peak at 500 mM NaCl (Figure 7).

Reduction of DNA and Endotoxins

Reduction of host cell derived impurities (DNA and endotoxins) in the first chromatography step had been evaluated in small scale for the first step. The reduction of DNA was typically in the range of 3-4 logs (Figure 3) and the reduction of endotoxins is in the range of 4,5–5 logs of magnitude (Figure 4).

Reduction of cells and particles

The reduction of cells and cell debris were in the order of 4–6 logs in the expanded bed adsorption experiments performed in small scale. In the GMP production campaign a 0.22 μm filter was mounted on the product outlet, resulting in no measurements of reduction of viable cells over the column. The level of CFU in the filtered eluent was less than 10/mL.

Yield

Initial experiments in small scale with expanded bed adsorption gave low yields (Figure 5) and only modest reduction of host cell derived impurities due to several technical problems. Several experiments had to be disrupted due to high pressure drop in the column. The main reason for the increase of pressure was an accumulation of DNA, aggregates and fines from the adsorbent under the net of the top adapter. In these

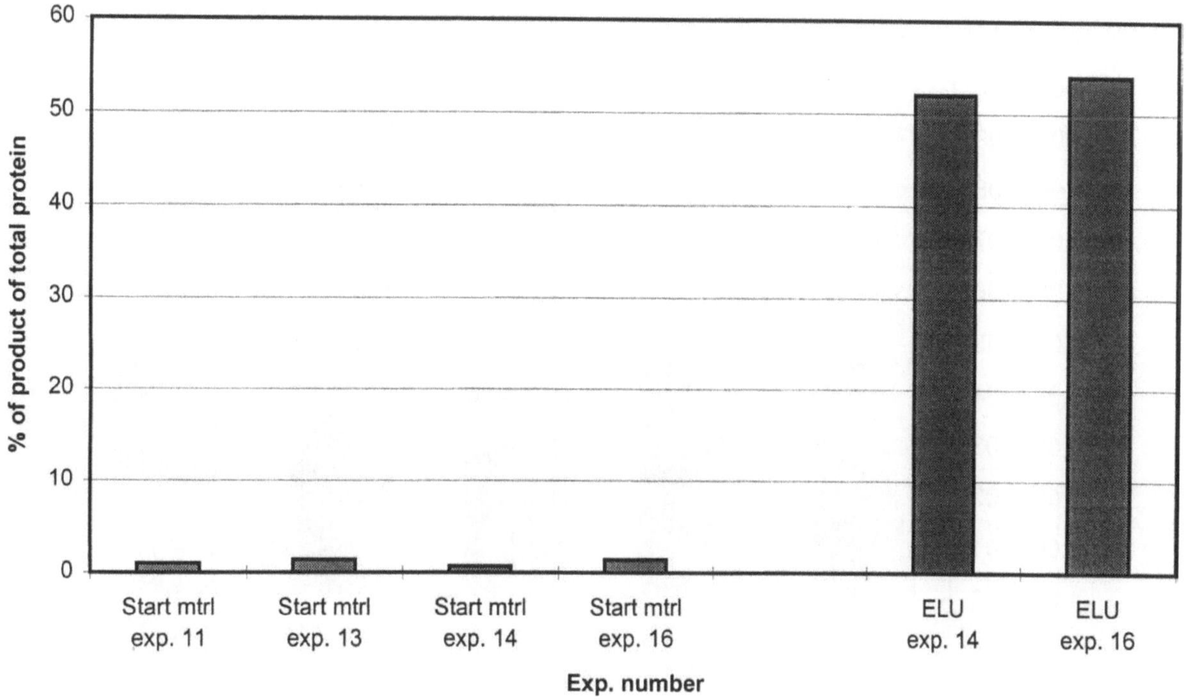

Figure 2. Total protein analysis (according to the Bradford technique) and concentration analysis of fusion protein (according to HPLC method) performed in the diluted homogenate and the eluent. In the graph the product concentration is represented as % of the total protein concentration.

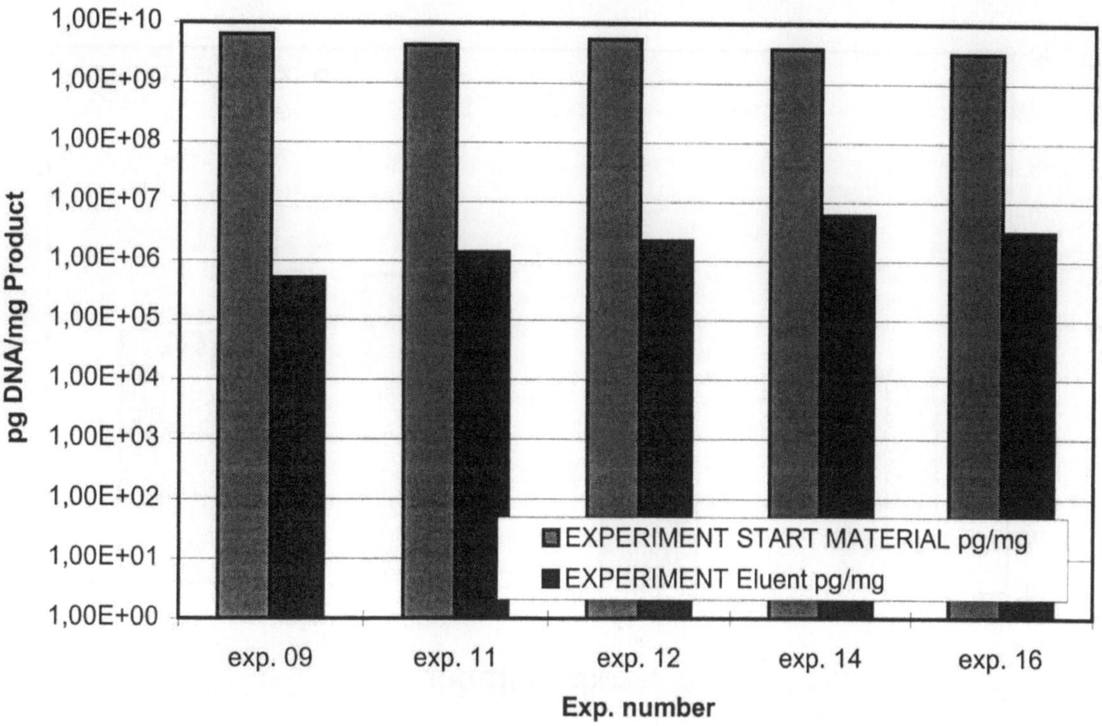

Figure 3. The reduction of DNA in a number of expanded bed experiments performed in small scale. The reduction of DNA (as measured with the threshold technique) was quite stable with a reduction of 3 to 4 logs of magnitude over the step.

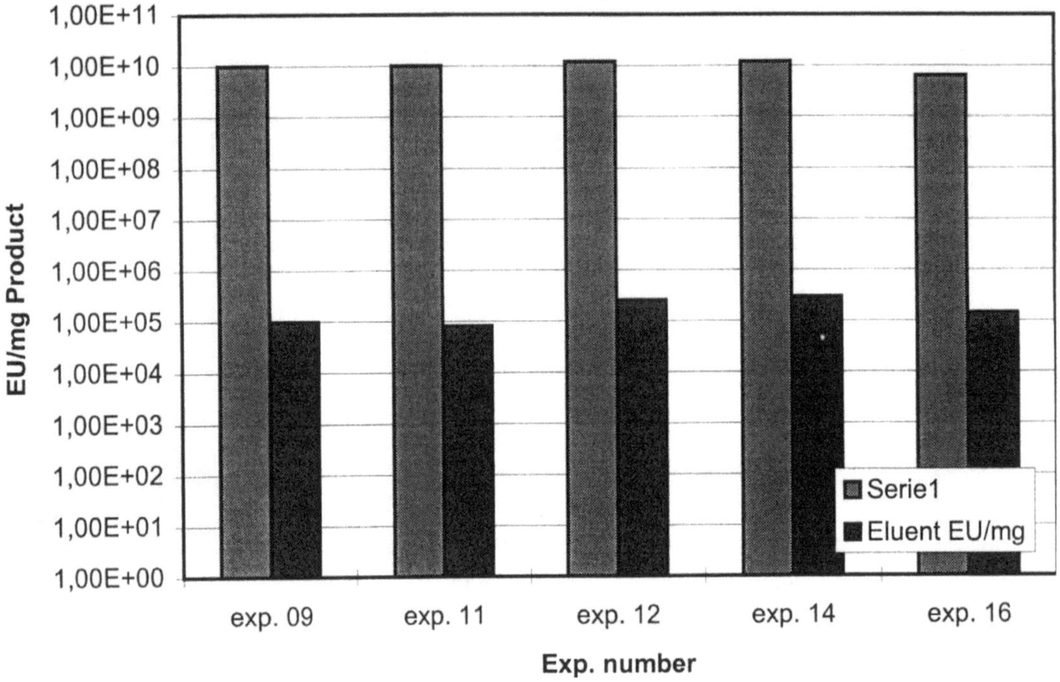

Figure 4. The reduction of endotoxins in a number of expanded bed experiments performed in small scale. The reduction of endotoxins (as measured with the LAL gel cloth technique) was quite stable with a reduction of approximate 5 logs of magnitude over the step.

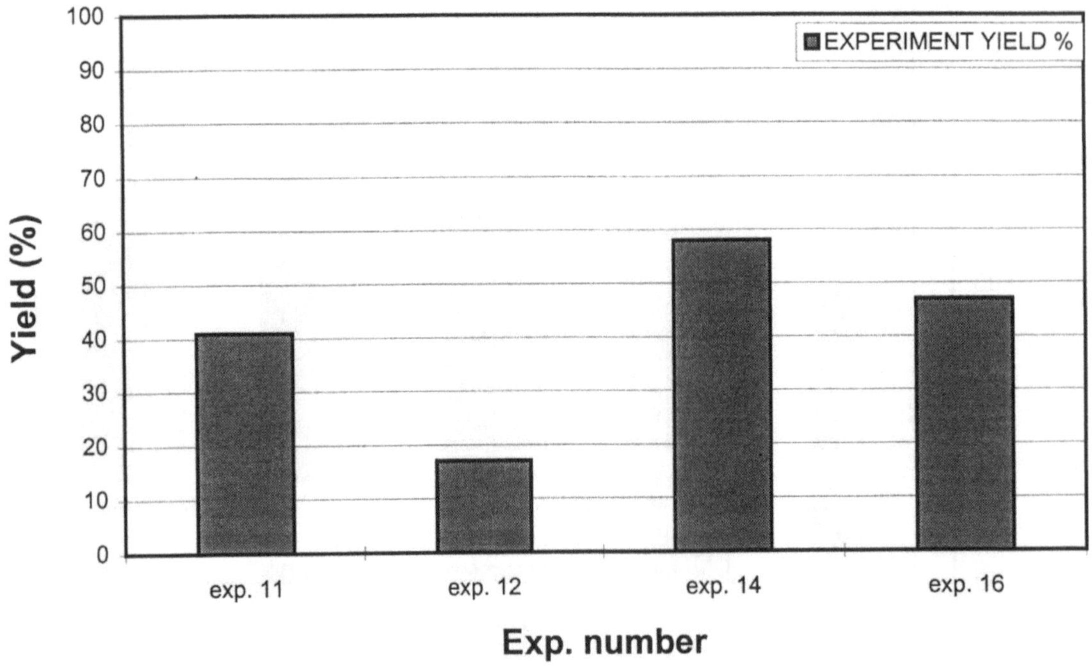

Figure 5. The yield (as measured with the HPLC technique) in the small scale experiments with expanded bed adsorption technology was in the range of 20 to 60%.

small scale experiments the fermentor broth was homogenized at 600 bar prior to dilution and condition for the capture step. The results from these runs indicate that the pressure was too low to efficiently break down the DNA molecules. By increasing the pressure and the number of cycles (from two to three) in the homogenizer the build-up under the adapter was reduced. By removing the top adapter net (and thereby also the distribution plate), no increase of pressure was noted during sample application.

The yield in the scaled-up GMP production process (Figure 6) was improved compared to the small scale results. Two factors that were important for the increase of recovery were the bed height and the dimension of the column. In the scaled-up process the packed bed height was 28 cm compared with 15 cm in the lab scale experiments. When the GMP process was scaled down and the packed bed height of the capture step was equivalent to the GMP process, yields above 70% were achieved (data not shown).

Wall effects, positioning of the column in a vertical position, fluctuations of the pump flow, temperature differences and densities seemed all to be more critical in small scale compared with pilot scale and can have influenced negatively on the yield in small scale.

Scale up and GMP production

The performance and reproducibility of the capture step were satisfactory in the GMP production campaign. The recovery was in the range 70–85% in the scaled-up method (Figure 6). All runs of the capture step in the GMP production campaign were performed using the same lot of resin. No reduced performance in terms of selectivity or capacity were noted during the campaign.

A typical chromatogram from the capture step in the scaled-up GMP production campaign is shown in Figure 7.

The reduction of host cell derived impurities in the step were stable in GMP production scale (Figure 8). The reduction of DNA was 3–4 logs, the reduction of ECP's was approx. 2 logs and the reduction of endotoxins was 4,5–5 logs of magnitude.

Discussion

The choice of a cation exchanger as chromatographic technique for the capture step in the process was a rather straightforward decision due to the high pI of the fusion protein (9.4) and the negative charge of nucleotides and Endotoxins at pH 8. Furthermore the pI for most ECP's (*E. coli* K12) are between 6 and 7 and less than 10% of the ECP's has a pI over 9 (Andrews et al., 1994). This together indicated that a cation exchanger performed at a relatively high pH had good prospects in reducing host cell derived impurities from the product fraction.

There were several advantages in using the expanded bed adsorption technique for the capture step. First; it was indicated that the fusion protein was unstable in the conditions existing after harvest and homogenization, the time consuming centrifugation step could, in that case, imply a loss of intact fusion protein and generate impurities difficult to separate downstream. Second; a centrifugation step is difficult to scale-up and scale-down in a reproducible manner. Third; the product is a toxic substance and the safety arrangements needed to centrifuge larger quantities of the fusion protein is considerable. Forth; by using expanded bed technology two unit operations could be transformed into one, resulting in higher overall yields and larger productivity.

Initial experiments with expanded bed adsorption resulted in low yield and only modest reduction of host cell derived impurities. This could be a result from a number of parameters influencing the performance of the EBA. The following parameters were pointed out as critical, the degree of homogenizing of the cells, the positioning of the column in an exact vertical position, the fluctuations of the pump flow, differences in temperature and densities. The pressure and number of passages in the homogenizer was one critical parameter that had to be optimized to ensure satisfactory performance of the following EBA. Normally above 90% of the intracellular product were free in the buffer after the first passage at 800 bar, but another two passages seemed to be needed to break down cells, cell debris and DNA further to achieve an acceptable performance on the following EBA column. If this is a result from reducing the particle size, the length of polynucleotides or the lower viscosity is still not investigated.

Conclusions

A capture step for a fusion protein was developed, scaled-up and used in production according to cGMP. The step fulfilled most of the goals that were set for the step early in the development process.

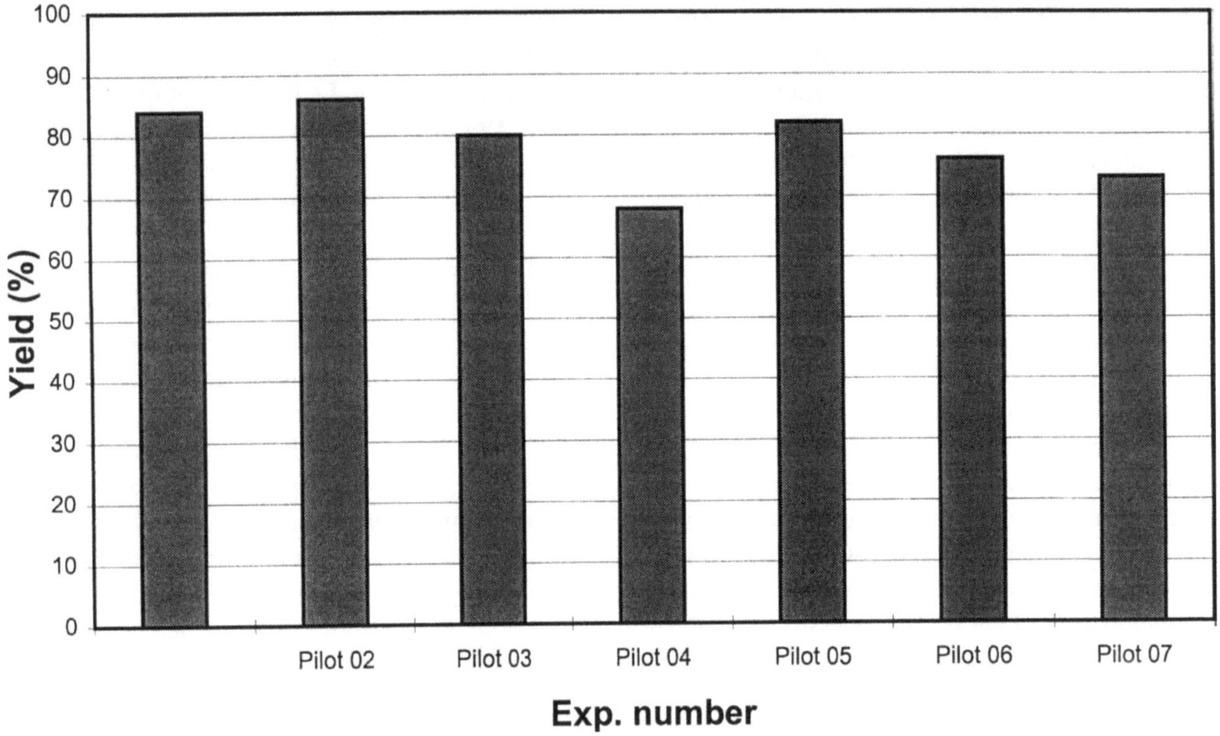

Figure 6. Yield of the capture step in the scaled-up method used in the production campaign according to cGMP. The yield was stable throughout the campaign with a yield of 70 to 80%.

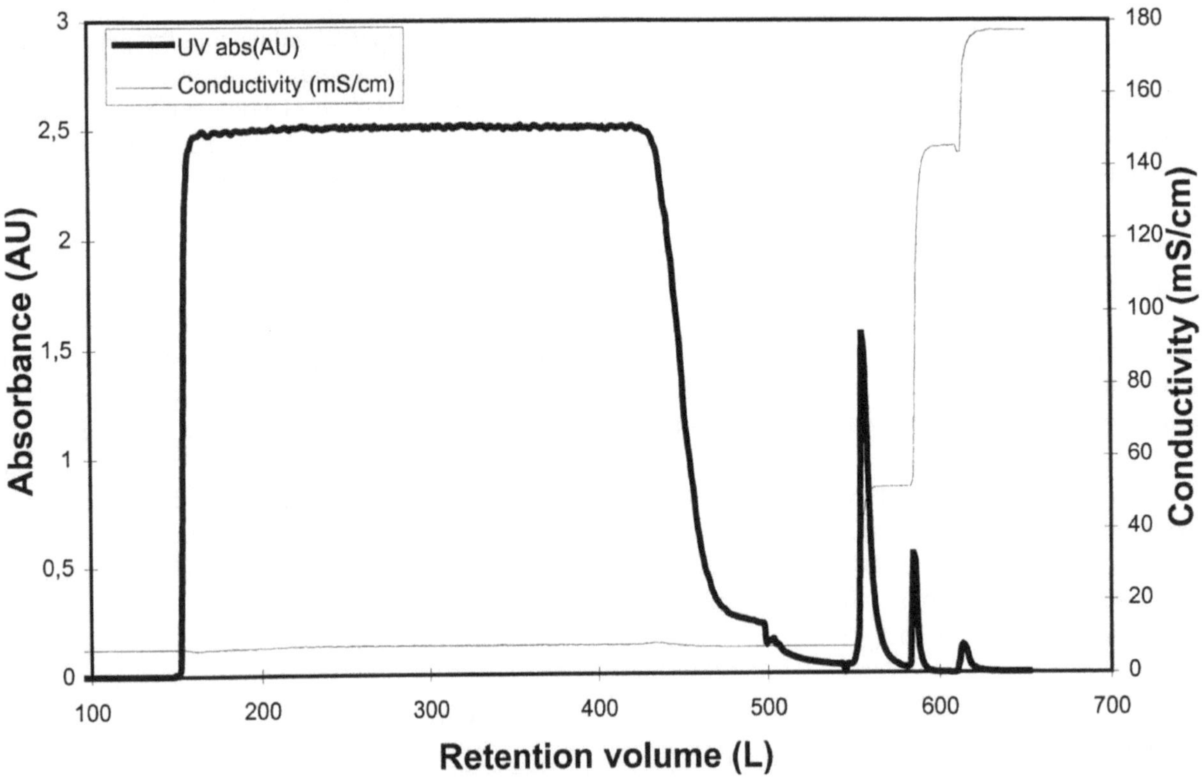

Figure 7. A typical chromatogram from the capture step in the scaled-up method used in the production campaign according to cGMP. The breakthrough of unbound material occurred after approx. 50 liters of diluted homogenate had been applied to the column. The bed was settled after the retention volume of 500 liters and the production fraction was collected at 550–580 liters of total retention volume.

DNA

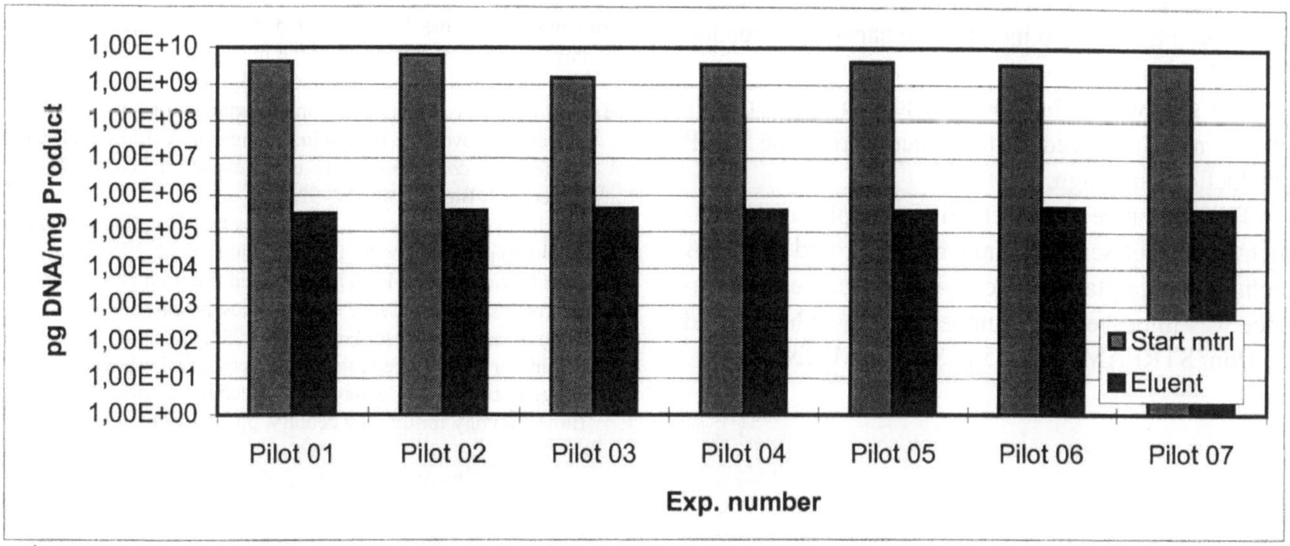

ECP's

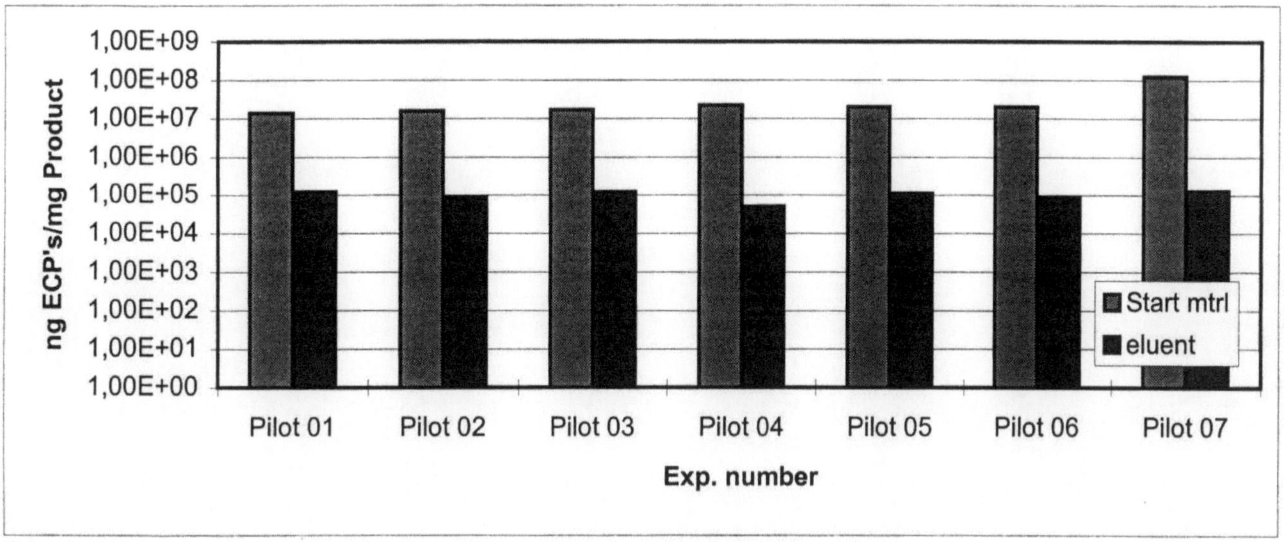

Endotoxins

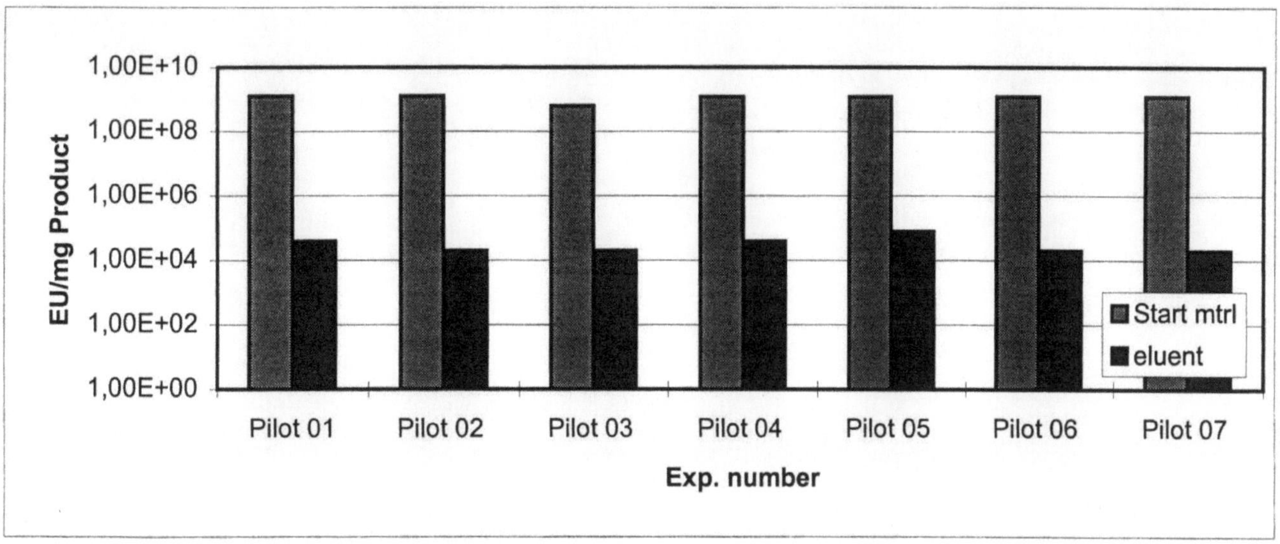

Figure 8. Reduction of host cell derived impurities over the capture step in the production campaign according to cGMP The reduction of ECP's were measured by the ELISA technique, the reduction of DNA was measured by the threshold technique method and the reduction of Endotoxins were measured by the LAL gel cloth method. The reduction of host cell derived impurities was stable throughout the campaign.

The capacity of the resin was one of the most critical issues in the development of the step and further investigation in how to increase the capacity is needed fulfill the goal of the step.

The step was stable and robust with no drop-outs or any noted reduced performance during the GMP production campaign.

The tolerance for fluctuations of physical parameters in pilot scale was much greater and the EBA technology is suitable for scale-up, in fact the performance was improved in a number of ways when scaled up from STREAMLINE 25 to STREAMLINE 200.

References

Andrews AT, Noble I, Keeratipibul S, and Asenjo, JA (1994) Physicochemical properties of the main matrix proteins of three important culture vehicles. Biotechnol. Bioeng., 44 29–37.

Barnfield Frej A-K, Hjorth R and Hammarström Å (1994) Pilot scale recovery of recombinant Annexin V from unclarified *Escherichia coli* homogenate using expanded bed adsorption, Biotechnol. Bioeng., **44** 922–929.

Chase HA (1994) Purification of proteins by adsorption chromatography in expanded beds. Trends in biotech. 12: 296–303.

Draeger MN and Chase HA (1991) Liquid fluidized bed adsorption of proteins in the presence of cells. Bioseparations 2: 67–80

Hedman P and Barnfield Frej A-K (1992) Adapting chromatography for initial large-scale protein recovery. In: Ladisch M and Bose A (eds) ACS Conference Proceedings Series, Harnessing Biotechnology for the 21st century. pp 271–274.

Bioseparation **8:** 229–235, 1999.
© 1999 *Kluwer Academic Publishers.*

On-line monitoring of glucose and/or lactate in a fermentation process using an expanded micro-bed flow injection analyser

M.P. Nandakumar[1], Arvind M. Lali[2] & B. Mattiasson[1]
[1]*Department of Biotechnology, Center for Chemistry and Chemical Engineering, Lund University, Lund, Sweden;* [2]*Chemical Engineering Division, Department of Chemical Technology, University of Mumbai, Mumbai, India*

Received 9 October 1998; accepted in revised form 16 February 1999

Key words: expanded bed, flow injection analysis, on-line analysis, glucose, lactate

Abstract

A novel flow injection biosensor system for monitoring fermentation processes has been developed using an expanded micro bed as the enzyme reactor. An expanded bed reactor is capable of handling a mobile phase containing suspended matter like cells and cell debris. Thus, while the analyte is free to interact with the adsorbent, the suspended particulate matter passes through unhindered. With the use of a scaled down expanded bed in the flow injection analysis (FIA) system, it was possible to analyse samples directly from a fermentor without the pretreatment otherwise required to extract the analyte or remove the suspended cells. This technique, therefore, provides a means to determine the true concentrations of the metabolites in a fermentor, with more ease than possible with other techniques.

Glucose oxidase immobilised on STREAMLINE was used to measure glucose concentration in a suspension of dead yeast cells. There was no interference from the cell particles even at high cell densities such as 15 gm dry weight per litre. The assay time was about 6 min. Accuracy and reproducibility of the system was found to be good. In another scheme, lactate oxidase was covalently coupled to STREAMLINE for expanded bed operation. With the on-line expanded micro bed FIA it was possible to follow the fermentation with *Lactobacillus casei.*

Introduction

A large variety of biosensors have been described in the literature (Scheller and Schubert, 1992). However, only a limited number have been utilised for on-line monitoring of fermentations and other similar applications (Esti et al., 1996; Scheper et al., 1996; Favier et al., 1997). One reason for this failure to fully exploit the powerful biosensor technology at industrial level has been the lack of suitable sampling and sample handling technologies (Mattiasson and Håkanson, 1993). However, attempts are being made to devise new techniques in this direction (Mandenius et al., 1984; Holst et al., 1988; Picque and Corrieu, 1992). It is obvious that the issues of sampling and sample handling constitute the real bottleneck when one attempts on-line sampling from a fermentor, using minimal precautions to keep the system sterile. A number of difficulties are encountered, that make it essential, either for the separation of cells and other suspended matter, or for the extraction of cell free analyte, or indeed, for both of these methodologies to be carried out.

Flow injection analysis (FIA) is, beyond doubt, a technique immensely suited for the analysis of metabolites appearing as a result of continuous sampling from a fermentor. In an attempt to impart certain specificity to the analysis, enzymatic reactions are combined with the FIA concept. There are two types of biosensors as applied to flow injection analytical systems. The first is the true sensor type with a membrane covered transducer (Buerk, 1993), and second has the immobilised enzyme reactor configuration (Shu et al., 1993). The latter has been used with a greater degree of success. The advantage with a separate immobilised enzyme reactor is that it is possible to use high loadings of enzyme in the reactor, thereby guaranteeing higher operational stability of the systems (Shu et al., 1994). Enzyme reactors used in FIA systems are packed bed reactors, wherein the enzyme is preferably covalently bound to an adsorbent matrix, which is in the form of beads. The matrix utilised is generally controlled pore glass or other inert polymer material. One limitation of such a system employing a packed bed enzyme reactor has been clogging of the bed as soon as viscous samples or samples containing suspended particulate matter are encountered. Such a situation requires pretreatment of the sample, preferably on-line on FIA to effect solid-liquid separation. Invariably, membranes are used for the desired filtration (Min et al., 1998). Membranes, however, are known to foul; more readily so with direct fermentation broths. Fouling progressively reduces flux and performance of the membrane, and therefore increasingly affects the measurements made on the system.

Expanded bed adsorption and chromatography has been introduced as a new technique that allows direct adsorption of target molecules from cell suspensions (Chase, 1994). The concept of expanded bed adsorption is based on the fact that in a bed of suspended adsorbent phase the unwanted suspended particulate matter, e.g. cells, passes around the adsorbent while the target molecules get adsorbed. Under similar conditions a packed bed would clog. The present work is an attempt to extend this very advantage of expanded bed operation, to flow biosensors, by replacing the traditional packed bed enzyme reactor on a FIA setup, with an expanded bed reactor. Typical packed bed reactors in FIA systems are a few hundred microlitres in volume. In this work, novel expanded micro beds have been designed and used in FIA assembly for analysing glucose and lactate, in samples drawn directly from fermentors along with the cells. It is further shown that such a system does indeed provide a technique for sampling and sample handling on a FIA system, and is free from the problems of sample prefiltration and dialysis, when applied as an on-line system on a fermentor.

Materials

Native STREAMLINE used as the designer expanded bed adsorbent phase was a generous gift from Amersham Pharmacia Biotech AB, Uppsala, Sweden. STREAMLINE is in the form of rigid cross-linked agarose gel on quartz core beads in the size range of 125 to 315 microns and has a specific gravity of 1.25. Glucose oxidase (E.C.1.1.3.4) and lactate oxidase (E.C. not available) were procured from Sigma Chemical Company, St. Louis, MO, USA. Tresyl chloride was obtained from Synthelec AB, Lund, Sweden. All other chemicals were of analytical grade.

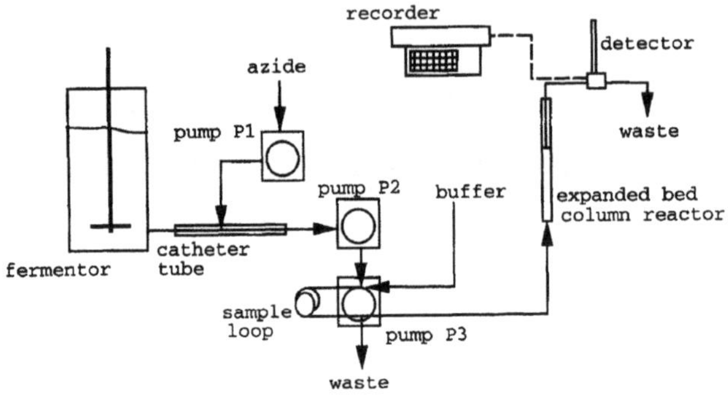

Figure 1. Schematic representation of the flow injection system with expanded bed reactor.

Experimental setup

The sampling unit, injection valve, flow systems and amperometric sensor for the off-line analysis of glucose, in absence and presence of dead yeast cells, and on-line monitoring of L-lactate during fermentation were set up as shown in Figure 1. The system comprised of a sample injection valve, an enzyme reactor (expanded bed column) and a detector (Clark oxygen electrode). A buffer was continuously pumped using a peristaltic pump (Alitea, Sweden). A similar pump was used for sample injection as described below.

For on-line analysis, samples (for lactate) were taken directly into the system, via a double lumen catheter (Håkanson et al., 1991). During this process, the metabolic activity of the cells was arrested by addition of sodium azide pumped at 0.3 ml/min by pump P1 through the outer shell of the catheter and mixed with the sample in the tip of the sampling unit. The sample was carried into the injection loop by the sample peristaltic pump P2 at a flow rate of 0.5 ml/min. When the sample loop was full, the sampling pump was stopped and the sample could be injected into the FIA using the FIA assembly pump P3. The main system worked at a controlled constant flow rate of 0.5 ml/min (unless stated otherwise). A controlled constant flow is important, as it not only gives reproducible flow dispersion and reactor conversion, but also a constant bed expansion in the expanded micro bed column. The assay cycle thus consisted of two steps; loading the sample into the injection loop and then injection of the sample into the carrier buffer. Depletion of oxygen by the oxidase in the reactor, was detected by the Clark electrode and recorded on a recorder or computer.

Off-line analysis using this setup was carried out for glucose in the presence or absence of dead yeast cells, by forcing samples into the sample loop, using a syringe. Glucose in the fermentation samples, in the presence of azide-killed *Lactobacillus casei*, was also estimated off-line, in the same way.

The enzyme reactor

Two vertical glass columns, specially fabricated by the Lund University workshop, were used as enzyme reactors. One had an inner diameter of 10 mm and the other 5 mm. Both could be used with a maximum bed height of 20 cm. Each end of the column was equipped with a movable adapter, also specially fabricated, to provide even flow distribution through a 100 micron stainless steel wire mesh. The columns, arranged vertically, could be operated both in packed bed and in expanded bed mode. As mentioned earlier, STREAMLINE was used as the adsorbent phase onto which the enzymes were immobilised.

Immobilisation of enzyme

Native STREAMLINE (5 ml settled gel) was activated with tresyl chloride according to established procedures (Nilson and Mosbach, 1980). Glucose oxidase was added in 10 ml of 0.2 M sodium phosphate buffer, pH 6.5 and gel activated with tresyl chloride (25 mg/ml gel). Coupling of the enzyme was carried out overnight, at 4 °C. The beads were then washed with water under suction with excess of 1 M NaCl and buffer to remove the unbound enzyme and then treated with 1 mM ethanolamine in buffer at pH 8.0 for one hour, in order to block the free activated sites on the gel. The gel was again washed with excess of 1 M NaCl and water and finally stored, refrigerated, in 0.2 M sodium phosphate buffer, pH 6.5 until use.

Lactate oxidase was also immobilised in a similar manner. The enzyme was added in 20 units per ml of settled STREAMLINE in a 0.2 M sodium citrate buffer and pH of 6.0. The same buffer was used for the storage and use of the immobilised matrix.

Methods of analysis

The expanded bed column was filled with STREAMLINE, carrying immobilised enzyme to a 3 cm settled bed height. Whilst the bed expansion studies were done both in 10 mm and 5 mm inner diameter columns, only the 5 mm column was used for off-line and on-line assays. This column, used at a flow rate of 0.5 ml/min, resulted in a uniform expansion of the bed to 6 cm. That is, a degree of expansion of 2 was obtained. The glucose oxidase expanded micro bed reactor was used to assay glucose, in the presence and absence of dead yeast cells. The FIA system was characterised with respect to sensitivity, speed and reproducibility. This study was performed by monitoring the effect of flow rate and sample size, on the response times (base line time and peak time).

Samples containing *Lactobacillus casei* cells, and glucose and/or lactate from the fermentor or otherwise, were injected into the FIA system through the loop described above. During its passage through the enzyme reactor, the substrate was consumed and the resulting decrease in oxygen was measured by the electrode.

After successful demonstration of the principle, that the cells did not indeed affect the glucose assay with the expanded bed reactor, the system was coupled to the lactate fermentor with the expanded micro-bed reactor, now containing the immobilised lactate oxidase. The fermentation system consisted of cultivation of *Lactobacillus casei*. The organism was grown on MRS broth (E. Merck, Germany) with a composition in g/l peptone, 10, meat extract 8.0, yeast extract, 4.0, glucose 20.0, di-potassium hydrogen phosphate, 2.0, Tween 80, 1.0, diammonium hydrogen citrate, 2.0, sodium acetate 5.0, magnesium sulfate, 0.2, manganese sulfate, 0.04. Prepared by dissolving the powdered media in distilled water and the final pH was adjusted to 6.3 with 0.1 N NaOH and autoclaved at 15 min 120 °C.

The fermentation was run in a 2 litre fermentor, with a working volume of 1 litre. A 16-hour-old culture was used as the inoculum, and fermentation was carried out at 37 °C at an agitation speed of 150 rpm. Samples were withdrawn every 30 min for off-line analysis of glucose, L-lactate and cell mass. Samples were also taken via the double lumen catheter for on-line FIA analyses.

Calibration of the system was done for glucose by off-line analysis of standard solutions. For on-line assays of L-lactate the calibration solution was added via the double lumen catheter which has earlier been shown to be successful (Kyröläinen et al., 1995).

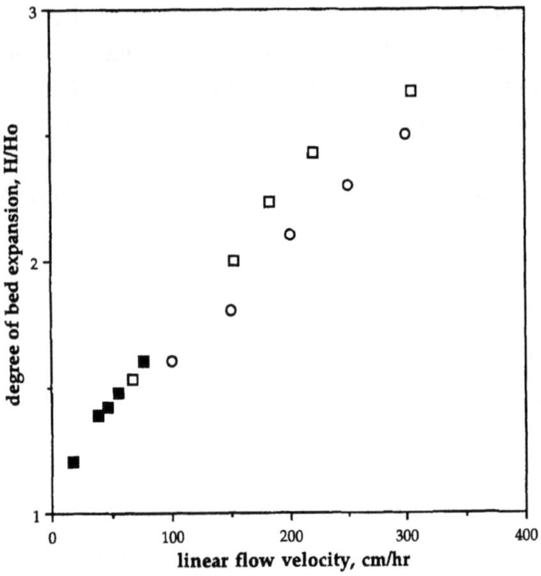

Figure 2. Degree of bed expansion vs. flow rate in 5 mm column (□); in 10 mm column (■) and that reported (Lali et. al., 1997) for 32 mm column (○).

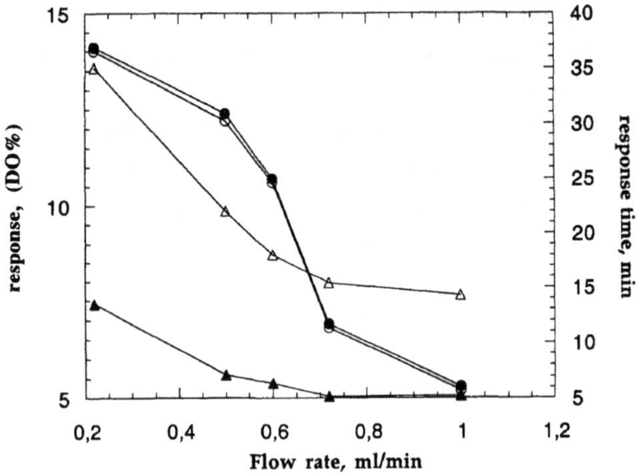

Figure 3. Effect of flow rate on response of the sensor; measuring conditions – 1% glucose analysed in the absence (•) and presence (○) of 1 g dry weight/l yeast cells. Time for peak maximum (▲) and baseline (Δ) at different flow rates.

Results and discussion

An expanded bed purification operation in bioprocessing has been shown to offer distinct advantages (Chase et al., 1994; Lali et al., 1997). The most important of these, in the present context, is the ability of expanded bed adsorption/chromatography to handle crude samples with suspended particulate material.

There is no existing information in the literature regarding operation of very small 'micro' solid-liquid fluidised bed reactors that will at most have an operating volume of 10 mL. Therefore two columns, one of 10 mm inner diameter and the other of 5 mm inner diameter were specially fabricated complete with the necessary adaptors and flow distributors. These adaptors and flow distributors were of the type used successfully elsewhere to operate a 3.0 cm diameter expanded bed (Lali et al., 1997).

Bed expansion in expanded micro beds

The expanded micro beds were analysed for wall effect on the bed expansion characteristics of STREAMLINE. It is known that settling behaviour of particles, and by analogy the expansion behaviour of a bed of monosized particles, is unaffected by the column walls so far, as the column diameter to particle diameter ratio is more than 100 (Kunii and Levenspiel, 1969). STREAMLINE particles with largest particle diameter of about 315 micron should then expand well without a significant wall effect in a column with a diameter of 31 mm and above. Thus both the columns used in this work were expected to exhibit wall effect; that is, expand more than reported in larger columns. Figure 2 shows the bed expansion behaviour of STREAMLINE in water, observed in the two columns, as well as that reported in a column of 32 mm diam. (Lali et al., 1997). Both 10 mm and 5 mm columns show somewhat larger bed expansion as compared to that reported for STREAMLINE DEAE in a 32 mm column, thereby positively indicating the presence of some wall effect. Larger bed expansion in smaller columns may also in part be due to a difference in the type of liquid distribution at the inlet adaptor plate. This was also perhaps the reason why bed expansion in the 5 mm column became erratic and the bed assumed disturbed and mixed behaviour above a degree of expansion value of 2. All the analysis runs in the following work were therefore restricted to flow rates that gave a degree of bed expansion of less than or equal to 2 in the 5 mm diam. column.

Glucose analysis in presence and absence of cells

Variation of the response signal with flow rate for glucose in clear solutions and in the presence of heat killed yeast cells (1 g dry weight/l) in the 5 mm expanded micro bed reactor with STREAMLINE-glucose oxidase is given in Figure 3. The yeast *Saccharomyces cerevisiae* were used as a model representing a particulate matter that easily could be introduced into the system. The sample glucose concentration was 1% (w/v) and a sample volume of 200 microlitre was used. The plot shows that the presence of cells does

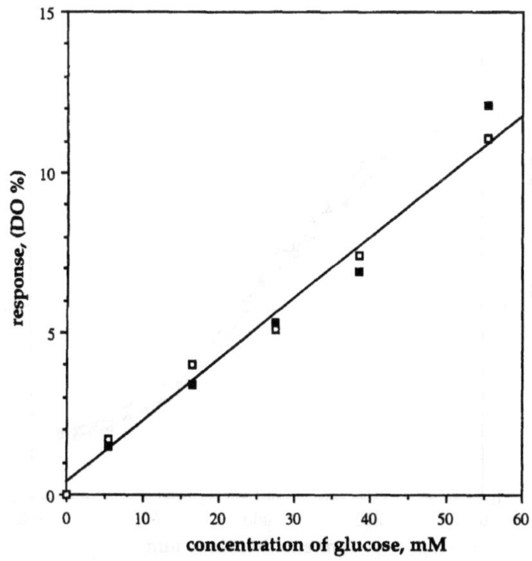

Figure 4. Calibration plot for glucose in the absence (■) and presence (□) of 1 g dry weight/l yeast cells; measuring conditions – 0.2 M sodium phosphate buffer pH 6.5, flow rate 0.5 ml/min, and 6 minute detection time.

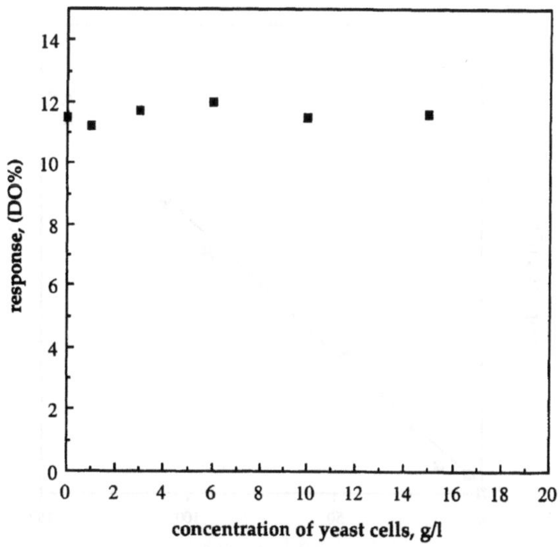

Figure 5. Effect of different concentration of yeast cells on response of the sensor; measuring conditions – 0.2 M sodium phosphate buffer pH 6.5, flow rate 0.5 ml/min and 6 minute detection time.

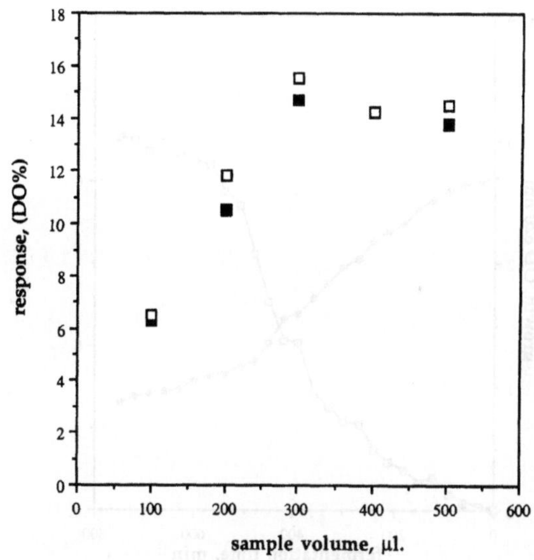

Figure 6. Expanded bed analysis – Effect of sample volume on response of the sensor in the absence (■) and presence (□) of 1 g dry weight/l yeast cells; measuring conditions – 0.2 M sodium phosphate buffer pH 6.5, flow rate 0.5 ml/min.

not affect the response. The plot also indicates that response time decreased with increasing flow rate, a fact noted with packed bed reactors as well. The plot also shows that for linear flow velocity of 0.5 ml/min, the response time remained constant (6 min), while the base line time became constant only above about 1 ml/min. It can be noted that these response times are much higher than normally observed for typical FIA systems. One also, however, does have to remember here that the analysis carried out applies well even in the presence of cells, and eliminates the need for cell (or other suspended matter) separation step/s. The absence of such a step must also increase the magnitude of the response, thereby rendering the present technique more sensitive at low analyte concentrations as compared to when separation steps involving dilutions are applied. Further, a sampling frequency of about 10 per hour in a fermentor system, should be adequate in most instances.

The calibration plot of glucose, analysed in the presence and absence of yeast cells, is shown in Figure 4. The reproducibility, measured separately, was found to be within ±6% for ten consecutive sample injections of 1% (w/v) glucose in the presence and absence of 1 g dry weight/l yeast cells. Further, variation over any one day was found to be below 10%. Glucose oxidase preparation on STREAMLINE was stable when stored at +4 °C in 0.2 M sodium phosphate buffer pH 6.5. An activity of more than 85% was retained after a period of thirty days.

The effect of heat killed yeast cell concentration on the response, recorded using a flow of 0.5 ml/min and response detection, at 6 min is shown in Figure 5. No influence of the cells was noted up to a cell concentration of 15 gm dry weight per litre.

Figure 6 shows variation of the response to a 1% (w/v) glucose concentration with changing sample size. The response increased up to a sample volume of 300 microlitre and then remained constant.

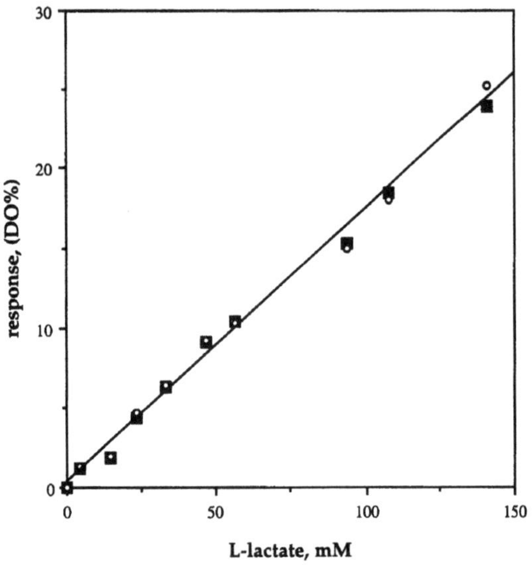

Figure 7. Calibration curve for L-lactate in the presence (○) and absence (■) of 1 g dry weight/l yeast cells; measuring conditions – 0.2 M sodium citrate buffer pH 6.0, flow rate 0.5 ml/min and 7 minute detection time.

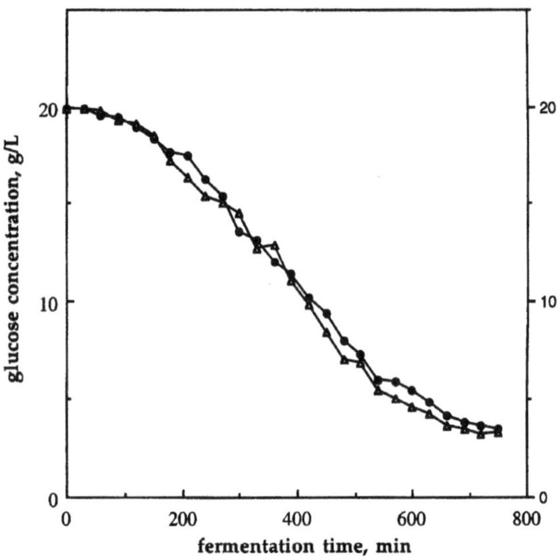

Figure 9. Profile of glucose consumption during lactic acid fermentation by *L. casei*. Off-line determination by expanded bed analyser (Δ), and by dinitro salicylic method (•). respectively.

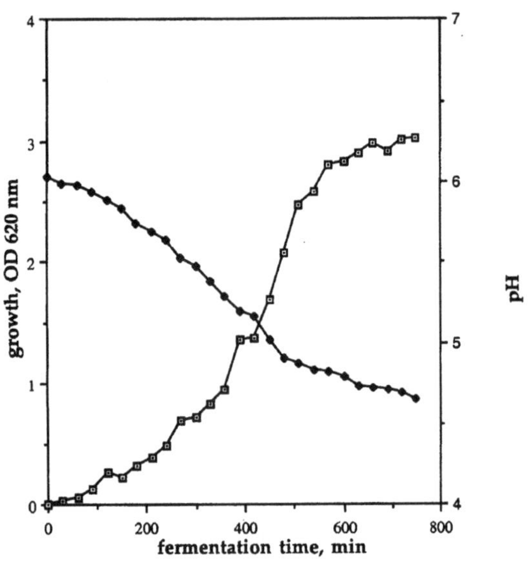

Figure 8. Profile of growth, OD at 620 nm (□), and pH (■) during fermentation by *L. casei* (for details see text).

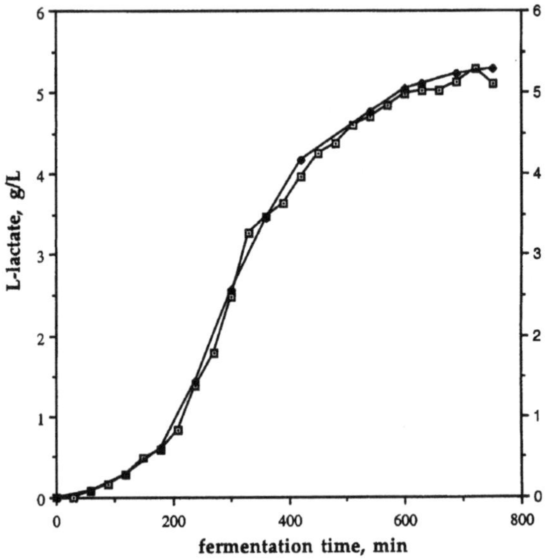

Figure 10. L-lactate determination during fermentation as determined by off-line HPLC (■), and by on-line expanded bed analysis (□), respectively.

On-line lactate analysis on a fermentor

Figure 7 shows the L-lactate analysis on the expanded micro bed FIA with STREAMLINE-lactate oxidase in the presence and absence of heat killed yeast cells (1 g dry/ml), and at a flow rate of 0.5 ml/min in 0.2 M pH 6.0 sodium citrate buffer and a detection time of 7 min. The accuracy and sensitivity of the method was considered satisfactory for the on-line application of the system on the fermentor.

Cultivation cycle of *Lactobacillus casei* was followed by off-line as well as by on-line analysis of glucose, L-lactate, biomass and pH. As mentioned in the methods section, samples were taken from the fermentor using a coaxial catheter. The inhibitor mixed with broth at the sampling point prevented further cell metabolism. Samples were loaded, using the sampling pump, into the injection loop of the micro-expanded bed FIA with STREAMLINE-lactate oxidase. Cell free samples were also taken and analysed for lactate on HPLC. Glucose was similarly analysed on the on-line STREAMLINE-glucose oxidase micro-expanded bed FIA as well as by off-line cell free dinitrosalicylic acid (DNSA) method (Miller, 1959). Cell mass was measured off line on a spectrophotometer at 620 nm. Figures 8, 9 and 10 show the results obtained.

A maximum cell density was observed after about 10 hours and this corresponded to a decrease in pH from 6 to 4.5 (Figure 8). Almost total glucose consumption was observed. The result from both off-line DNSA glucose analysis and the expanded micro bed FIA glucose analysis were in good agreement, as indicated by Figure 9. L-lactate measurements on off-line HPLC and expanded micro bed FIA were also highly consistent (Figure 10).

Conclusions

Results reported above clearly demonstrate the usefulness of the expanded micro bed FIA system for on-line monitoring of fermentation processes. The technique presented is robust and sensitive enough and extends the use of an expanded bed operation concept to FIA analysis, of samples containing suspended particulate matter, like cells. It is expected that this new approach will permit more direct and accurate monitoring of broth components and metabolites in fermentations, leading to implementation of better fermentation control strategies.

Acknowledgements

This project was supported by the Swedish National Board for Technical and Industrial Development (NUTEK), The SIK (The Swedish Foundation for Food Science and Biotechnology) and to some extent by The Swedish International Development Aid Organisation (Sida/SAREC).

References

Buerk DG (1993) Biosensor: Theory and Applications. Technic Publishing Company, USA.

Chase HA (1994) Purification of proteins by adsorption chromatography in expanded beds. Trends Biotechnol. 12: 296–303.

Esti M, Messia M C, La Notte E, Lembo P, Compagnone D and Palleschi G (1996). Curd-ripening evaluation by flow injection analysis of L-lactic acid with an electrochemical biocell during mozzarella cheese manufacture. J. Agri. Food Chem. 10: 3102–3107.

Favier JP, Bicanic D, Helander P and van Iersel, M (1997) The optothermal approach to a real time monitoring of glucose content during fermentation by brewers yeast. J. Biochem. Biophys. Meth. 34: 205–211.

Håkanson H, Nilson M and Mattiasson B (1991) General sampling system for sterile monitoring of biological processes. Anal. Chim Acta. 249: 61–65.

Holst O, Håkanson H, Miyabayashi A and Mattiasson B (1988) Monitoring of glucose in fermentation processes using a commercial glucose analyzer. Appl. Microbiol. Biotech. 28: 32–36.

Kunii D and Levenspiel O (1969) Fluidisation Engineering. John Wiley and Sons, N.Y.

Kyröläinen M, Håkanson H and Mattiasson B (1995) Online calibration of a computerised biosensor system for continuous measurement of glucose and lactate. Biotech Bioeng. 45: 122–128.

Lali A, Kaul R, Galaev I Yu and Mattiasson B (1997) Purification of L-lactate dehydrogenase from porcine muscle by expanded bed affinity chromatography. Isolation and Purification 2: 289–300.

Mandenius CF, Dannielsson B and Mattiasson B (1984) Evaluation of a dialysis probe for continuous sampling in fermentors and complex media. Anal. Chim. Acta 163: 135–142.

Mattiasson B and Håkanson H (1993) Sampling and sample handling – crucial steps in process monitoring and control. Trends in Biotech. 11: 136–142.

Miller GL (1959) Use of dinitrosalicylic acid reagent for determination of reducing sugar. Anal. Chem. 31: 426–428.

Min RW, Rajendran V, Larsson N, Gorton L, Planas J and Hahn Hägerdal B (1998). Simultaneous monitoring of glucose and L-lactic acid during a fermentation process in an aqueous two-phase system by on-line FIA with microdialysis sampling and dual biosensor detection. Anal. Chim. Acta. 366: 127–135.

Nilsson K and Mosbach K (1980) *p*-Toluenesulfonyl chloride as an activating agent of agarose for the preparation of immobilised affinity ligands and proteins. Eur. J. Biochem. 112: 397–402.

Picque D and Corrier G (1992) Performance of asceptic sampling devices for online monitoring of fermentation processes. Biotech. Bioeng. 40: 919–924.

Scheller F and Schubert F (1992) Biosensors. Elsevier Science Publishers B.V., Amsterdam, The Netherlands.

Scheper TH, Hilmer JM, Lammers F, Muller C and Reinecke M (1996) Biosensors in bioprocess monitoring. J. Chromatogr. A 725: 3–12.

Shu HC, Håkanson H and Mattiasson B (1993) D-Lactic acid in pork as freshness indicator monitored by immobilised D-lactate dehydrogenase using sequential injection analysis. Anal. Chim. Acta, 283: 727–737.

Shu HC (1994) D-lactic acid analysis using sequential injection analysis and amperometric biosensor. Ph.D. thesis, Lund University (ISRN LUTKDH/TKBT-94/1020-SE): 19–26.

Address for correspondence: Prof. B. Mattiasson, Center of Chemistry and Chemical Engineering, University of Lund, P.O. Box 124, S-22100 Lund, Sweden. (Fax no: 46 222 4713; e-mail: Bo.Mattiasson@biotek.lu.se).

Bioseparation **8:** 237–245, 1999.
© 1999 *Kluwer Academic Publishers.*

Binding assays in heterogeneous media using a flow injection system with an expanded micro-bed adsorption column

Bo Mattiasson* & M. P. Nandakumar
*Department of Biotechnology, Center for Chemistry and Chemical Engineering, Lund University, P.O. Box 124, S-221 00 Lund, Sweden *Address for correspondence. Fax +46-46-222 47 13, E-mail: Bo.Mattiasson@biotek.lu.se*

Received 9 October 1998; accepted in revised form 29 December 1998

Key words: expanded micro-bed, flow injection ELISA, immunoassay

Abstract

Competitive binding assays have been performed in flow injection systems. To further increase the versatility of the system, and to enable it to deal with samples containing particulate matter, the adsorption step was designed as an expanded bed column. Immunochemical quantification of human serum albumin was chosen as a model system to use for the development of the technology. A competitive ELISA was set up using peroxidase labelled HSA as competing ligand. The introduction of the expanded bed immunosorption column made the system tolerant to samples containing suspended particulate matter. The analytical outcome is very similar to that from the packed bed system even though more time is required for each assay cycle. The capability of the system was tested by addition of increasing amounts of yeast cells. The results clearly indicate that the system is suitable e.g. for process monitoring of fermentations.

Introduction

Immunochemical binding assays are available in a broad spectrum of configurations (Maggio, 1980; Ngo, 1988). Most development has been devoted to discrete assays, but in recent years an increasing interest has been seen in the application of immunochemical binding assay principles in the construction of immuno-based biosensors (Nakamura et al., 1992; Ngo, 1987). Today, there is a range of different potential possibilities for the construction of immunobiosensors (Mattiasson et al., 1977; Collins and Janata, 1982; Daniels et al., 1988; Mattiasson and Miyabayashi, 1988, Puchades et al., 1992). One of the earlier constructions which has been developed for quick assays has been Flow-ELISA, which is a combination of flow injection analysis (FIA) and enzyme linked immunosorbent assay (ELISA) (Mattiasson et al., 1977, Mattiasson and Larsson, 1987; Lim et al., 1990; Kusterbach et al., 1990; Miller, 1989). Such assays were developed and applied successfully for process monitoring and control (Nilsson et al., 1991, 1992, 1994). Flow-ELISA as well as FIA with reactors containing immobilized enzyme are all susceptible to clogging problems. The area of application has therefore been restricted to analysis of clean samples. This means that sample pretreatment has become increasingly important. Dialysis and other membrane based treatments have proven successful for analysis of low molecular weight compounds (for review see Mattiasson and Håkanson, 1993). When dealing with macromolecules, it is more complicated to introduce efficient, fast and reliable separation methods. Sampling and sample handling may be separated such that sampling is done without further precautions other than retarding the metabolic activity in the sample. This can be done by using the double lumen catheter (Håkanson et al., 1991).

The introduction of expanded bed reactors offers a new possibility since such systems have in large scale applications been proven to allow passage of particulate matter through the column without interferring with the performance (Chang and Chase, 1995; Lali et al., 1996). An expanded bed is characterised by having a flow of liquid from below, upwards, thereby expanding the bed as compared to the situation in packed bed mode. Furthermore, the preparation of beads is heterogeneous, such that there are differences in size and density. The larger and denser beads will be enriched towards the bottom of the column. Since the bed is expanded, there will be free liquid between the particles, and therefore particulate matter in the sample may pass. Owing to the heterogeneity of the beads in the column, a stabilised expanded bed is attained. This will give less back mixing as compared to what could be expected from a conventionally fluidized bed. In the present paper, we are reporting on an expanded bed, antibody based column for the determination of human serum albumin, in the presence of cells without any sample pretreatment using the flow ELISA principle.

Materials and Methods

Chemicals

Horseradish peroxidase type II (HRP) was purchased from Sigma Chemical Co., St Louis, Mo, USA. Human serum albumin and rabbit antibodies to human serum albumin were purchased from Dakopatts (Glostrup, Denmark). Native Streamline® was a generous gift from Pharmacia Biotechnology AB, Uppsala, Sweden. The preparation contains spherical beads with a mean diameter of 125–315 mM and a mean density of 1.25. Tresyl chloride was obtained from Synthelec AB, Lund, Sweden. All other chemicals were of analytical - reagent grade.

Immobilisation of antibodies on native Streamline®

Native Streamline® was activated with tresyl chloride according to established procedures (Nilsson and Mosbach 1980). Rabbit antibodies against HSA were added in 10 mL of 0.2 M phosphate buffer pH 7.4 containing tresyl chloride activated gel (10 mg per mL of settled gel) and coupling proceeded overnight at +4 °C. The beads were then washed under vaccum with excess of 1 M NaCl to remove unbound antibodies, followed by washing with 0.2 M phosphate buffer pH 7.4 and then treated with 1 mM ethanolamine pH 9 for 1 h, in order to block the remaining groups in the gel. The gel was again washed with excess of 1 M NaCl and water and finally suspended in 0.2 M phosphate buffer pH 7.4 and refrigerated until use.

Preparation of human serum albumin conjugates with horseradish peroxidase (HRP)

Preparation of human serum albumin conjugates with HRP was carried out according to Nilsson et al. (1992).

Preparation of expanded bed antibody column

The expanded bed antibody reactors used in these analytical applications were specially designed glass columns, with an inner diameter of 0.5 cm and a height of 20 cm. Each end of the column was equipped with a movable adaptor, being specially constructed to give an even flow and to distribute particulate materials evenly in the column. The adaptors had a distributor with a 1 mm dia single hole and then a wire mesh (mesh 100 microns). The column can be operated both in packed bed and expanded bed mode. The column was evaluated with regard to expansion properties for a bed of native Streamline® at different flow rates (Nandakumar et al., 1999).

Streamline® (400 μL) with immobilised rabbit antibodies was mixed with native Streamline® (this enables dilution) and packed in a glass column of 0.5 x 20 cm. This expanded bed column was connected to a Flow injection system (FIA), similar to the system described previously by Nilsson et al (1993), but the present system was operated manually (Figure 1). This mode of operation was chosen since it was found suitable during the development of the system. As we have described earlier (Nilsson et al., 1993), flow-ELISA can be computer controlled, and will duly be set up with these systems once they have been optimized.

The concept of diluting a bed with immobilized antibodes with native gel has been tried before when developing packed bed immunoassays (Borrebaeck et al., 1978). It was then demonstrated that some of the aberrations that are observed with a too high concentration of antibodies could efficiently be eliminated. The concentration of the ligand (in this case the antibody) is critical only from the point of view that the amount of binding sites must be a limiting factor, otherwise there will not be any competition in the binding situation. It is from this point of view of less interest to determine the binding capacity of the antibody column since the system is deliberately operated far from equilibrium.

The flow system used contained tubings of Teflon with an inner diameter of 0.5 mm. The total length of tubings in the system was 30 cm. The expanded bed column was as described mainly of a dia of 0.5 cm, but some columns with double the diameter were also

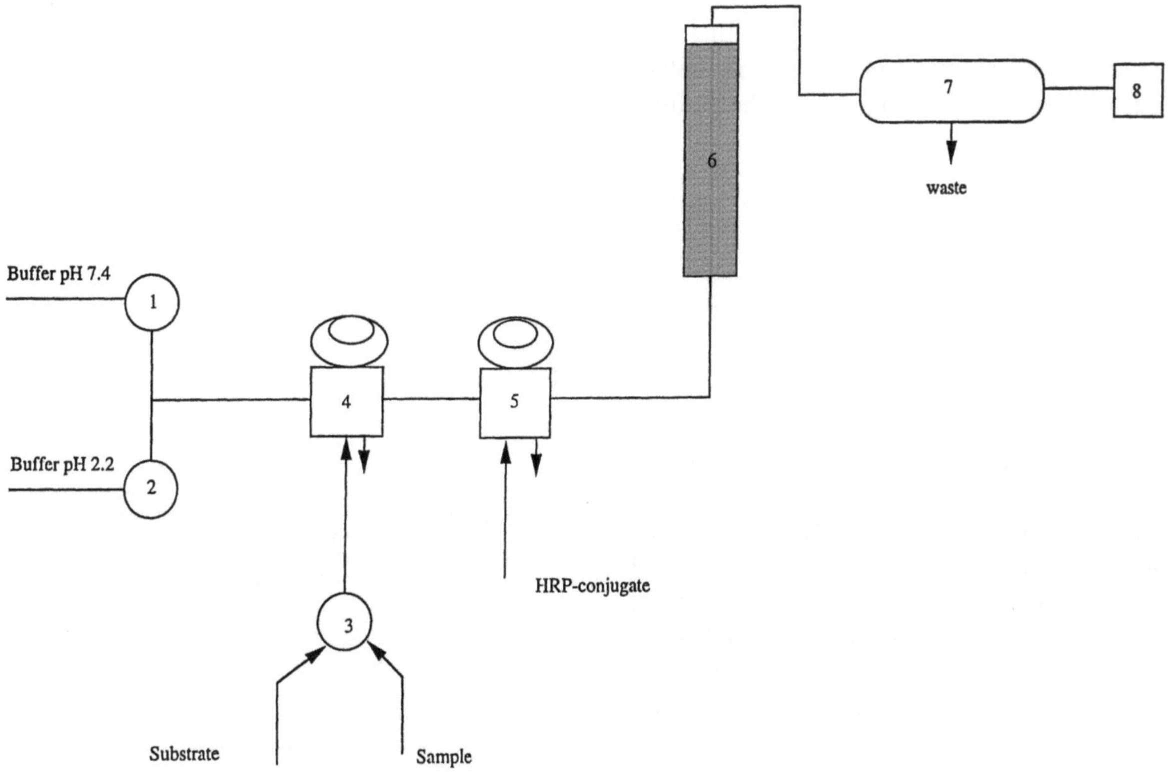

Figure 1. Schematic presentation of the experimental set-up used in the studies. 1-3, Three way valves; 4 and 5, injection valves; 6 = expanded bed column; 7 = spectrophotometer; 8, recorder.

studied initially. The length of the bed, when in the expanded state was 60 mm, and in packed bed mode was 30 mm, i.e. the expansion factor was 2. The total volume of the bed in expanded mode was 1.2 mL. The flow cell placed in the spectrophotometer had a dead volume of 30 μL and a light path of 10 mm. The FIA system consisted of two peristaltic pumps (Alitea, Sweden). The first was used for pumping peroxidase conjugated HSA and substrate, and the second for pumping the sample, 0.15 M sodium phosphate buffer pH 7.4 (reaction buffer) and glycine-HCl buffer pH 2.2 (dissociating buffer). A spectrophotometer with a flow cell was used (Shimadzu UV 120-02) for the spectrophotometric determination of cells (suspended material under the conditions used) and the reaction catalyzed by the labelling enzyme used in the assay. The noise level of this instrument was 0.001 to 0.003 abs. units. All valves and pumps were controlled manually. A sample loop was used for introducing the sample into the flow system. Normally a sample volume of 100 μL was used.

A sample containing HSA and cells (heat killed yeast cells) was introduced and passed through the expanded bed antibody column. A fraction of the HSA in the sample was bound by the antibodies in the column. After this, a pulse of HSA–HRP conjugate was injected through the 500 μL injection valve and passed through the column. The amount of conjugate was kept constant. The conjugate was bound to antibodies not occupied by the HSA. The column was then washed with the carrier buffer and then a pulse (75 μL) of substrate solution (130 mg of phenol, 16.5 mg 4-amino antipyrine and 105 μL perhydrol per 100 mL of 0.15 M sodium phosphate buffer pH 7.4, 0.5 M NaCl) was injected. When passing the sample containing cells through the expanded bed column, the cells which have a low density will pass through the column first and give an absorbance reading in the spectrophotometer at 540 nm which is then followed by a second absorbance reading given by the coloured product formed by the enzymatic reaction (between substrate and the HRP in the bound conjugates). The registered absorbance comprised of one peak given by the cells and one by the enzyme activity from the bound conjugates. The next step in the assay cycle involved elution of the HSA and conjugates from the column with a 3 min pulse of pH 2.2 dissociating buffer. A range of standard solutions of HSA were assayed according to this scheme and a calibration curve constructed from these results.

The flow rate of the carrier buffer and eluting buffer were normally 0.5 mL/min unless otherwise stated. The residence time of the sample inside the column was approximately 150 sec.

Result and discussion

Choice of column dimensions

Columns with an inner diameter of 0.5 cm were used throughout all the experiments presented here. When using columns with small diameters, the wall effects will be more pronounced as compared to when preparative columns are used. The performance of expanded beds in two different columns (0.5 and 1.0 cm in dia, respectively) were evaluated and found to behave according to expectations with regard to expansion during operation at different flow rates. This was valid in a range of flow rates that have been used in the studies discussed here.

The immunosorbent

Antibodies immobilised to Streamline were used. From studies on Sepharose-immobilized antibodies in Flow-ELISA it has been shown important to keep a low concentration of antibodies as long as competitive binding assays are carried out. A high concentration (above 1 mg/ml) will influence the sensitivity of the assay since then larger amounts of native and labelled antigen are needed for creating the competitive binding situation. It has earlier been shown that if a too high amount of antibodies have been immobilised, it is possible to dilute the antibody-Sepharose with native Sepharose, thereby reducing the unfavourable situation with regard to the competitive binding reaction. When operating with too high concentrations of antibodies, a "low concentration hook" in the calibration curve has been observed. This disappears upon dilution of the immunosorbent with inert matrix. One should also stress that only a minor fraction of the available antibodies are occupied in the binding reaction. The exposure of antibodies to antigen is carried out far from equilibrium, and then the dynamic responses are of interest, but not the total capacities. Contact time for a typical assay in packed bed mode with a column volume of approx. 100 μL is in the order of eight seconds, and for the expanded bed at least 2 minutes.

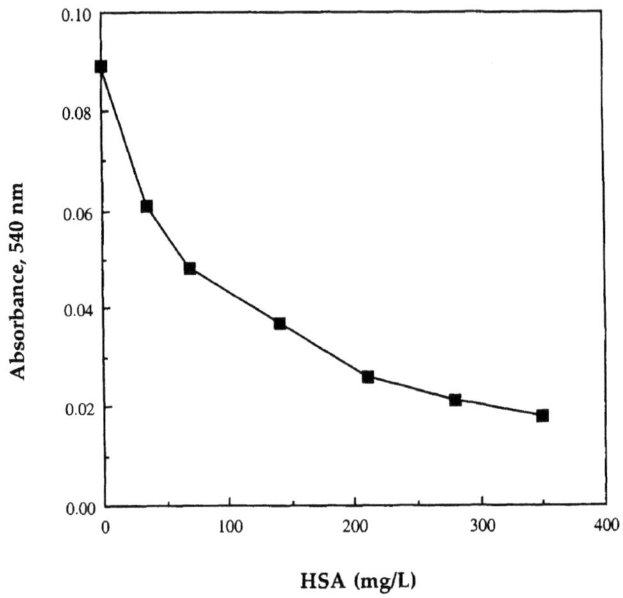

Figure 2. Calibration curve obtained after assaying standard solutions containing varying amounts of HSA and a constant amount of enzyme-labelled HSA.

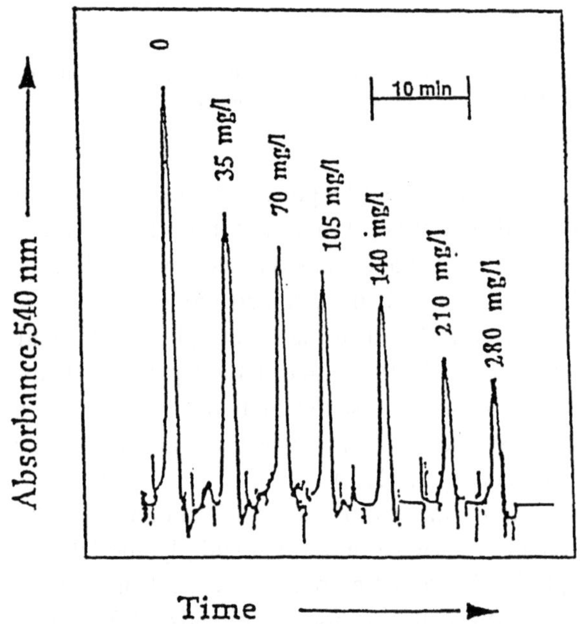

Figure 3. Typical peaks registered on a chart recorder from the Flow-ELISA when analysing samples containing varying amounts of HSA. No particulate matter was introduced in these assays.

Calibration curve

A typical calibration curve for different concentrations of HSA is presented in Figure 2 and the peaks registered in the chart recorder in Figure 3. In the competitive binding assay the species to be measured in a sample, competes with a constant level of the same species which is labelled with an enzyme, on binding to a limiting number of binding sites in the immunosorbent. The calibration curve has the typical shape of a competitive binding assay with lower readings at higher concentrations of the target molecule. Thus, the region with the best reproducibility is the steepest part i.e. the lower concentration region. The calibration curves show that a maximum concentration of 350 mg/L of HSA can be determined using the expanded bed antibody column with a detection time of 14 min (peak maximum time) and it returned to the base line after 12 min. This time needed for an assay cycle is more than of an immunoassay carried out using a packed bed antibody column (Nilsson et al 1992).

The reproducibility of the assay of 100 mg HSA/L in the presence of yeast cells was determined by repeating an assay 7 times within one day, and the SD was found to be ±4%. In absence of cells the S.D. was ±3%. Assays carried out at different days were done by correcting each day for decrease in binding capacity of the immunosorbent. Therefore, between-day-variations were not measured.

Dispersion in the system varied for the soluble substances and the particulate matter. The suspended cell particles gave more dispersion and caused some tailing. Inside the column the sample moves as expected in an expanded bed with no back-mixing as judged by the behaviour that is according to the Rickardson–Zaki plot (Lali et al., 1996; Nandakumar et al., 1999).

HSA determination in the presence of cells

Different concentrations of heat killed yeast cells ranging from 0.1 g/L to 1 g/L (dry weight) were prepared in the presence of 100 mg/L of HSA, injected in the system and the response checked. Since the system was operated in expanded bed mode, the yeast cells passed through the column without binding to the immunosorbent. However, HSA in the sample got bound to the antibodies. This means that the cells were washed out of the system prior to addition of substrate to the labelling enzyme. The presence of cells in the sample did not interfere with the determination of HSA, as was shown by analyzing samples containing 100 mg/L of HSA and varying amounts of yeast cells. The maximum peak height for the yeast cells appeared at 6.3 min and the signal returned to baseline before a second peak from the immunochemical determination of HSA was registered. The latter reached a maximum in 14 min and finally reached baseline after 25 min. The time for an assay may be reduced further by e.g. operating with support materials with higher density. Such material is now under investigation.

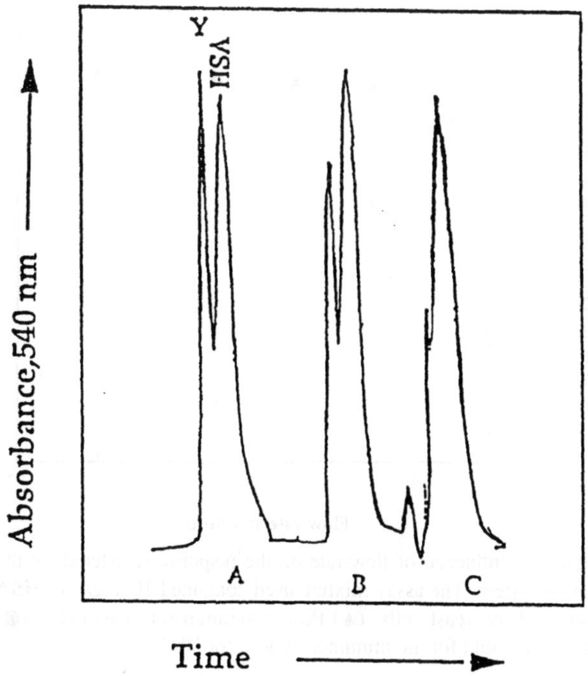

Figure 4. Typical peaks registered upon analysis of mixed samples containing both yeast cells and HSA. The content of HSA was kept constant at 100 mg/L while the yeast content was varied from 0.1 to 0.7 g/L.

The absorbance registration yields a double peak. When operated under slightly different conditions two separate peaks were registered, but it transpired to take too long for each assay cycle, and therefore the addition of reagents to HRP was done before the yeast cells had fully left the system. Typical peaks registered in the chart recorder are presented in Figure 4. This shows that it was not possible to detect a separate yeast peak at concentrations lower than 0.1 g/L. However, the separation of the two peaks was more distinct as the concentration of yeast cells increased.

Dead yeast cells were chosen in this study to represent particles in the flow. The use of dead cells provided a preparation that stays constant over time. This was preferred as a standardised addition to the system was desired. We have, subsequently also used viable cells as well as cell homogenates in the flow system and no interferences with the binding assays were observed. Furthermore, when analysis of a real system is due to take place, the cells will be inhibited/killed prior to entering the flow system. This is done using a coaxial catheter (Håkanson et al., 1991). Hence, the heat killed yeast cells are better able to simulate conditions of real analyses.

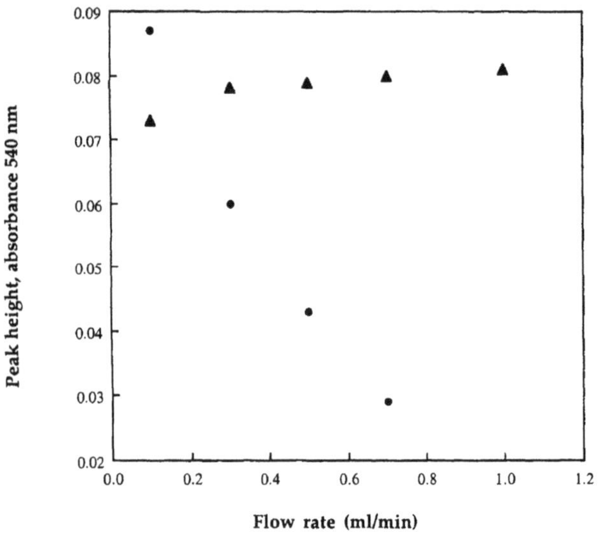

Figure 5. Influence of flow-rate on the response registered by the sensor system. The assay mixture used contained 100mg/L of HSA and 1 g/L of yeast cells. (▲) Peak maximum for yeast cells; (●) peak maximum for the immunoresponse for HSA.

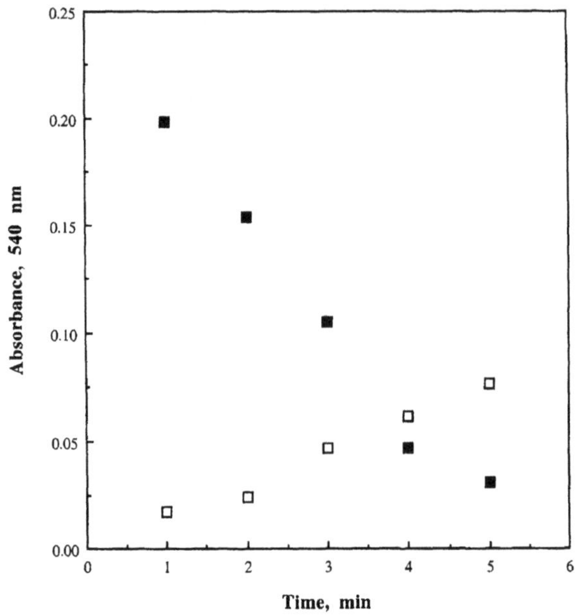

Figure 6. Effect of addition of substrate for the marker enzyme as a function of time elapsed since the peroxidase labelled antigen was introduced into the flow system. The experiments were carried out with 100 mg/L of HSA in the presence of yeast cells. Operating flow-rate 0.5 mL/min. (■) Peak maximum from immunoassay, (□) peak maximum for yeast cells.

Binding dependence of the flow rate

The influence of flow-rate on the immunochemical binding reaction as well as the enzymatic reaction, were studied. This was done by varying the flow-rate and keeping all other parameters constant.

The immunosorbent column discussed here is of the expanded bed type, where the possibility of mixing is higher and the contact time of the sample with immobilised antibody, are expected to be longer than in a comparable packed bed reactor. Furthermore, a change in flow-rate in an expanded bed system influences the system in at least two different modes. The contact time for a certain liquid segment decreases since the liquid passes through the immunosorbent quicker. However, an increased flow rate will also result in a more pronounced expansion of the bed. This latter effect will contradict the former observation to some extent. Figure 5 shows the outcome of such a study, where the flow rate during the binding step of the operation cycle was varied. It is clear that the flow-rate is a very important factor. When using an expanded bed antibody column, the response obtained at a lower flow rate was higher than that obtained at higher flow-rates. Similar results were observed in packed bed operation when the sample pulse flow-rate was varied. However, the variations were not as drastic as observed in the case of expanded bed columns. Possible reasons may be that the slow dispersion of sample and the conjugates in the expanded bed column at lower flow rate, followed by the addition of substrate, results in reactions between substrates and enzyme in both the bound and free conjugates. However, at higher flow-rates a reduction in peak height was observed indicating that the free conjugate had been washed out of the system. At a flow-rate of 0.5 mL/min, the binding of antigen followed by conjugates is expected to be uniform with a similar contact time for both. The substrate addition was carried out at the same flow rate and the results from the assay indicated that free conjugate was never exposed to the substrate pulse.

The lower flow-rates influence both the response and the time for analysis. Which flow-rate that one wants to use is a matter of sensitivity in the assay *vs* time comsumption. In this case we chose to run all further experiments at 0.5 mL per min.

Influence of time for addition of substrate

From the results above it was found that the mixing of substrate with the free conjugates in the flow system may create problem in the analysis. Thus, for evaluating this interference, the time for substrate addition, after the addition of conjugates varied. This was done at constant flow-rate (0.5 mL per min). The results are depicted in Figure 6.

An increase in response was significant when the substrate was added one minute after the addition of conjugates and decreased when the substrate was ad-

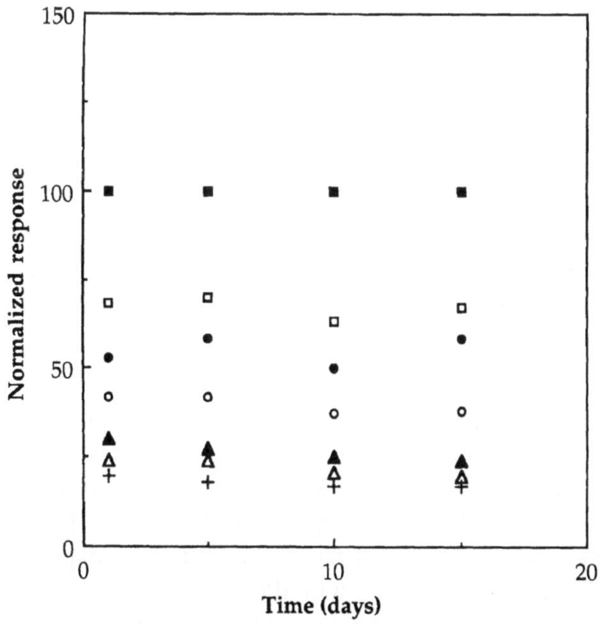

Figure 7. Operational stability of the flow-ELISA system monitored by assay of standard solutions of HSA. The value with no free antigen was set = 100%, and all other values were referred to that value. Recalibration with regard to the capacity of the immunoadsorbent was done once a day. (■) 0 mg/L, (□) 35 mg/L), (●) 70 mg/L), (○) 140 mg/L, (▲) 210 mg/L, (△) 280 mg/L, (+) 350 mg/L.

ded after a period exceeding 4 min. A further increase in time did not influence the response. This clearly indicates that early addition of substrate will result in reaction with free conjugates, and this will give higher readings, but less reliable. Later addition, results in reaction with only bound conjugates resulting in an accurate quantification of the bound conjugates and thereby giving a value useful for determining the concentration of antigen. The time of addition of substrate was fixed to 4 min in all subsequent experiments. The yeast cells present in the sample gave a peak that was registered independent of substrate addition.

The time of the peak maximum for the immunochemical reaction for quantifying HSA and for the yeast cells present in the sample were also noted (Figure 6). The yeast cells were registered at a fixed time whereas the immunochemical binding reaction varied with the time for substrate addition.

Operational stability of the immunoassay

The binding capacity of the column decreased slightly for each assay cycle due to denaturation caused by the low pH buffer used in the dissociating step and by the exposure to the substrate solution used for quantifying the amount of bound labelling enzyme. The peaks registered when measuring standard solutions for constructing the calibration curve gradually decrease as the capacity of the immunosorbent decreases. Thus, a calibration curve measured on a fresh column was much higher than a calibration curve measured on the same column after it had been used for a while. The shape of the curve remained the same throughout the lifetime of the column but the amplitude decreased gradually (Borrebaeck et al., 1979; Nilsson et al., 1993). Thus, if the readings were normalised with regard to capacity, such that the reading from assays with no free antigen present was set equal to 100%, then congruent calibration curves were obtained. A similar result of loss of column binding capacity and maintained, normalised, calibration curves were reported for flow ELISA operated by using a packed bed with immobilized antibody. It can be seen quite clearly from Figure 7 that the readings of fixed concentrations stayed constant after the values had been normalised. The data shown in Figure 7 were generated by measuring samples containing heat-killed yeast cells. Live cells would not be allowed into the system since living cells will attach to the walls of the flow system and form a kind of bioreactor. It has been shown earlier that such complications destroy analyses of low molecular weight compounds that are also substrates for the cells (Mandenius et al., 1981). In order to cope with this problem, sampling may be carried out with the concomitant addition of an inhibitor (Håkanson et al., 1991). The results presented in Figure 7 represent more than 150 assays of samples containing heat killed yeast cells.

Biofouling may be a potential threat to the performance of the analytical system. We do not, at present, have a clear picture about the impact of biofouling on this system. However, the system is based on use of an excess of immobilised antibodies used under non-equilibrium conditions. The capacity of the column decreases during repeated assay, but there is no information available concerning how much can be ascribed to denaturation and how much is due to fouling of the system. From earlier studies it is known that even if the readings from calibration assays loose in absolute values, they stay constant when normalised with respect to the capacity of the column (Mattiasson et al., 1977; Nilsson et al., 1993).

From downstream processing it is a well known fact that fouling may disturb the performance of the separation. In the analytical case, one can foresee less serious disturbances since the sample volumes are small, the time for contact is relatively short and each of them is followed by a regenerating step. If neces-

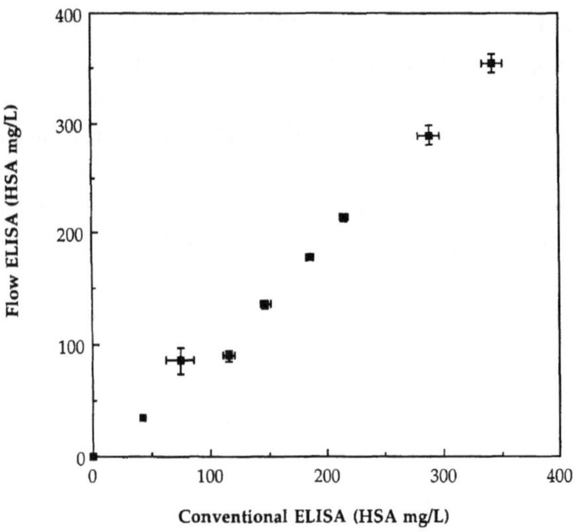

Figure 8. Human serum albumin, as measured by the present method vs the conventional ELISA technique. Each point represents the average of three assays with each method.

sary it would be possible to add a special reagent to improve the cleaning of the column between the assays. If however, fouling takes place, then a reduced capacity will be detected. As long as the calibration curve is valid, i.e. as long as there are enough active antibodies to give a calibration curve, it will be relatively easy to compensate for the disturbances by simply normalising the values to the capacity value for the column (Nilsson et al., 1993).

Comparison of results from analyses on HSA by conventional flow ELISA and expanded bed flow ELISA.

Figure 8 shows a comparison of data obtained by using conventional ELISA and flow expanded bed ELISA, respectively. A variation of 5–10% was observed in case of flow ELISA, possibly due to the errors incurred during manual operation of the system. This may be eliminated or at least substantially reduced by complete automation of the system.

The differences in time needed for an assay cycle are at present too large, and a development is going on to speed up the expanded bed Flow ELISA. At present the cycle time for the expanded bed mode of operation is approx. 25 min and for the packed bed mode approx. 6 min.

How far one can shorten the cycle time is too early to speculate about, and it is thus difficult to compare the future areas of application of the two. It is as far as can be seen today no reason to use an expanded bed when particulate free liquid samples of low viscosity are to be analysed. In many other cases it may be realistic to use the expanded bed concept provided the speed requirement is met.

Conclusions

The results reported here for development of a flow injection based expanded bed immunoassay, are the first of its kind. The analytical system can be used for determination of antigens in competitive binding assays, in the presence of suspended materials, like cells. The system operated with an expanded bed of immunosorbent. Interferences were minimised by using an adsorbent with low non-specific binding and the expanded mode of operation that allows particulate matter to pass through the bed. It is expected that this new approach will allow a more direct monitoring of specific proteins and other analytes in the presence of suspended materials like cells. The expanded bed mode of operation eliminated the needs for separation, prior to binding assays. This omits the inevitable bottleneck in previous technologies. The time gained by eliminating the separation step is lost at present due to the extra time required by the expanded bed mode of operation compared to a corresponding packed bed. However, the use of an automated process over a manual separation step helps to improve the reproducibility of the assay. This is essential when using the signals for process monitoring and control. One of the main direct advantages of using the expanded mode is that several different steps are integrated in the analytical procedure.

Acknowledgements

This project was supported by The National Swedish Board for Technical Development (NUTEK) and The Swedish Foundation for Food Science and Biotechnology (SIK). The cooperation with Pharmacia Biotechnology AB took place under the programme of The Swedish Competence Center for Bioseparation.

References

Borrebaeck C, Börjesson J and Mattiasson B (1978) Thermometric enzyme linked immunosorbent assay in continuous flow systems: Optimization and evaluation using human serum albumin as a model system. Clin. Chim. Acta 86: 267–278.

Chang YK and Chase HA (1995) Development of an expanded bed technique for an affinity purification of G6PDH from unclarified yeast cell homogenates. Biotechnol. Bioeng. 48: 355–366.

Collins S and Janata J (1982) A critical evaluation of the mechanism of potential response of antigen polymer membrane to the corresponding antiserum. Anal. Chim. Acta 136: 93–99.

Daniles PB, Deakon JK, Eddowes MJ and Pedley DG (1988) Surface plasmon resonance applied to immunosensing. Sensors Actuators 15: 11–18.

Håkanson H, Nilsson M. and Mattiasson B. (1991) A general sampling system for sterile monitoring of biological processes. Anal. Chim. Acta 249: 61–65.

Kusterbeck AW, Wemhoff GA, Charles PT, Yeager DA, Bredehorst R, Vogel CW and Ligler FS (1990) A continuous flow immunoassay for rapid and sensitive detection of small molecules. J. Immunol. Methods 135: 191–197.

Lali A, Kaul R, Galaev I Yu and Mattiasson B (1997) Purification of L-lactate dehydrogenase from crude homogenate of porcine muscle by expanded bed affinity chromatography (EBAC). Isolation and Purification 2: 289–300.

Lim CS, Miller JN and Bridges JW (1980) Automation of an energy transfer immunoassay by using stopped flow injection analyses with merging zones. Anal. Chim. Acta 114: 183–189.

Maggio ET (ed.) (1980) Enzyme-immunoassay. CRC-Press, Boca Raton Fl, USA.

Mandenius C-F, Danielsson B and Mattiasson B (1981) Process control of an ethanol fermentation with an enzyme thermistor as a sucrose sensor. Biotechnol. Letters 3: 629–634.

Mattiasson B, Borrebaeck C, Sanfridsson B and Mosbach K (1977) Thermometric enzyme linked immunosorbent assay: TELISA. Biochim. Biophys. Acta 483: 221–227.

Mattiasson B and Håkanson H (1993) Sampling and sample handling - crucial steps in process monitoring and control. Trends Biotechnol. 11: 136–142.

Mattiasson B and Larsson KM (1987) Flow injection enzyme immunoassay - a quick and convenient binding assay. Proc. 4th Europ. Congr. Biotechnol. Amsterdam, The Netherlands, June 14–19, 1987. Eds. OM Neijssel, RR van der Meer and K Ch AM Luyben. vol 4, pp 517–522.

Mattiasson B and Miyabayashi A (1988) Registration of the streaming potential over an affinity column in a continuous flow system as a means to biospecifically quantify proteins. Anal. Chim. Acta 213: 79–89.

Miller JN (1989) Flow injection analysis using antibodies and other receptor proteins. Anal. Proc. 26: 317–318.

Nakamura RM, Kasahara Y and Rechnitz RA (eds) (1992) Immunochemical assays and biosensor technology for the 1990s. American Soc. for Microbiology, Washington, USA.

Nandakumar MP, Lali A and Mattiasson B (1999) On-line monitoring of a fermentation process using an expanded bed micro bed flow injection analyser. Bioseparation, this volume.

Ngo TT (ed.) (1987) Electrochemical sensors in immunological analysis. Plenum Press, New York.

Ngo TT (ed.) (1988) Nonisotopic immunoassay. Plenum Press, New York.

Nilsson K and Mosbach K (1980) *p*-Toluenesulfonyl chloride as an activating agent of agarose for the preparation of immobilised affinity ligands and proteins. Eur. J. Biochem. 112: 397–402.

Nilsson M, Håkanson H and Mattiasson B (1991) Flow injection ELISA for process monitoring and control. Anal. Chim. Acta 249: 163–168.

Nilsson M, Håkanson H and Mattiasson B (1992) Evaluation of a sequential competitive binding assay. Proc. Control Quality 597: 383–389.

Nilsson M, Mattiasson G and Mattiasson B (1993) Automated immunochemical binding assay (flow-ELISA) based on repeated use of an antibody column placed in a flow-injection system. J. Biotechnol. 31: 381–394.

Nilsson M, Vijayakumar AR, Holst O, Schornack C, Håkanson H and Mattiasson B (1994) Integrated operation of fermentation, downstream processing and on-line process monitoring in cultivation of *Bacillus amyloliquifaciens* for production of α-amylase. J. Ferment. Bioeng. 78: 356–360.

Puchades R, Maquieira A, Atienza J and Montoya A (1992) A comprehensive overview on the application of flow injection techniques in immunoanalysis. Critical Reviews in Analytical Chemistry 23: 301–321.

Bioseparation **8:** 247–254, 1999.
© 1999 *Kluwer Academic Publishers.*

Use of a micro-expanded bed containing immobilised lysozyme for cell disruption in flow injection analysis

M. P. Nandakumar, Anita Tocaj & B. Mattiasson
Department of Biotechnology, Center for Chemistry and Chemical Engineering, Lund University, P.O. Box 124, S-221 00 Lund, Sweden

Received 9 October 1998; accepted in revised form 13 April 1999

Key words: cell disruption, ultrasonication, immobilised lysozyme, flow injection analysis

Abstract

A method for cell disruption in Flow Injection Analysis (FIA) systems has been developed. The principle involves on-line cell disruption by means of immobilised lysozyme followed by an ultrasonic treatment. In order to avoid flow problems in the analytical system, the lysozyme was immobilised to Streamline® that was used in an expanded bed in the flow system. Samples of suspensions of *Micrococcus lysodeikticus* were treated and the success of the treatment was evaluated in terms of released protein and as a decrease in the optical density at 450 nm. The new technology offers a powerful tool in flow injection analyses for quantification of intracellular compounds. The concept of integration, i.e. combining cell disruption with handling of cell debris and assay procedure in one continuous flow process facilitates its use and increases the probability of reaching reproducible and reliable results.

Introduction

Fermentation monitoring has been facilitated by the introduction of biosensors and flow injection systems. These techniques in combination with proper sampling and sample handling procedures makes it possible to monitor on-line or at least with high frequency both low molecular weight substances and macromolecules, provided they are excreted to the medium. The information that can be gained is far superior to what has been available previously and also to a large extent, compared to what is still used in many places. However, all the reactant pools in the fermentation broth reflect events in the interior of the cells, and it would of course be very attractive to have a reading of the intracellular concentration of a certain metabolite. To reach that goal, new analytical devices need to be developed. Two different approaches can be applied: the non-invasive (e.g., Fiechter and Sonnleitner, 1994) and the invasive one. The first seems most attractive, even if there is still a long way to go before methods will be available for a range of the more important metabolites. The use of fluorescence gives a good estimate of the NADH/NAD^+ ratio (Sonnleitner, 1991) and NMR has been used to monitor phosphorylated intermediates in the glycolysis (Lohmeier-Vogel et al., 1986). However, there is, to date, no generally applicable technology for non-invasive monitoring. The other alternative would be to disintegrate the cells and then monitor the concentrations of various metabolites. This has been the established routine for off-line analysis for many years. However, it is characteristically time-consuming, inaccurate and has poor reproducibility.

It has been repeatedly proven that the use of Flow Injection Analysis (FIA) is a very powerful technique when dealing with reactions that have to be handled under non-equilibrium conditions (Mattiasson and Borrebaeck, 1978; Ruzicka & Hansen, 1988). Automation of the different steps in the process enables greater accuracy (Nilsson et al., 1993) than that using skilled personel in manually operated procedures. Thus, cell disruption is with its inherent dynamic behaviour very suitable for integration in FIA systems. One problem is, however, that bioassay-FIA so far has been based either on the use of open tubular systems or on using packed bed reactors. The use of expanded beds makes it possible to purify proteins in fermentation broths without an initial cells or cell debris separation process (Chase, 1994). The recent development in our laboratory concerning the use of expanded beds for analytical purposes in conjunction with flow injection analysis additionally offers an interesting possibility for handling particulate containing material, such as cell containing fermentation broths (Mattiasson and Nandakumar, 1999; Nandakumar et al., 1999).

The present paper describes the development of an on-line cell disruption procedure adapted to FIA using an expanded bed with immobilised lysozyme followed by ultrasonic treatment in a flow-through cell.

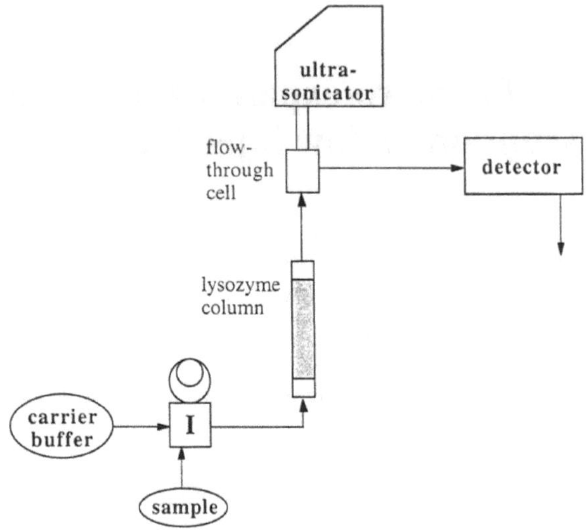

Figure 1. Schematic presentation of the complete flow-system.

Materials and methods

Preparation of beads

Materials

Native Streamline® used for immobilisation of enzyme was a generous gift from Pharmacia Biotechnology AB, Uppsala, Sweden. Chicken egg white lysozyme, lyophilised *Micrococcus lysodeikticus* cells were obtained from Sigma Chemical Company (St. Louis, Mo. USA). Tresylchloride used for activation of the support was obtained from Syntelec AB (Lund, Sweden). All other chemicals were of analytical grade.

Immobilisation procedure

Immobilisation of lysozyme was carried out according to an established procedure (Nilsson and Mosbach, 1980). Ten ml of settled gel of native Streamline® was activated with tresylchloride and suspended in 0.2 M sodium phosphate buffer pH 7.0. Lysozyme was added to the activated gel at a concentration of 25 mg of enzyme per ml of gel. Coupling of the enzyme was carried out overnight at 4 °C. The beads with immobilised enzyme were then washed successively with an excess of water, 1 M NaCl and 0.2 M sodium phosphate buffer pH 7.0, to remove unbound enzyme. The enzyme beads were then treated with 1 mM ethanolamine in sodium phosphate buffer (pH 8) for 1 h at room temperature in order to block the free activated sites on the gel. The gel was washed again with an excess of water and 1 M NaCl and finally stored in 0.2 M sodium phosphate buffer pH 7.0 until use. All the washing procedures were carried out under reduced pressure. The residual activity of immobilised lysozyme was determined by comparing the specific activity of the immobilised lysozyme to that of the added free enzyme.

Characterisation studies using immobilised lysozyme – batch mode

Batch digestions were carried out in cell suspensions with 0.5 g/L *Micrococcus* cells in 0.2 M sodium phosphate buffer pH 6.4 at 25 °C. The amount of immobilised enzyme was varied by adding known amounts of beads with immobilised lysozyme to the cell suspensions. The contact time was investigated by removing small amounts of sample, from a 50 mL cell suspension containing 1 g (wet weight) beads, at fixed time intervals. The operational stability was investigated by repeatedly adding fresh cell suspension to the beads and washing them between each addition.

Instrumentation

Flow system

A flow system, as depicted in Figure 1, was used. The continuous flow passes through an expanded bed where the immobilised lysozyme is held, via a flow-through ultrasonicator unit to a detector. The cell containing samples were introduced into the flow system via a sample loop, with an injection volume of 100 μL. In such assays it is important to have a constant and accurate flow, and for this purpose a peristaltic pump (Alitea, Sweden) was used. The on-line detector used was a spectrophotometer for OD_{450} measurements. The protein concentration was determined by Bradford /BioRad assay after removing the cell debris by filtration.

Expanded bed column

A column with an inner diameter of 0.5 cm was used. The Streamline® with immobilised lysozyme was kept in the column between two flow adapters. The flow adapters were supplied with special arrangements to secure a homogenous distribution of the liquid at the entrance of the column (Nandakumar et al., 1999). The flow was introduced from the bottom. The distance between the adapters was slightly more than double that of the height of the sedimented gel (in this case 4.5 cm) in the column. On initiation of flow through the bed, first packed bed behaviour is observed and when the flow rate increases, the bed starts to expand. Under standard conditions, expansion to approximately twice the packed bed height was used.

Expansion of the lysozyme-Streamline® bed

The column was placed vertically and the flow was varied. The settled bed volume was registered prior to pumping buffer through the system and then the bed heights at different flow-rates were compared to that of the settled one.

Ultrasonication

The Ultrasonicator UP 400 S (Dr. Hielscher GmbH, Stahnsdorf, Germany) was equipped with a flow through cell (D 14/22) provided with a cooling system. The titanium probe that fits into the flow through cell has a diameter of 14 mm. The ultrasound was pulsed in cycles of one second and the on/off time was chosen to be 0.5 s each. The sound intensity can be varied between 20 and 100% of the maximal value; which for the S14 D probe is 105 W/cm^2.

Analytical methods

Lyophilised *M. lysodeikticus* cells were suspended in 0.2 M sodium phosphate buffer, pH 6.4, and used for evaluation of all disruption methods in terms of released protein and in optical density (OD_{450}) readings. *Micrococcus lysodeikticus* cells are frequently used as a substrate in lysozyme activity determinations (e. g., Datta et al., 1973; Nakamura et al., 1991; Lee and Ku, 1994; Chen and Chen, 1996).

Cell lysis.

The decrease in optical density at 450 nm was taken as a measure of cell lysis. In batch experiments the cell samples were diluted with 0.9% NaCl at OD_{450} values exceeding 1.0.

Table 1. OD_{450} readings and protein released from the cells after treatment with different cell disruption methods

Cell disruption method (off-line)	OD_{450}	protein (mg/mL)
Blank (no disruption)	5.1	0.08
Freeze and thaw	4.1	0.11
Lysozyme	1.6	0.31
Lysozyme + EDTA	1.5	0.33
Lysozyme + urea	1.5	0.32
Freeze and thaw + lysozyme	1.4	0.34
Lysozyme + EDTA+ urea	1.3	0.39
Ultrasonication	1.1	0.42
Lysozyme + ultrasonication	0.6	0.64
Lysozyme + EDTA + ultrasonication	0.6	0.67

Table 2. OD_{450} readings and protein released from the cells after treatment with the on-line cell disruption methods

Cell disruption method	OD_{450}	protein (mg/mL)
Immobilised lysozyme	1.9	0.29
Immob. lysozyme + ultrasonication	1.1	0.53

Reference methods of cell disruption

A range of different disruption methods was investigated and the amount of released protein as well as OD_{450} was compared. *Micrococcus lysodeikticus* was suspended in 0.2 M sodium phosphate buffer, pH 6.4, at a concentration of 1 g cell dry weight (cdw)/L for this study. The different methods used, together with the results, are shown in Table I. The experimental procedures were according to those described in the relevant literature (Gray et al., 1972; Lee et al., 1979; Wang et al., 1979; Fish et al., 1983; Caldeira and Cabral 1994).

Results and discussion

An efficient, quick and reliable method to analyse the intracellular content of certain products or metabolites in samples from bioreactors is highly desirable. Proteins produced by a microorganism, can be found in the cytoplasm, in the periplasmic space or excreted to the exterior. However, in the latter case, there will also be at any given time, some molecules in the two other locations. In such cases, analysis of the content of the cell-free medium will often give a fairly good estimate. However, under conditions where the product is located in the cytoplasm, in the periplasmic space or even adsorbed to the cell surface an erroneous result will be obtained. Clearly there is a need for additional steps prior to analysis.

A gram-positive organism, *Micrococcus lysodeikticus*, was chosen as it is routinely used as a substrate for lysozyme activity measurements and is thus a favourable target in the first disruption study. Furthermore, by choosing a gram positive organism, any concern about protein enrichment in the periplasmic space is obsolete.

The present paper reports on the developement of a FIA system for monitoring of intracellular compounds in biotechnological processes. This assay technique is becoming increasingly popular as it offers high reproducibility, high reliability and is easy to use (Mattiasson and Danielsson, 1996; Ruzicka and Hansen, 1988). Thus, such a method would be attractive provided it is possible to integrate the pretreatment steps as indicated above in order to release the compounds that are to be analysed. As the cells are disrupted their intracellular content is released and particulate matter (cell debris) will also be found in the sample. Usually the cell debris has to be removed prior to analysis and that is often done by means of centrifugation or filtration in off-line analyses. If, however, a continuous integrated process is to be developed it would be a great advantage to be able to carry out the analysis without removing the cell debris. We have recently presented the use of an expanded bed integrated in flow injection assay systems both for enzymatic analysis (Nandakumar et al., 1999) and in immunochemical binding assays (Mattiasson and Nandakumar, 1999) in the presence of cells. It should thus be possible to perform the assays in the presence of cell debris and the challenge is then the disruption step *per se*.

Cell disruption studies

A range of different cell disruption methods are described in literature: osmotic shock (Wang et al., 1979), chemical lysis (Gray et al., 1972; Lee et al., 1979), enzymatic lysis (Fish et al., 1983), sonication (Caldeira and Cabral, 1994). At this stage it deserves discussion of a definition of a successful disintegration. How should the cell disruption process be evaluated? When is 100% of the intracellular protein released and how is that quantified? As shown in Table I, we took the stand that by comparing a broad range of disruption methods, a value that was equal to the best of established methods was set as an indication of success. A method suitable for being integrated in a flow system must be quick and convenient and require only small volumes of sample. An obvious candidate is enzymatic lysis of cells which has been reported in preparative applications to be efficient and mild with regard to the fate of the cell content (Andrews and Asenjo, 1987). Lysozyme has been used for cell lysis and it has been reported to be efficient for gram-positive cells. One problem though is how to apply the enzyme. If the enzyme is to be added in free solution, fairly large amount of enzyme has to be used at high costs. This may be acceptable as long as an inexpensive enzyme like lysozyme is utilised, but it would be attractive to develop a technology for the use of other, more expensive enzymes. Thus, to use immobilised enzymes was an obvious choice. However, even if there are reports that immobilised lysozyme has been efficient in killing/permeabilising bacterial cells (Datta et al., 1973, Mattiasson, 1977; Crapisi et al., 1993) the foreseen flow problems when pumping a cell suspension through the bed makes such an alternative, less attractive. Alternative modes of operation with the immobilised enzyme are at hand (Chen and Chen 1996; Lee and Ku, 1996). As the expanded bed concept has proven efficient in binding assays (Mattiasson and Nandakumar, 1999) as well as in enzyme assays (Nandakumar et al., 1999), it was decided to evaluate its performance with immobilised lysozyme.

Batchwise disruption studies using immobilised lysozyme

The effect of immobilised lysozyme on *Micrococcus* cells is dependent on the amount of enzyme available and also on the time for contact between the cells and the enzyme. Figures 2A and 2B, show the effect of these two parameters. Initially the lytic action increases with increasing amounts of lysozyme added (Figure 2A). The data fit a saturation type of kinetics. The contact time in this study was 15 min. The effect of increasing contact time showed that above 6 min, the effect levels off towards a constant value (Figure 2B). The bead concentration was in this case 20 mg/mL and was used directly after immobilisation. When integrating the lytic step into a flow system, some characterisation studies had to be done. The release of protein and decrease in OD_{450} upon repeated exposure of *Micrococcus* to the beads was determined and in Figure 3 it can be seen that a slight denaturation of the immobilised enzyme (or at least loss of enzyme activity) takes place. This effect on the lysozyme

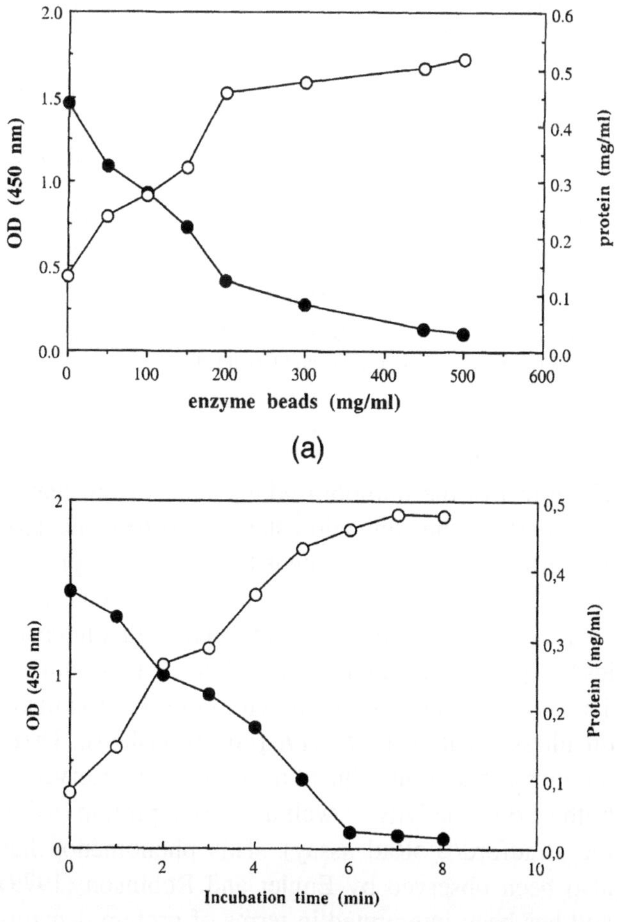

Figure 2. Lysis of *Micrococcus* cells using lysozyme immobilised to Streamline® in batch operations. The outcome is measured in terms of released protein (○) and OD_{450} (●); (A) effect of the amount of immobilised lysozyme added (2 ml cell suspension was used for lysis studies); (B) effect of incubation time.(50 ml cell suspension was used for lysis studies).

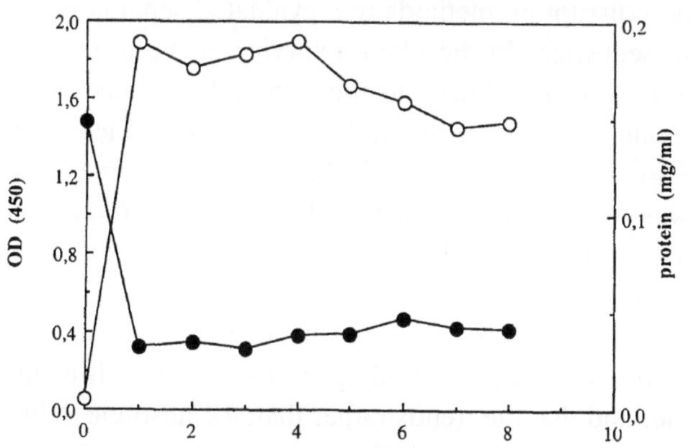

Figure 3. Operational stability of lysozyme immobilised to Streamline® Shown in terms of released protein (○) and OD_{450} (●).

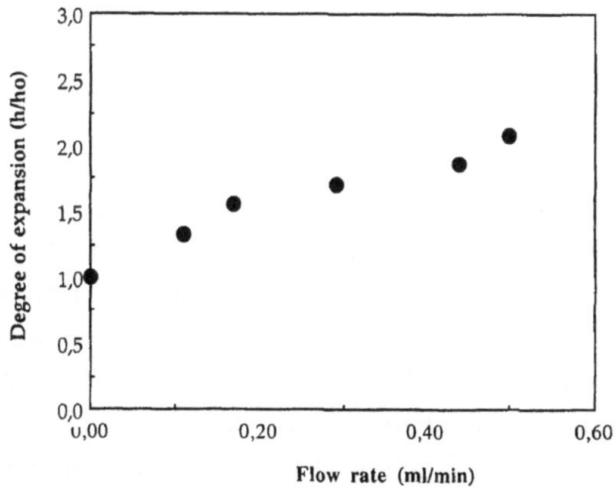

Figure 4. Expansion of the lysozyme-Streamline® bed as a function offlow-rate. Degree of expansion (h/ho).

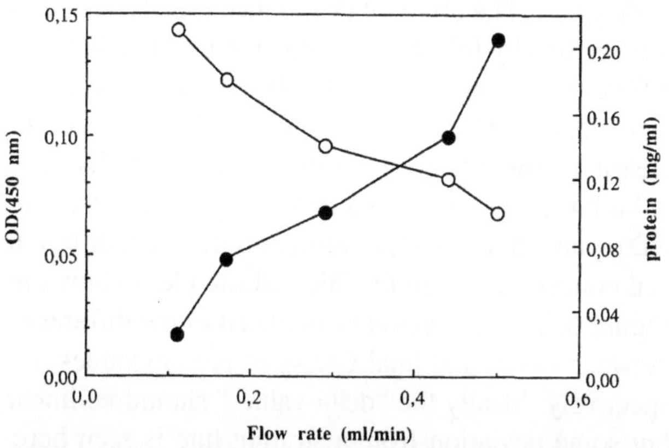

Figure 5. Lysis of cells in an expanded bed reactor as a function of flow-rate. Shown in terms of released protein (○) and OD_{450} (●).

could also be due to shielding of the enzyme by non-specifically adsorbed material to the enzyme-support. This was not further studied in this investigation.

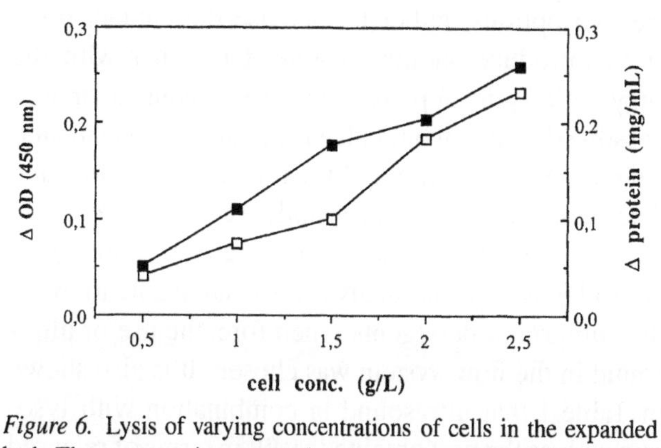

Figure 6. Lysis of varying concentrations of cells in the expanded bed. The outcome is presented as the amount of protein released (final – initial) Δ protein (□) and as the decrease in optical density (initial – final) ΔOD_{450} (■).

Evaluation of the expanded bed

The importance of the contact time is well-known and was also shown in Figure 2B when a batch process was studied. In a flow system the contact time is determined by the reactor size and the flow rate. The flow-rate does not only determine the contact time when an expanded bed is used, but is also an important parameter that influences the behaviour of the expanded bed. The bed height as a function of the flow-rate is shown in Figure 4. Previous studies have shown that when using columns of this small size the degree of expansion should not exceed 2 (Nandakumar et al., 1999). In this case it limits the flow rate to ~0.5 mL/min as shown in Figure 4. The effect of flow-rate on the lytic action was studied by injecting a fixed amount of cells using different flow rates, i.e. with different retention times in the reactor. The results are shown in Figure 5. An increase in flow-rate gives a decreased lysis, as observed in both OD_{450} and in protein readings. The change in OD_{450} and in released protein as a function of different cell concentrations in the injected sample is shown in Figure 6. The changes are calculated as the difference between initial and final OD_{450} or protein values, respectively. Ideally the "delta-values" should be linear but some deviation from a straight line is seen here. However the correlation coefficients are satisfactory, 0.96 for Δprotein and 0.98 for ΔOD_{450}. It should in this context be stressed that one can not expect that all absorbance will disappear as the disruption process approaches completion as the cell debris will also scatter light and contribute to an absorbance reading.

It is clear from these studies that the *Micrococcus* cells are heavily influenced by the action of lysozyme, but not completely lysed. This was also the case when using lysozyme in free solution (see Table I). There are two options, either to increase the contact time, or to introduce another treatment together with the enzymatic lysis step. An increase in contact time is possible, but that would also mean an increase in analysis time. The combined treatment of enzymes and detergents in order to permeabilise cells is often used. However, when the released intracellular compounds are to be used in bio-analysis, it is advisable to avoid the addition of detergents. Therefore, the use of ultrasound in the flow-system was chosen. It is also shown in Table I that ultrasound in combination with lysozyme gives the best obtained result in terms of released protein and in decreased OD values for those methods tested.

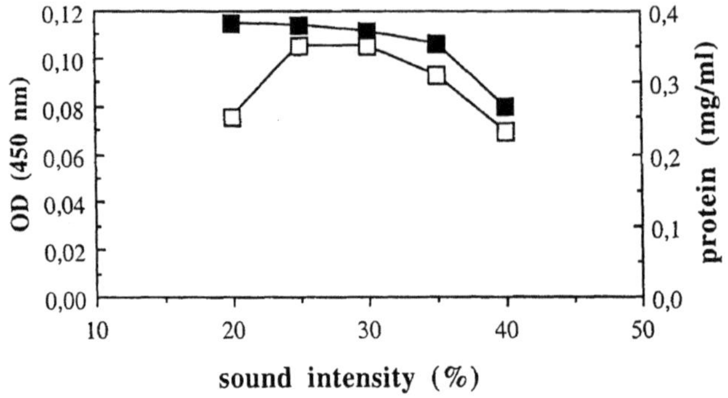

Figure 7. Effect of sound intensity on *Micrococcus* cells shown interms of released protein (□) and OD_{450} (■).

Ultrasonication

The effect of the immobilised enzyme was evaluated and similary, the sonication step has to be evaluated. The effect of the sound intensity on released protein and on OD_{450} is seen in Figure 7. As the intensity increases, more protein is released, but at higher intensities less protein is seen in the absorbance measurement. This is consistent with the previous observation made on ultrasonication of *E. coli* (not published). There the increase and subsequent decrease was observed in both enzyme activity as well as in total protein (using the Bradford/BioRad assay). This phenomenon has also been observed by Engler and Robinson (1979) and has been interpreted in terms of protein denaturation which leads to a lower reading of soluble protein in the sample.

Combining lysozyme and ultrasound in the flow-system

The lysozyme treatment was combined with ultrasound and investigated by continuously pumping a cell suspension through the flow-system. The impact of the two disruption methods was evaluated separately and in sequence. In the blank experiment the ultrasonicator was turned off and the immobilised beads were replaced by native Streamline®. It is quite clear from Figure 8a and 8b and Table 1 that lysozyme alone was less efficient than the ultrasound alone and that a combination of the two was by far the best method for disintegration. This is seen both as OD_{450} and as protein released. It should be mentioned that these evaluation methods both give a rather rough estimate, and that the trends rather than the absolute values should be considered. To further investigate the system, a more specific method for determination of the target molecule should be implemented. The results obtained in this work, however, clearly indicate that the concept of combining enzymatic and ultrasonic

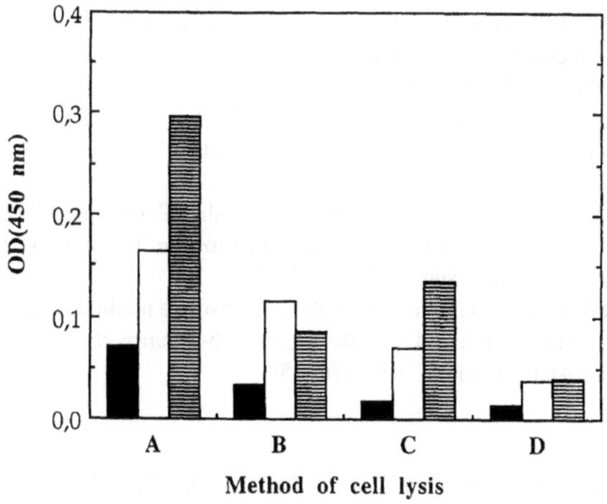

(a)

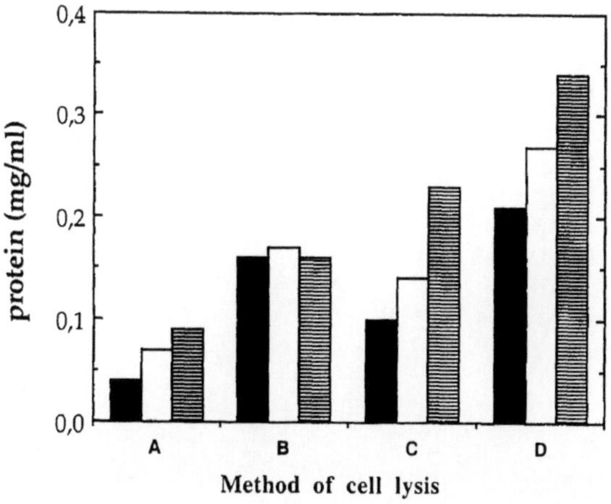

(b)

Figure 8. a & b. The effect of the different disruption methods in the flow system. A: without lytic treatment. B: ultrasonication. C: immobilised lysozyme. D: immobilised lysozyme plus ultrasonication. Results are presented for three different cell concentrations. 0.5:(■) 1.0: and (□) 2.0 g/L(⊟) a) in terms of OD_{450} b) in terms of released protein (mg/mL).

treatment in a flow-system is a promising technique for studying intracellular compounds.

Concluding remarks

An efficient and fast method to disrupt cells in an integrated fashion together with analysis, will offer good possibilities to quantify compounds and to measure parameters that earlier were not registered as frequently. The value of quantifying intracellular proteins, nucleic acids or even low molecular metabolites may open new possibilities to a better understanding of cellular metabolism and physiology. Such studies are under way.

Acknowledgements

This project was supported by The National Swedish Board for Technical and Industrial Development (NUTEK), The Swedish Foundation for Food Science and Biotechnology (SIK), The Swedish Agency for Research Cooperation with Developing Countries (Sida/SAREC), The Trygg Hansa Alf Åkerman Foundation, The Royal Physiographic Society in Lund and The Helge Ax:son Johnson Foundation.

References

Andrews BA and Asenjo JA (1987) Enzymatic lysis and disruption of microbial cells. Trends Biotechnol. 5: 273–277.

Caldeira JCG and Cabral JMS (1994) Extraction of a steroid 1,2 dehydrogenase by sonication of *Arthrobacter simplex* cells. Bioseparation 4: 271–278.

Chase H (1994) Purification of proteins by adsorption chromatography in expanded beds. Trends Biotechnol. 12: 296–303.

Chen J-P and Chen Y-C (1996) Improvement of cell lysis activity of immobilised lysozyme with reversibly soluble-insoluble polymer as carrier. Biotechnol. Techniques. 10: 749–754.

Crapisi A, Lante A, Pasini G and Spettoli P (1993) Enhanced microbial cell lysis by the use of lysozyme immobilised on different carriers. Process Biochem. 28: 17–21.

Datta R, Armiger W and Ollis DF (1973) Lysis of *Micrococcus lysodeikticus* by lysozyme covalently immobilised on cellulose and polyacrylamide. Biotechnol. Bioeng. 15: 993–1006.

Engler CD and Robinson CW (1979) New method for measuring cell-wall rupture. Biotechnol. Bioeng. 21: 1861–1869.

Fiechter A and Sonnleitner B (1994) Non-Invasive concepts in metabolic studies. Adv. Microbial Physiol. 36: 145–180.

Fish NM, Harbron S, Allenby DJ and Lilly MD (1983) Oxidation of n-alkenes: Isolation of alkane hydroxylase from *Pseudomonas putida*. Eur. J. Microbiol. Biotechnol. 17: 57–63.

Gray PP, Dunnill P and Lilly MD (1972) Kinetics of β-galactosidase production by a constitutive mutant of *Escherichia coli*. J. Ferment Technol. 50: 381–387.

Lee C-H, Tsang SK, Urakabe R and Rha CK (1979) Disintegration of dried yeast cells and its effect on protein extraction, sedimentation property, and viscosity of the cell suspension. Biotechnol. Bioeng. 21: 1–17.

Lee C-K and Ku M-C (1994) Lysis of *Micrococcus lysodeikticus* cells by lysozyme covalently immobilised on the lumen of hollow fibres. Biotechnol. Techniques 8: 193–198.

Lohmeier-Vogel E, Hahn-Hägerdal B and Vogel HJ (1986) Phosphorus-31 NMR studies on maltose and glucose metabolism in *Streptococcus lactis*. Appl. Microbiol. Biotechnol. 25: 43–51.

Mattiasson B (1977) The use of coimmobilized lysozyme as a bactericide in enzyme columns: a step towards the design of self-sterilising enzyme columns. Biotechnol. Bioeng. 19: 777–780.

Mattiasson B and Danielsson B (1996) Flow injection analysis (FIA) in combination with biosensors. In: RF Taylor and JS Schultz eds. Handbook of Chemical and Biological Sensors. IOP Publishing Ltd. Bristol and Philadelphia. pp 533–552.

Mattiasson B and Borrebaeck C (1978) Non-equilibrium, isokinetic enzyme immunoassay of insulin using reversibly immobilised antibodies. In: B. Pal. ed. Enzyme labelled immunoassay of

hormones and drugs. Walter de Gruyter & Co. New York. pp 91–105.
Mattiasson B and Nandakumar MP (1999) Binding assays in heterogeneous media using a flow injection system with an expanded bed adsorption column. Bioseparation, this volume.
Nakamura S, Kato A and Kobayashi K (1991) New antimicrobial characteristics of lysozyme-dextrane conjugate. J. Agric. Food Chem. 39: 647–650.
Nandakumar MP, Lali A and Mattiasson B (1999) On-line monitoring of fermentor using expanded bed flow injection analyser. Bioseparation, this volume.
Nilsson K and Mosbach K (1980) p-Toluenesulfonyl chloride as an activating agent of agarose for the preparation of immobilised affinity ligands and proteins. Eur. J. Biochem. 112: 397–402.
Nilsson M, Mattiasson G and Mattiasson B (1993) Automated immunochemical binding assay (flow-ELISA) based on repeated use of an antibody column placed in a flow-injection system. J. Biotechnol. 31: 381–384.
Ruzicka J and Hansen EH (1988) Flow injection analysis. 2nd ed. John Wiley & Sons, Inc., New York, USA.
Sonnleitner B (1991) Dynamics of yeast metabolism and regulation. Bioproc. Eng. 6: 187–193.
Wang DIC, Cooney CL, Demain AL, Dunhill P, Humprey AE and Lilly MD (1979) Fermentation and Enzyme Technology. John Wiley & Sons, New York. pp 238–310.
Whitaker JR and Granum PE (1980) An absolute method for protein determination based on differences in absorbance at 235 and 280 nm. Anal. Biochem. 109: 156–159.

Address for correspondence: Fax +46-46-222 47 13 e-mail: Bo.Mattiasson@biotek.lu.se

Bioseparation **8:** 255–267, 1999.
© 1999 *Kluwer Academic Publishers.*

Flow injection analysis of intracellular β-galactosidase in *Escherichia coli* cultivations, using an on-line system including cell disruption, debris separation and immunochemical quantification

Anita Tocaj, M.P. Nandakumar, O. Holst & B. Mattiasson
Department of Biotechnology, Center for Chemistry and Chemical Engineering, Lund University, P.O. Box 124, S-221 00 Lund, Sweden

Received 9 October 1998; accepted in revised form 1 April 1999

Key words: beta galactosidase,immunoassay, micro-expanded bed, on-line disintegration

Abstract

A continuous integrated process for on-line quantification of intracellular components has been developed. By applying the concept of expanded micro-beds in a flow injection system it was possible to first perform on-line cell disintegration followed by an on-line binding assay for quantification of a reporter protein (β-galactosidase) from the cell interior. The disintegration process involved the use of an expanded bed with immobilised lysozyme followed by ultrasonic treatment in a flow-through cell. The cell debris does not interfere in the binding assay as it is carried out in an expanded bed. The time for an assay cycle is at present approx. 35 min. This integrated system can be used for quantification of proteins down to at least 10^{-7} mol/L.

Introduction

To fully understand and control fermentation processes all essential process variables should be determined, preferentially on line. Analysis of important variables such as biomass, substrate concentration, metabolic by-products and of course intracellular enzyme activities or product concentrations are often carried out off-line. The use of flow injection analysis (FIA) has made it possible to monitor almost any component with a much higher sample frequency than using off-line methods. Provided an on-line cell disruption process could be set up, it would be possible to design an analytical system for monitoring intracellular compounds during cultivation processes. Intracellular compounds have earlier been monitored on-line during downstream processing (Recktenwald et al., 1985) and systems have been developed for monitoring in cultivation processes (Ahlmann et al., 1986; Blankenstein and Kula, 1991; Kracke-Helm et al., 1991; Valero et al., 1992; Steube and Spohn, 1994). Several different disruption methods have been used and one important factor to consider is the interference of particulate matter originating from the disrupted cells. An on-line cell disintegration system designed for disruption of grampositive organisms, using a combination of enzymatic and ultrasonic treatment, has been developed in our laboratory (Nandakumar et al., 1999b). That method involves the use of immobilised lysozyme in an expanded bed of analytical size. The use of an expanded bed facilitates the contact between lysozyme and cells compared to a packed bed. We have also demonstrated the possibility of using expanded beds in conjunction with FIA for enzymatic analysis (Nandakumar et al., 1999a) as well as for immunoassays (Mattiasson and Nandakumar, 1999).

There are two major ways to quantify proteins using immunochemical binding assays, i.e., either via a competitive binding assay or using the sandwich assay. When the target molecule is an enzyme a third possibility arises. The enzyme is bound to the column by means of immunochemical binding and then the enzyme activity is determined by adding its specific substrate. This latter method will only give a value for the active enzyme molecules while the two former will give both active and denatured (but still immunocompetent) enzyme content.

In the present paper we address the issue of disintegration of gram-negative cells using a combination of enzymatic and ultrasonic treatment and subsequent quantification of a specific intracellular compound.

Materials and methods

Preparation of expanded beds

Materials

Lysozyme (E.C. 3.2.1.17, L6876) from chicken egg white, β-galactosidase from *E. coli* (E.C. 3.2.1.23, G5635) and polyclonal mouse antibodies against *E. coli* β-galactosidase (G4644) were all obtained from Sigma Chemical Co. (St. Louis, MO). Native Streamline® was a generous gift from Pharmacia Biotechnology AB (Uppsala, Sweden). Tresyl chloride was obtained from Synthelec AB (Lund, Sweden). All other chemicals used were of analytical grade.

Immobilisation procedure

Native Streamline® was activated with tresyl chloride according to the established procedure (Nilsson and Mosbach, 1980). Coupling of the enzyme (lysozyme) to the activated gel was carried out overnight at 4 °C in a 0.2 M sodium phosphate buffer pH 7.0. The antibody (anti-β-galactosidase) was coupled using the same conditions except for the buffer pH which was 7.2. The amount of enzyme or antibody added per ml settled gel was 25 or 20 mg, respectively. The beads with immobilised protein were washed with excess water, 1 M NaCl and 0.2 M sodium phosphate buffer (pH 7.2 or pH 7.0 as above) and then treated with 1 mM ethanolamine in sodium phosphate buffer pH 8 for 1 h at room temperature. The gel was again washed with excess 1 M NaCl and water and finally stored in 0.01 M sodium phosphate buffer (pH 7.2 or pH 7.0) containing 100 mM NaCl and 1 mM $MgCl_2$ and 0.02% (w/v) sodium azide until further use. The final coupling yield for antibody was 87% and for lyzozyme was 83%, respectively.

Expanded bed columns

The columns used were specially designed for operating expanded beds in analytical applications (for details see Nandakumar et al., 1999a). Each end of the column has a movable adapter which enables operating in an expanded as well as in a packed bed mode. The inner diameter is 0.5 cm and the total length is 20 cm.

Beads with immobilised antibodies (500 μL) were diluted with (1.0 mL) native Streamline® and one gram (wet weight, sucked gel) of the beads were packed in the column.

Instrumentation

A schematic set-up for the FIA system is outlined in Figure 1. It consists of two major parts. In the first part the cells are disrupted by means of enzymatic digestion, followed by ultrasonic treatment. In the second part the activity of the intracellular enzyme is detected.

The cell disruption part consists of a sample injection valve, a column with immobilised lyzozyme and an ultrasonicator, UP 400 S (Dr. Hielsher GmbH, Stansdorf, Germany). The second part consists of the column with immobilised anti-β-galactosidase antibody, a substrate injection valve and a spectrophotometer equipped with a flow-through cuvette for detection. Two peristaltic pumps (Alitea, Sweden) are used, one for carrier buffer and one for dissociation buffer. All the pumps and valves are operated manually. The FIA system was operated as described in Table 1.

Ultrasonication. The ultrasonicator was equipped with a flow-through cell with cooling. The titanium probe for the flow-through cell has a diameter of 14 mm and gives a maximal sonication intensity of 105 W/cm^2. For off-line sonication a micro-probe with a diameter of 3 mm and a maximal sonication intensity of 420 W/cm, was used. The sonication intensity is variable between 20 and 100% of the maximal intensity for each probe. The ultrasound was pulsed in cycles of a second. The on/off time is selected by adjusting the cycle time. In these studies the cycle time was chosen to 0.5 s (0.5 s on, 0.5 s off) and sonication intensity was set to 20%. The sonication time in the flow-system is determined by the flow-rate through the ultrasonicator flow-through cell.

Calibration with β-galactosidase The Flow ELISA system was calibrated for β-galactosidase activity under operating conditions. Standard solutions of pure β-galactosidase were injected in the carrier stream. Upon passage through the antibody column it was affinity bound. Substrate was injected and the colour developed, due to enzymatic reaction, was measured

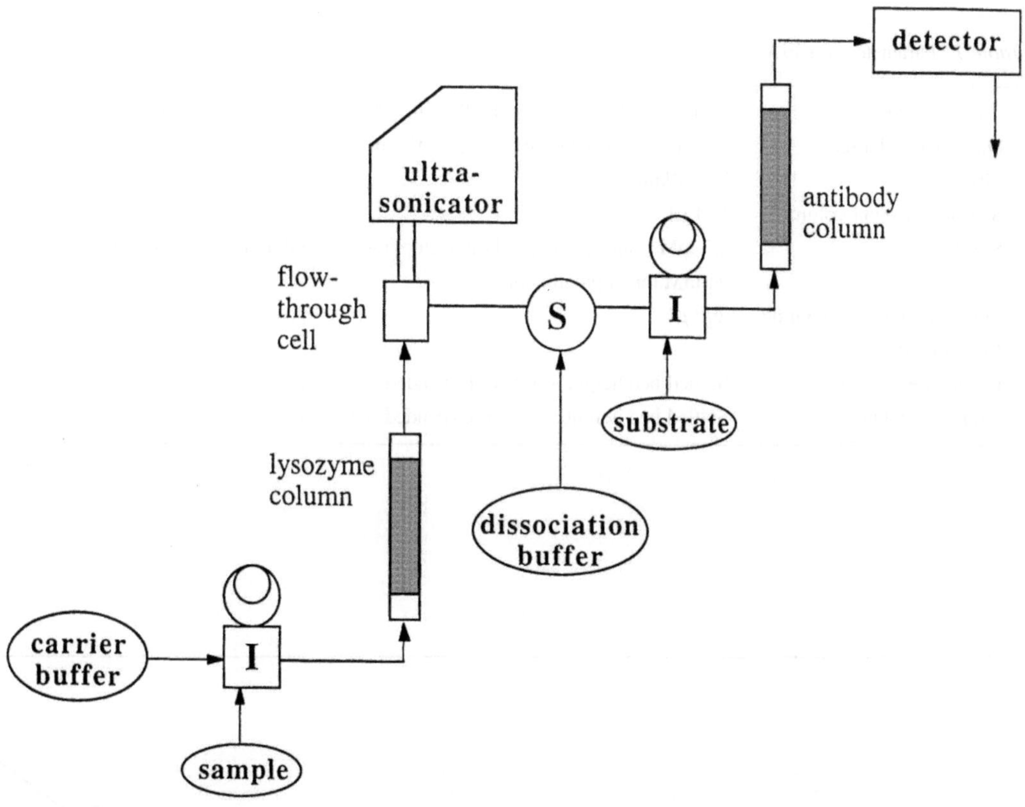

Figure 1. Schematic set up of the analytical system. I: injectors, S: switch valve.

and a calibration curve constructed (see Figure 2). The capacity of the system was measured.

Cultivation conditions

Microorganism and media

The microorganism used was *E. coli* TOPP1 (Stratagene, La Jolla, CA, USA).

The media consisted of (in g/L): lactose, 8; $(NH_4)_2SO_4$, 2.0; KH_2PO_4, 1.6; $Na_2HPO_4 \cdot 2H_2O$, 6.6; $(NH_4)_2$-H-citrate, 0.5. Trace element solution (Holme et al., 1970), 2.0 ml/L, $MgSO_4$ (1 M) 2.0 ml/L, and Adekanol (antifoam) 0.05 ml/L was also added.

Shake-flask cultivations. Shake-flask cultivations were carried out in baffled shake-flasks in the same media as above, except that the lactose was replaced with glucose. β-galactosidase production was induced after 6 h of cultivation by adding lactose (5 g/L).

Inoculum preparation. The inoculum was prepared from a working stock (1 ml) in 100 ml of the same media as above. It was grown in a baffled shake-flask (1 L) for 10 h, at 30 °C, on a rotary shaker.

Batch cultivations. The cultivations were carried out in a 3 L (working volume 2.0 L) bioreactor (Chemoferm FLC-B-3, Hägersten, Sweden), equipped with control loops for pH and temperature. Data logging and dissolved oxygen (DO) control were carried out using a control system, SattLine from Alfa Laval Automation AB (Malmö, Sweden). The temperature was 30 °C and the pH was kept at 7.0 by titration with 12.5% ammonia. Dissolved oxygen (DO) was measured with a galvanic electrode calibrated to 100% at air saturation at 30 °C. The DO was controlled at 30% by means of a PID controller (from Satt Lines control library) manipulating the stirrer speed. Supplementary lactose (30 ml 300 g/L) was added as the initial lactose was depleted and then repeated twice upon lactose depletion.

Sampling and sample treatment. Samples were withdrawn through a rubber septum at the bottom of the bioreactor, using evacuated blood collecting tubes, 3 ml Venoject (Terumo, Madrid, Spain).

Samples, including cells, for off-line determinations of β-galactosidase activity were kept frozen at −20 °C before being subjected to further disruption.

Table 1. Parameters for FIA

Carrier buffer	0.01 M Na-Phosphate, pH 6.9, with 100 mM NaCl and 1 mM $mgCl_2$.
Dissociation buffer	0.2 M Glycine-HCl, pH 2.2, 0.5 ml/min
Flow rate	0.5 ml/min
Sample injection volume	100 μl
Substrate	2.3 mM *o*-nitrophenyl-β-D-galactopyranoside(ONPG), in carrier buffer with 94 mM mercaptoethanol.
Substrate injection volume	200 μl
Bed height:	
Lysozyme column	Settled bed height – 5.0 cm Expanded – 9.8 cm
Antibody column	Settled bed height – 4.5 cm Expanded – 9.0 cm

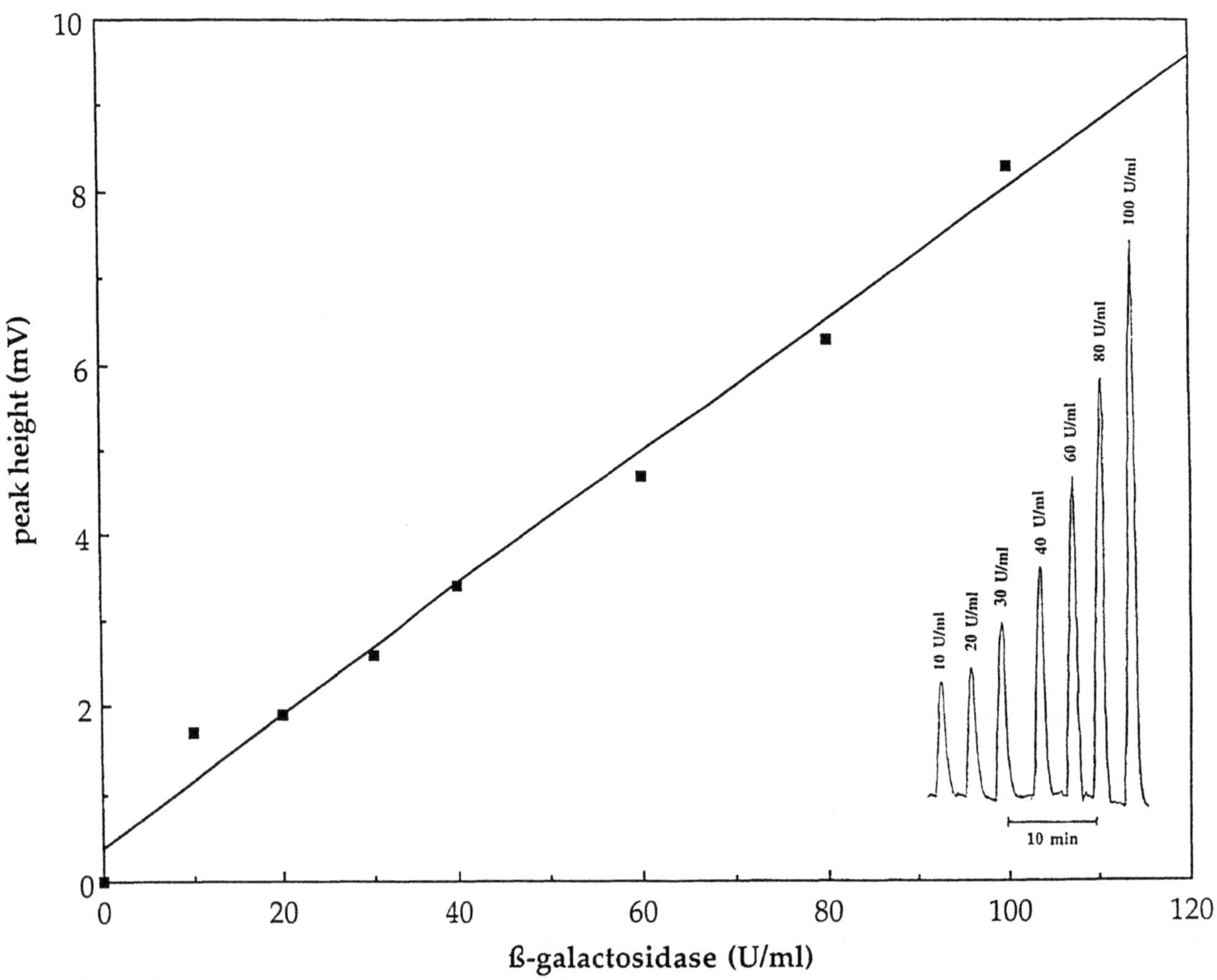

Figure 2. Calibration curve for β-galactosidase. As an insert is shown the peaks registered when increasing amounts of betagalactosidase were added as pulses in the flow system. The operating conditions were defined in Table 1.

Table 2. Result of β-galactosidase activity released from the cells after treatment on-line with different cell disruption methods

Cell disruption method (on-line)	β-galactosidase activity (U/ml)
Lysozyme	13
Ultrasonication	3900
Lysozyme + ultrasonication	4300

Samples (10 hour old culture at late exponential phase, OD at 620 nm was 15.0) were taken after induction by lactose during cultivation.

Samples used for enzymatic determination of lactose were centrifuged at 15.000**g* for 1 min and then the supernatant was kept frozen at −20 °C. Prior to lactose analysis the supernatant was heated to 80 °C for 15 min and then centrifuged at 15.000**g* for 2 min.

Analytical methods

β-galactosidase activity. The β-galactosidase activity was determined using the FIA system described above and compared with results from the off-line analysis according to Miller et al. (1972) (with modifications in terms of pH (at pH 6.9 instead of pH 7.5) and concentration of mercaptoethanol (94 mM was used instead of 4.5 M) and the volume of the reaction mixture was doubled (2 ml)). The activity was determined following the enzymatic hydrolysis of 14 mM *o*-nitrophenyl-β-D-galactopyranoside (ONPG) in 0.01 M sodium phosphate buffer, pH 6.9, containing 1 mM $MgCl_2$, 100 mM NaCl and 94 mM mercaptoethanol, at 25 °C. The released *o*-nitrophenol gives a yellow colour which was determined spectrophotometrically at 420 nm. The molar extinction coefficient for *o*-nitrophenol (ONP) is 3.1 $mM^{-1} \cdot cm^{-1}$.One unit of enzymatic activity hydrolyses 1 μmole of ONPG per minute at 25 °C.

Optical density (OD). The OD was measured at 620 nm. Samples were diluted with 0.9% NaCl at OD values exceeding 0.5.

Lactose. Lactose concentration was determined enzymatically using a test kit (no.176 303) from Boehringer Mannheim GmbH (Mannheim, Germany).

Off-line cell disruption methods. The samples were initially frozen-and-thawed prior to further disruption. The concentrations of the different chemicals used were: toluene, 5%; sucrose, 20% in 0.01 M sodium

Table 3. Result of β-galactosidase activity released from the cells after treatmentwith different cell disruption methods

Cell disruption method (off-line)	β-galactosidase activity (U/ml)
Freeze and thaw (F&T)	230
(F&T) + lysozyme	250
(F&T) + toluene	440
(F&T) + osmotic shock (sucrose)	460
(F&T) + EDTA	510
(F&T) + lysozyme + EDTA	780
(F&T) + urea + EDTA	1550
(F&T) + urea + EDTA + lysozyme	1600
(F&T) + toluene + EDTA	1600
(F&T) + ultrasonication	3930
(F&T) + lysozyme + ultrasonication	4450
(F&T) + lysozyme + EDTA + ultrasonication	4800

Samples (10 hour old culture at late exponential phase, OD at 620 nm was 15.0) were taken after induction by lactose during cultivation. 3 ml of the samples were used for lysis experiments.

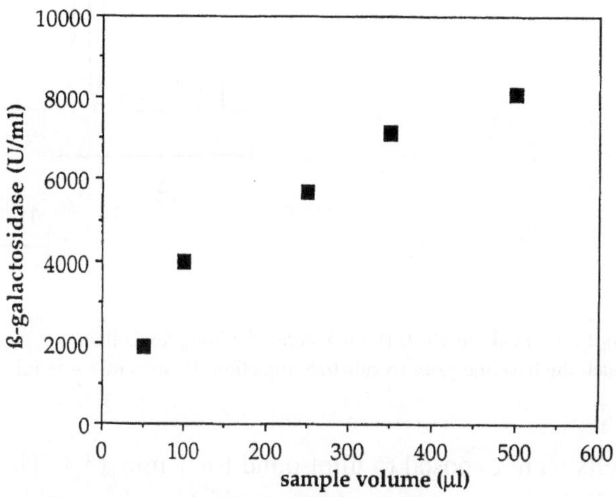

Figure 3. Effect of sample volume on β-galactosidase response(4000 U/ml of betagalactosidase detected when 100 μl sample volume injected in the FIA system was used for comparison). Experimental details are given in Table 1.

phosphate buffer pH 7.0; EDTA, 10 mM; urea 3 M. Lysozyme concentration was 10 mg/ml. The samples were incubated at room temperature for 30 min together with the chemical or enzymatic agents. Ultrasonication was carried out in Eppendorf tubes placed on ice using a micro probe (3 mm dia.). The ultrasonicator was run for 30 s, then switched off for 30 s. This was repeated for total time of 5 min. The cycle time was set to 0.5 s and the intensity to 20%. Thus, the

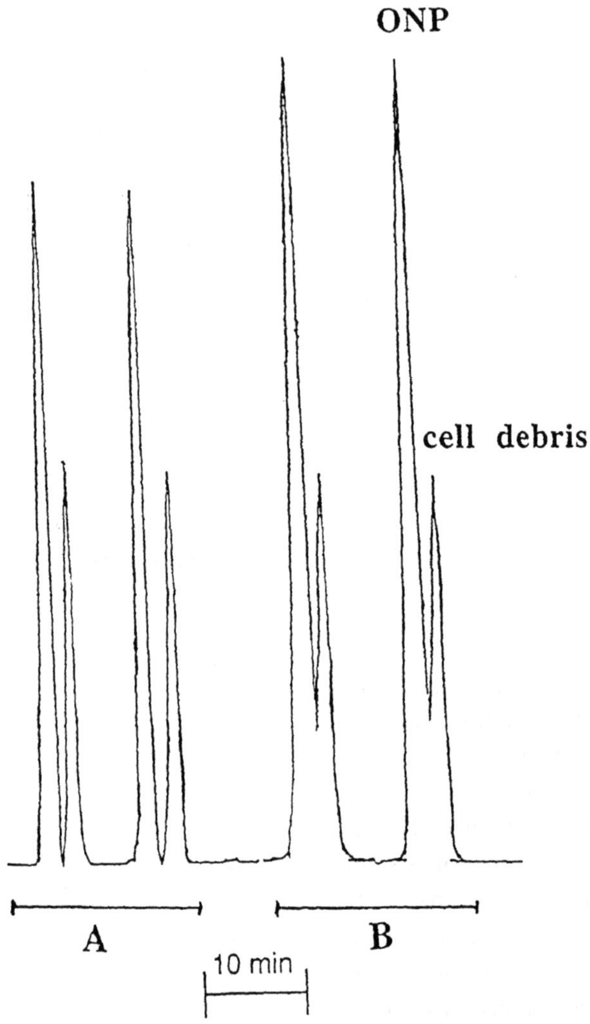

Figure 4. Peak height (mV) registered for samples containing β-galactosidase as well as cell debris. A: the cell debris peak was allowed to reach the baseline prior to substrate injection. B: substrate was injected before the cell debris peak had reached the base line.

cells were exposed to ultrasound for 1 min 15 s. The cell debris was removed by centrifuging the samples at 15.000*g for 2 minutes and the supernatants were analysed for β-galactosidase activity.

Results and discussion

The general approach to measure the concentration of an intracellular compound in a cultivation process is to take a sample, disrupt the cells, remove the cell debris and then carry out the analysis. This procedure involves a number of steps that are both labour- and time-consuming. Furthermore, a range of manual steps will lead to lower reproducibility and more assays will thus be needed. For process-control the analytical results must be available with as little time delay as possible. Integration and automation of sampling, sample treatment and assay procedure would meet both accuracy and speed requirements.

It was shown earlier that flow injection analysis (FIA) offers many advantages when it comes to handling of samples. These include application of reproducible amounts of a liquid sample, maintaining a constant time of contact between a sample and the analytical equipment and operation of complex systems also far from equilibrium in a reproducible way. Thus, FIA is a good choice when a complex assay procedure is going to be computerised and run in an automatic mode.

From earlier studies it is known that non-equilibrium assays can be performed in flow-injection

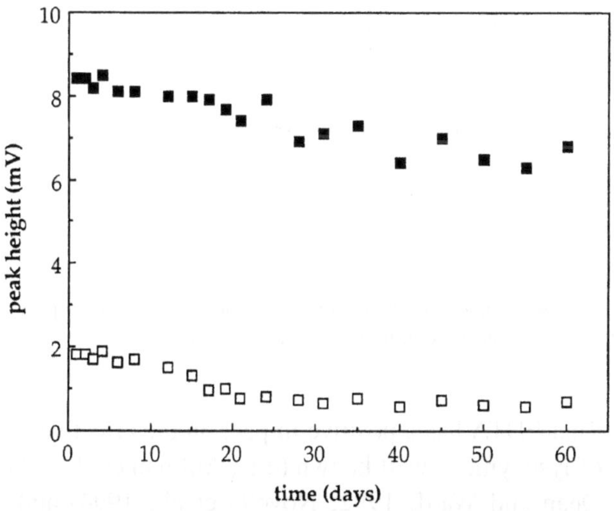

Figure 5. Operational stability of immobilized anti-β-galactosidase antibodies expressed as ability to trap enzyme activity from a standardized pulse passing through the column. (■) 100 U/ml, (□) 10 U/ml.

systems (Mattiasson et al., 1978) and in an automated way (Nilsson et al., 1993). The use of expanded beds in analytical flow systems makes it possible to carry out analysis in cell containing samples. This method has successfully been used for analysis of media components and of extracellular products in fermentation processes (Nandakumar et al., 1999a) as well as for immunochemical binding assays (Mattiasson and Nandakumar, 1999). For measurements of intracellular components in flow-systems dialysis membranes have been used to separate the cell debris from the analyte (Ahlman et al., 1986; Blankenstein and Kula, 1991; Steube and Spohn, 1994). Other workers have measured intracellular compounds in the presence of cell debris (Kracke-Helm et al., 1991; Valero et al., 1992). This is possible when the target molecule is an enzyme so that the sample can be injected and mixed with the substrate in the tubing and thus no enzyme reactor in the FIA system is needed. The sample has to be diluted to minimise disturbances from the cell debris. The combination of enzymatic and ultrasonic treatment in a flow system, using immobilised lysozyme in an expanded micro-bed, has earlier proven useful for disrupting grampositive cells (Nandakumar et al., 1999b). In this study the integration of the cell disrupting system described by Nandakumar et al., (1999b) together with immunochemical binding in expanded micro-beds for determination of intracellular components in gram-negative *E. coli*, was evaluated.

E. coli was chosen as it is an important organism for the production of recombinant proteins in a number of industrial processes. A reporter enzyme, β-galactosidase (Schupp et al., 1995), was selected, as the main issue was to evaluate the integrated concept.

Cell disruption using ultrasound

An important question in cell disruption studies is how to quantify the effectiveness of the disintegration process. Several different methods are used and have been described in the literature. Among these, are the analysis of soluble protein in the supernatant (Engler and Robinson, 1979; Sauer et al., 1989) and methods based on optical density measurements (Kuboi et al., 1995). Other authors have characterised disruption processes by the particle size distribution (Agerkvist and Enfors, 1990). Very often the success of a disruption process is described by the release of the target molecule. This is obviously an interesting parameter as the accessibility of the target molecule is a key issue in purification studies as well as in analytical applications. Ultrasonic treatment is often used as a reference method when evaluating chemical and/or enzymatic disruption methods and the protein or enzyme content found after ultrasonic treatment is set to 100%. How the ultrasonication is carried out is usually not well described. Cell disruption, by means of ultrasound is caused by the collapse of gas bubble cavities created with high-intensity sound-waves in the liquid. As the cavities collapse, extremely high pressures are developed. A correct way to describe the use of ultrasound would be to give the sound intensity that the cells have been exposed to and the duration of exposure. Information on the length of pauses and on cooling conditions are also of interest. Ultrasonication is quite a harsh method which may damage the released proteins or enzymes at prolonged exposure.

Enzymatic cell disruption

Lysozyme splits the β-1,4-glycosidic bond of the peptidoglycan, responsible for the mechanical strength, in the bacterial cell wall. In gram-negative organisms the peptidoglycan layer is very thin compared to that present in gram-positive organisms. In gram-negative microorganisms the peptidoglycane is surrounded by one outer and one inner membrane. The outer membrane of *E. coli* is relatively permeable to small molecules but not to enzymes or other large molecules (Brock and Madigan, 1988). In order to use lysozyme for disrupting gram-negative bacteria, the outer membrane should be removed or permeabilised. That can be achieved by treating the cells with detergents

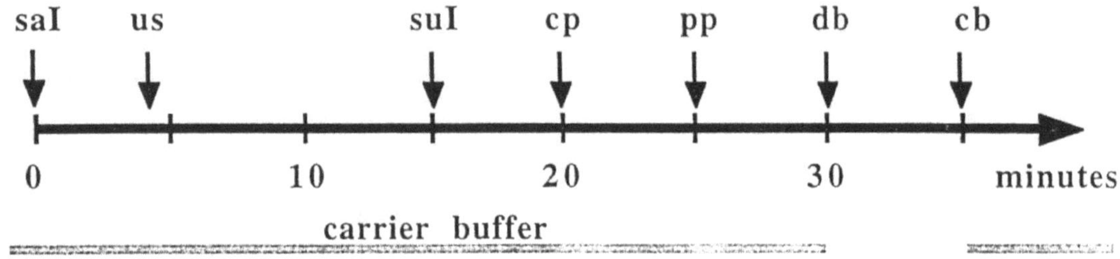

Figure 6. The sequence of events during an assay cycle. saI, sample injection; us, sample reaches the ultrasonicator; suI, substrate injection; cp, cell peak maximum registered; pp, product peak (ONP) maximum registered; db, dissociating buffer; cb, carrier buffer.

(Andrews and Asenjo, 1987), by adding EDTA and lyzozyme (Ahlman et al., 1986; Dean and Ward, 1992; Novella et al., 1994) or for certain strains freezing will have the same effect (Neidhardt et al., 1990). To freeze the cells in the on-line system was not practical and the use of detergents is not recommended due to possible interferences in the analytical assay. EDTA additions caused precipitates which clogged the tubings as well as the flow-through cuvette, and could not be used in the on-line system. Therefore, the system was tested without adding any EDTA in the carrier buffer and we found that lysozyme was still contributing to the release of β-galactosidase. There is EDTA present in the trace element solution which could contribute to the effect of lysozyme. It should be mentioned that by immobilising a more specific enzyme the effectiveness of the enzymatic disruption step would increase. These enzymes are not as well investigated as lysozyme, but have been reviewed by e.g. Andrews and Asenjo (1987).

Evaluation of cell disrupting methods

It is a difficult task to determine when the cell disintegration process is complete. From earlier studies, both in other laboratories (Pandolfe and Kinney, 1983; Engler, 1985) and by ourselves it is not possible to state that the cells have been completely disrupted and that the target enzyme is completely unaffected by the disruption. However, in order to get a better idea about the effectivity of different disintegration techniques, a range of methods were applied. The effect of the two on-line methods, individually and in combination, is found in Table 2. For off-line methods the sample was kept frozen and subsequently thawed prior to disruption by the other methods. The results are found in Table 3. Lysozyme pre-treatment together with ultrasonication increased the activity released by approximately 10% compared to ultrasonication alone. That EDTA has a positive impact on the effectiveness of lysozyme is well known (e.g., Ahlman et al., 1986; Dean and Ward, 1992; Novella et al., 1994) and it is clearly seen in Table 2. The best method (among the tested ones) in terms of released β-galactosidase activity would be lysozyme/EDTA treatment followed by ultrasonication. As it was not possible to add EDTA in the flow-system we chose the second best method, lysozyme treatment followed by ultrasonication to be used in the rest of this study.

Characterisation of the antibody column

The analytical expanded bed contains immobilised antibodies directed towards the target protein, β-galactosidase. The binding assay was set up to bind the target enzyme to the immunosorbent and then a specific substrate (ONPG) was added in order to quantify the bound activity. The adsorbent was characterised in the sense that well defined amounts of β-galactosidase were injected into the system and the amount of enzyme activity trapped was quantified. Typical peaks and the corresponding calibration curve are found in Figure 2. The sensitivity of the assay will to some extent be determined by the volume of sample injected. Similar studies have earlier been carried out in detail for packed bed mode of operation (Nilsson et al., 1993). In Figure 3 the relation between the activity registered and the sample volume applied, is shown.

It was shown previously, that cells present in the sample did not interfere with the binding (Mattiasson and Nandakumar, 1999). To make sure that cell debris did not interfere, assays were done on samples containing disintegrated cells and β-galactosidase and the appropriate time for substrate addition was investigated. In Figure 4 the peaks originating from cell debris and ONP are shown. It was found that the analytical result, calculated as the β-galactosidase activity bound to the column, was not dependent on absolute

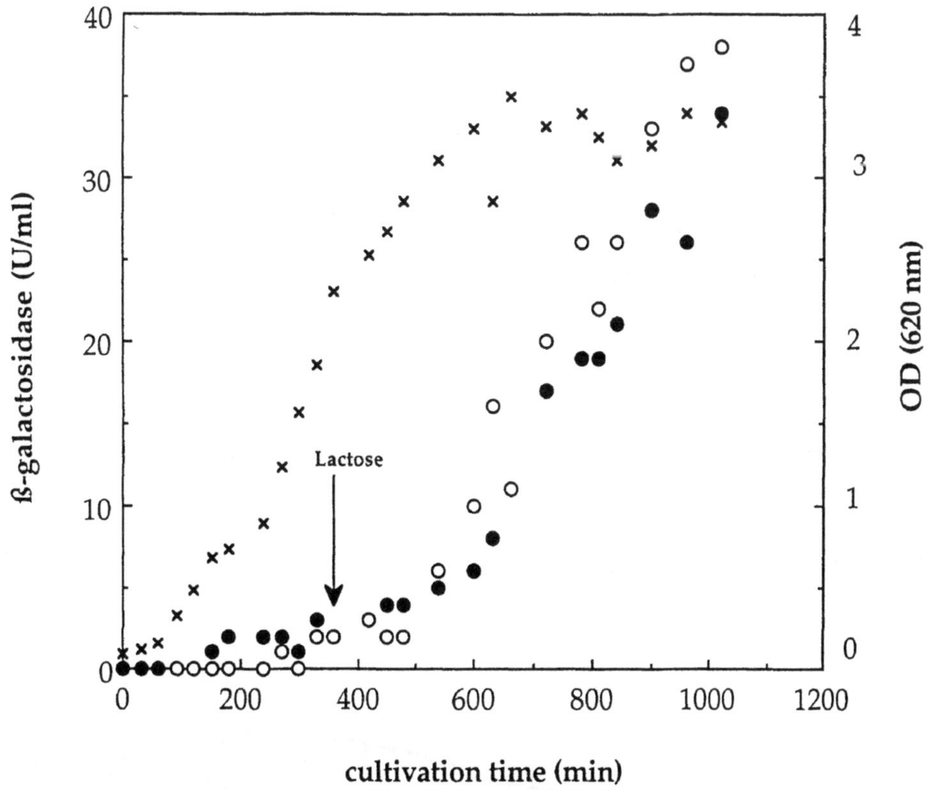

Figure 7. Monitoring of a shake-flask experiment. On-line (•) and off-line (o) β-galactosidase activity (U/ml of culture broth) and OD (x) measurements. At the arrow the inducer lactose was added.

peak separation. That means that the substrate could be added before the absorbance peak caused by the cell-debris reached the base-line and thereby the time needed for an assay could be reduced.

The stability of the immunosorbent preparation was studied by running a number of repetitive assays. Two standard solutions of enzyme (10 and 100 U/mL) were used. As can be seen from Figure 5 there is a slight decrease per incubation over a period of 60 days. This can most likely be attributed to the dissociation conditions damaging the immobilised antibodies. It should be stressed that during this period the same preparation of immobilised antibodies was used for more than 400 assays. As there is no competitive binding involved, an increase in operational stability can be achieved by simply adding more antibodies to the system. Still, the stability, as shown in Figure 5, is sufficient in most cases. The stability is to a large extent dependent on the molecular properties of the immobilised entity. In a study by Nilsson et al. (1992) using immobilised Concanavalin A it was shown that the stability in that assay was rather poor. This was due to the instability of Concanavalin A whereas in systems with immobilised antibodies good stability was obtained.

Experimental set-up of the complete system

The experimental set-up is shown in Figure 1. The system has two sequentially coupled expanded bed columns. In the first expanded bed the cells are subjected to enzymatic digestion by lysozyme covalently bound to Streamline® and then the sample enters the ultrasonicator for further disruption. In the second expanded micro-bed the target protein is trapped to an immunosorbent and when the substrate is injected the enzymatic activity is determined. The analytical sequence is shown in Figure 6. The carrier buffer is passed through the system continuously, except for the short period when the dissociation buffer is passed through the antibody column. The dissociation buffer is introduced at the switch valve (see Figure 1) and does not pass through the lysozyme column.

The degree of expansion is controlled by the flow rate. This was studied previously (Nandakumar et al., 1999b) and the result from that study was applied here.

Shake flask experiments

The complete, integrated system was tested and compared with off-line analysis of β-galactosidase in a

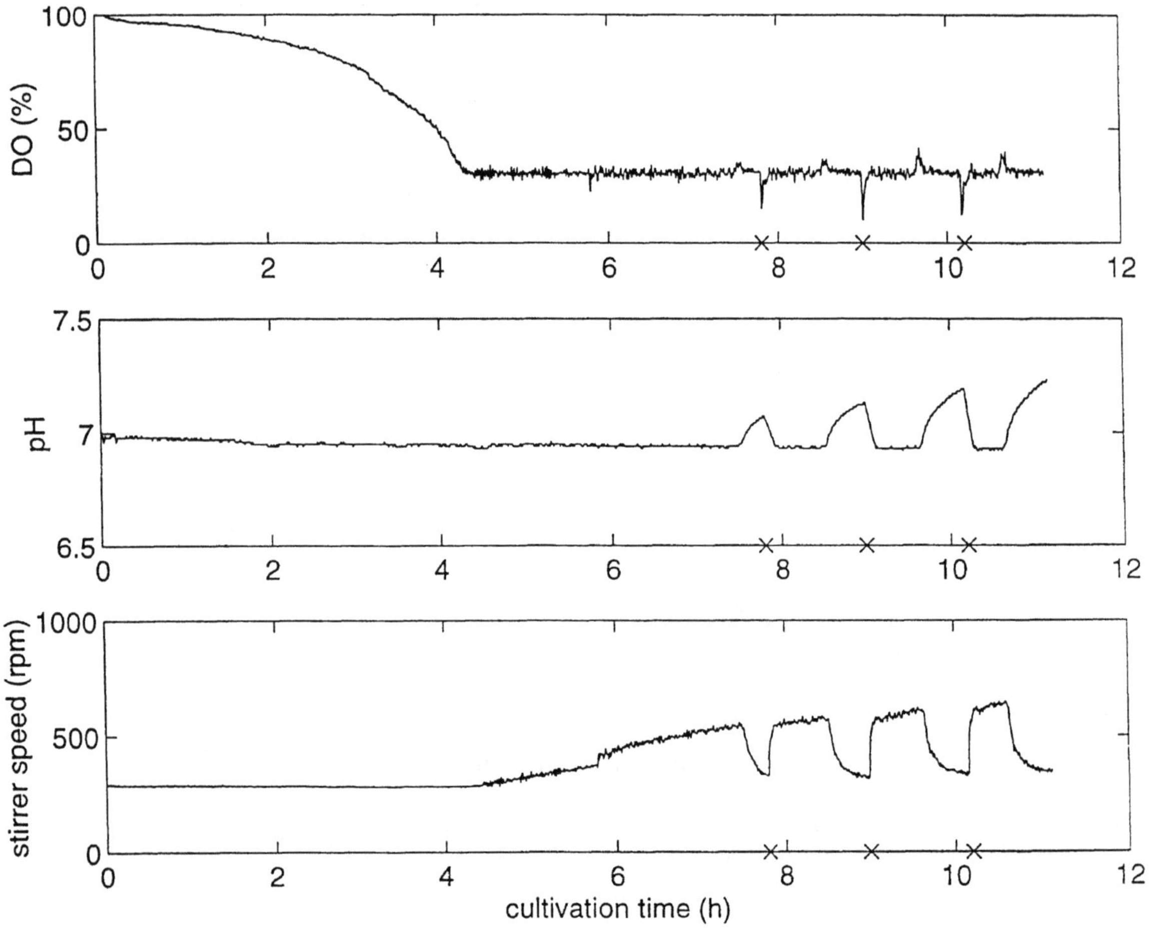

Figure 8a. Monitoring of a batch cultivation in a 3L fermenter. The experimental details are given in the text. (a) Logged signals and (b) analytical data for batch cultivation. Lactose additions are indicated with an (x).

shake-flask cultivation of *E. coli*. The culture was initially grown on glucose and then β-galactosidase production was induced by adding lactose. Figure 7 shows the growth curve as OD_{620} together with readings of β-galactosidase (U/ml of culture broth) on-line as well as off-line. The absence of enzyme activity registered before induction indicates that the system works. Once β-galactosidase production has been initiated there is a good correlation between on-line and off-line enzyme activities. The fact that off-line and on-line methods often give different results in absolute values, is well known and has also been observed in other systems (e. g., Kracke-Helm et al., 1991). Whether the small difference observed here was due to a more efficient disintegration in the off-line case and thereby better release of the enzyme, or whether it is an effect of shielding of the antibody is not clear. The likliehood of shielding of the antibody binding sites by material non-specifically bound to the adsorbent taking place seems implausible as there is a surplus of binding capacity in the column. The immobilised antibodies were supplied as a fairly small volume of derivatised Streamline®. In order to make the expanded immunosorbent more suitable for trapping the target protein, the immunosorbent beads were mixed with native Streamline® to increase the time of passage through the column. This way of 'diluting' immobilised antibody preparations has been shown to be efficient when dealing with competitive immunoassays (Borrebaeck et al., 1978).

On-line measurements during batch cultivation

The system was finally tested during a fermentation of *E. coli* grown on lactose in a labscale bioreactor. The logged process signals, together with the corresponding analytical results are found in Figure 8. The depletion of lactose could be observed in all three (DO, pH, stirrer speed) process signals. As the carbon source is utilised the need for oxygen is decreased

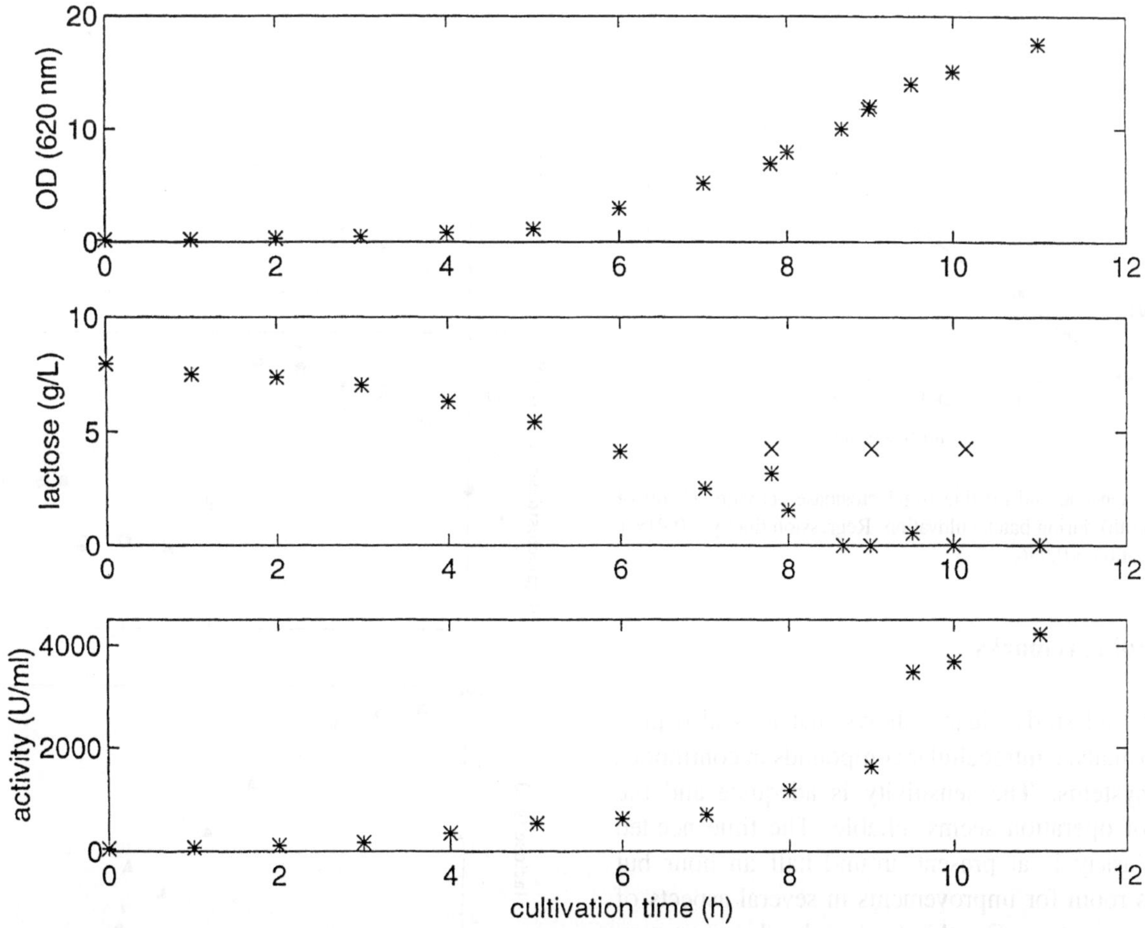

Figure 8b.

which is clearly indicated by the decreasing stirrer speed at 7.5 hours. Upon lactose addition (indicated by x), the stirrer speed increased in order to keep the DO at the set-point (30%). The pH rise and fall observed together with lactose depletion and addition, respectively, is likely to be a result of the utilisation of other carbon sources such as aminoacids or acetate, present in the broth. The correlation between on-line and off-line values of β-galactosidase activity during the batch cultivation is shown in Figure 9. The regression line indicates that the on-line and off-line values have good correlation. In Figure 10b where the β-galactosidase activity per OD unit is shown, one advantage of using on-line FIA systems compared to off-line results becomes obvious. Namely that the on-line values seems much more reliable and do not fluctuate as much as the off-line do. In that same Figure it is also possible to detect why the need for on-line measurements of intracellular compounds is vital, and is indicated by the differences of enzyme content in the cells at different stages of the cultivation. Initially the cells build up a pool of the enzymes, needed for substrate utilisation, and the growth rate (shown as ln OD in Figure 10a) is not at μ_{max}. As the growth rate increases the β-galactosidase content in the cells does not increase any further. At the end of the cultivation as lactose is added as pulses (shown as (Δ) in Figure 10c) the intracellular concentration seems to increase slightly. This indicates the need for on-line analysis of intracellular products to be used as a control signal for substrate addition. Using optimised substrate addition, the productivity could be increased. There is another possible explanation for the apparent increase in enzyme activity in the late phase of the cultivation. That is, that cells in the stationary phase, are more sensitive to the disruption methods applied and therefore more intracellular compounds are released. The effect of the disruption methods on cells from different cultivation stages was not investigated in this study.

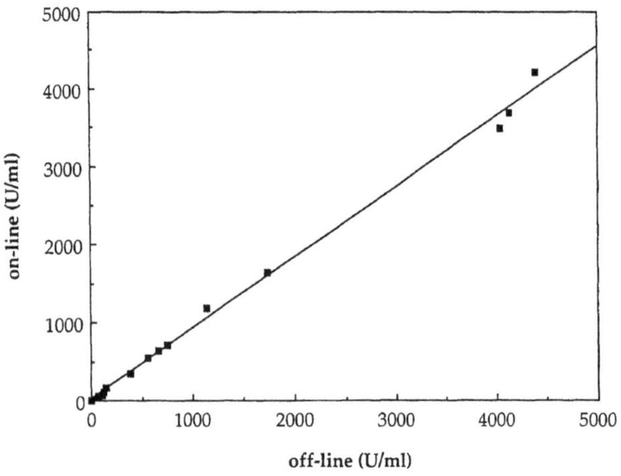

Figure 9. On-line and off-line β-galactosidase activities (U/ml of culture broth) during batch cultivation. Regression line: y = 0,91x + 20, correlation = 0,996.

Concluding remarks

The present study clearly shows that it is also possible to analyse intracellular compounds in continuous assay systems. The sensitivity is adequate and the mode of operation seems reliable. The time needed for an assay is at present around half an hour but there is room for improvements in several aspects of the assay system. For this further development may be necessary concerning, e.g., adsorbents and disruption procedures. The sensitivity was not a major focus in this study, so sensitivity could probably be improved further. The technology presented here, i.e., the combination of enzymatic and ultrasonic cell disruption methods, offers great possibilities for measuring intracellular compounds found in the different compartments of the cell. That will most certainly add to our knowledge about the organisms we are using in the production stages and hopefully also in a longer-term perspective improve conditions for efficient process control.

Acknowledgements

This project was supported by The National Swedish Board for Technical Development, The Swedish Foundation for Food Technology and Biotechnology, the Trygg-Hansa Alf Åkerman Foundation and the Helge Ax:son Johnsson Foundation. The support from Amersham Pharmacia Biotech AB was given under the programme of The Swedish Competence Center for Bioseparation.

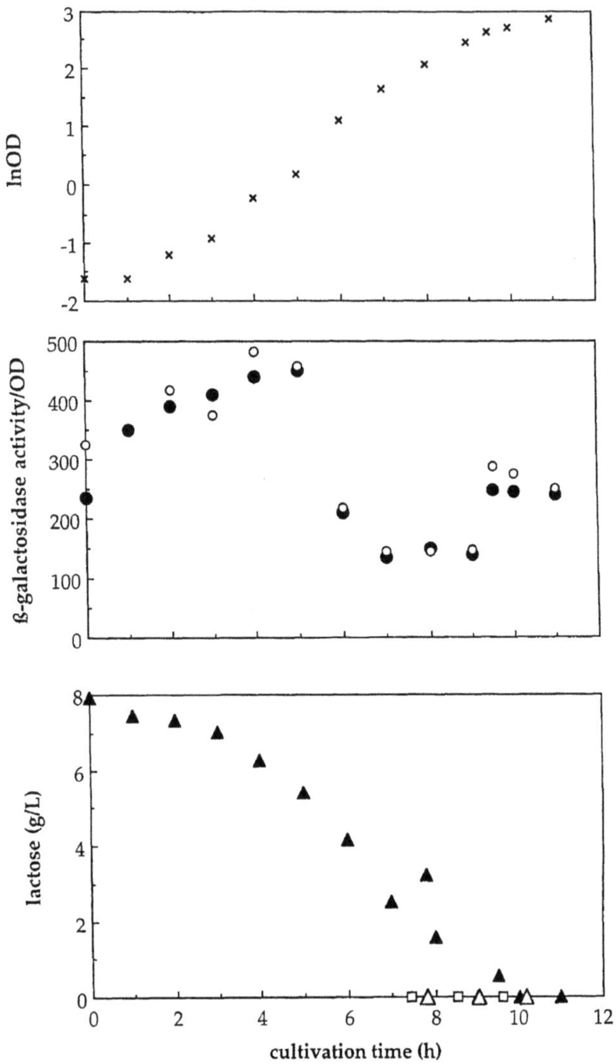

Figure 10. (a) ln OD versus cultivation time. (b) enzyme content in the cells, shown as β-galactosidase activity per OD unit, during the cultivation. On-line (●) and off-line (○) determination of β-galactosidase activity. (c) lactose concentration in g/L. Lactose depletion is indicated with (□) and additions with (△).

References

Agerkvist I and Enfors S-O (1990). Characterisation of *E. coli* cell disintegrates from a bead mill and high pressure homogenizers. Biotechnol. Bioeng. 36: 1083–1089.

Ahlmann N, Niehoff A, Rinas U, Scheper T and Schügerl K (1986). Continuous on-line monitoring of intracellular enzyme activity. Anal. Chim. Acta 190: 221–226.

Andrews BA and Asenjo JA (1987). Enzymatic lysis and disruption of microbial cells. Trends. Biotechnol 5: 273–277.

Blankenstein G and Kula M-R (1991). Cell permeabilization as a tool for measurement of intracellular enzyme activity in a flow-injection system. Anal. Chim. Acta 248: 371–378.

Borrebaeck C, Börjesson J and Mattiasson B (1978). Thermometric enzyme linked immunosorbent assay in continuous flow systems: Optimization and evaluation using human serum albumin as a modelsystem. Clin. Chim. Acta 86: 267–278.

Brock TD and Madigan MT (1988). Biology of microorganisms. Prentice-Hall International, Englewood Cliffs, New Jersey.

Dean CR and Ward OP (1992). The use of EDTA or polymyxin with lysozyme for the recovery of intracellular products from *Escherichia coli*. Biotechnol. Tech. 6: 133–138.

Engler CR (1985). Disruption of microbial cells. In: Moo-Young M, Humphrey AE and Cooney CL (eds) Comprehensive Biotechnology, Vol. 2. Pergmon, Oxford, pp. 305.

Engler CD and Robinson CW (1979). New method for measuring cell-wall rupture. Biotech. Bioeng. 21: 1861–1869.

Holme T, Arvidsson S, Lindholm B and Pavlu B (1970). Shake-flaskcultivation of bacteria. Process Biochem. 5: 62–66.

Kracke-Helm H-A, Brandes L, Hitzmann B, Rinas U and Shügerl K (1991). On-line determination of intracellular β-galactosidase activity in recombinant *Escherichia coli* using flow injection analysis (FIA). J. Biotechnol. 20: 95–104.

Kuboi R, Umakoshi H, Takagi N and Komasawa I (1995). Optimal disruption methods for the selective recovery of β-galactosidase from *Escherichia coli.* J. Ferment. Bioeng. 79: 335–341.

Mattiasson B and Nandakumar MP (1999). Binding assays in heterogenous media using a flow injection system with an expanded bed micro-bed adsorption column. Bioseparation, this volume.

Mattiasson B, Svensson K, Borrebaeck C and Kronvall G (1978). Non-equilibrium enzyme immunoassay of gentamycin. Clin. Chem 24: 1770–1773.

Miller JM (1972). In Experiments in molecular genetics. pp 352–355. Cold Spring Harbour Laboratory, Cold Spring Harbour, New York.

Nandakumar MP, Lali A and Mattiasson B (1999a). On-line monitoring of fermentor using expanded micro-bed flow injection analyser. Bioseparation. (*In press*).

Nandakumar MP, Tocaj A and Mattiasson B (1999b). Analysis of cell containing samples from fermentors using an integrated samples from fermentors using an integrated approach involving sampling, cell desintegration and analysis in a continuous flow injection system containing a micro-expanded enzyme column. Bioseparation, this volume.

Neidhardt FC, Ingraham JL and Schaechter M (1990). Physiology of the bacterial cell: A molecular approach. Sinauer Assosiates, Inc. Publishers, Sunderland, Massachusetts.

Nilsson M, Håkansson H and Mattiasson B (1992). Evaluation of direct versus competitive binding assays in flow injection systems. Proc. Control Qual. 4: 37–45.

Nilsson M, Mattiasson G and Mattiasson B (1993). Automated immunochemical binding assay (flow-ELISA) based on repeated use of an antibody column placed in a flow-injection system. J. Biotechnol. 31: 381–384.

Nilsson K and Mosbach K (1980). p-Toluenesulfonyl chloride as an activating agent of agarose for the preparation of immobilized affinity ligands and proteins. Eur. J. Biochem. 112: 397–402.

Novella IS, Fargues C and Grévillot G. (1994). Improvement of the extraction of penicillin acylase from *Escherichia coli* cells by a combined use of chemical methods. Biotechnol. Bioeng 44: 379–382.

Pandolfe WD and Kinney R (1983). Recent developments of the understanding of Homogenisation parameters, Paper presented at 1983 Summer National Meeting, AIChE, Denver, CO.

Recktenwald A, Kroner K-H and Kula M-R (1985). On-line monitoring of enzymes in downstream processing by flow injection analysis (FIA). Enzyme Microb. Technol. 7: 607–611.

Sauer T, Robinson CW and Glick BR (1989). Disruption of native and recombinant *Escherichia coli* in a high-pressure homogenizer. Biotechnol. Bioeng 33: 1330–1342.

Schupp JM, Travis SE, Price LB, Shand RF and Keim P (1995). Rapid bacterial permeabilization reagent useful for enzyme assays. BioTechniques 19: 18–20.

Steube K and Spohn U (1994). On-line monitoring of intracellular activities with flow-injection analysis. Anal. Chim. Acta 287: 235–246.

Valero F, Lafuente FJ, Solà C, Benito A, Vidal M, Cairó J and Villaverde A (1992). Simultaneous on line monitoring of intracellular β-galactosidase activity and biomass using flow injection analysis in *Escherichia coli* batch fermentations. Biotech. Technol. 6: 213–218.

Address for correspondence: B. Mattiasson, Department of Biotechnology, Center for Chemistry and Chemical Engineering, Lund University, P.O. Box 124, S-22100 Lund, Sweden. Tel: 46-222-8264; Fax: 46-222-4713; E-mail: bo.mattiasson@biotek.lu.se

Key word index

Author index

MIX
Papier aus verantwortungsvollen Quellen
Paper from responsible sources
FSC® C105338

If you have any concerns about our products,
you can contact us on
ProductSafety@springernature.com

In case Publisher is established outside the EU,
the EU authorized representative is:
Springer Nature Customer Service Center GmbH
Europaplatz 3, 69115 Heidelberg, Germany

Printed by Libri Plureos GmbH
in Hamburg, Germany